The Politics of Life Itself

Series Editor
Paul Rabinow

A list of titles in the series appears at the back of the book

The Politics of Life Itself

Biomedicine, Power, and Subjectivity in the Twenty-First Century

Nikolas Rose

PRINCETON UNIVERSITY PRESS
PRINCETON AND OXFORD

Published by Princeton University Press, 41 William Street, Princeton, New Jersey 08540
In the United Kingdom: Princeton University Press, 3 Market Place, Woodstock, Oxfordshire OX20 1SY

ISBN-13: 978-0-691-12190-1 (alk. paper)
ISBN-10: 0-691-12190-7
ISBN-13 (pbk.): 978-0-691-12191-8 (alk. paper)
ISBN-10 (pbk.): 0-691-12191-5

Library of Congress Control Number: 2006932070

British Library Cataloging-in-Publication Data is available

This book has been composed in Sabon with Futura display.

Printed on acid-free paper. ∞

pup.princeton.edu

Printed in the United States of America

10 9 8

Contents

CONTENTS

Acknowledgments

The ideas presented here have been developed over several years in dialogue and collaboration with many others. I would particularly like to acknowledge the support given by the members of the BIOS research network, initially established from Goldsmiths College, University of London, and to thank my colleagues at Goldsmiths who were involved, notably Mariam Fraser and Monica Greco. I would also like to thank Monica, Ilpo Helen, and Mariana Valverde, who helped organize the BIOS Vital Politics conference held at the LSE in September 2002—an event that brought together an outstanding array of talented researchers on social aspects of the biosciences. At the London School of Economics and Political Science (LSE), it has been my privilege to work in the very supportive research community of the BIOS Centre for the Study of Bioscience, Biomedicine, Biotechnology and Society, and to be able to attend the many stimulating seminars and symposia that BIOS has hosted. I would like to thank Pat Spallone for her vital input into the early life of BIOS, and to Sarah Franklin who has given her energies so fully to the development of this work at the LSE: Sarah has also generously given her time to read drafts of chapters, to discuss them with me in detail, to lend me books and papers, and guide me to literature that I would otherwise have missed. I wish I had been able to follow up more of the leads she gave me.

My thinking has constantly been provoked by the research students and postdoctoral fellows with whom I have worked, and I would particularly like to thank Carlos Novas, who not only demanded that I keep up with his prodigious reading and with the empirical demands of research, but also was so generous with his concepts. More recently, thanks to Filippa Cornelliussen, Chris Hamilton, Annette V. B. Jensen, Kirstin Klein,

Linsey McGoey, David Reubi, Ayo Wahlberg, and Scott Vrecko for keeping me more or less up to speed. I would also like to thank my brother, Steven Rose, for a lifetime of debate over many of these issues: my involvement with the biopolitical began when, influenced by my enjoyable experience washing equipment in his lab, I went to Sussex University to read biology and learned my genetics at the feet of John Maynard Smith. Over the years I have discussed issues of essentialism, reductionism, genetics, neuroscience, and the politics of biology both with Steven and with my sister-in-law Hilary Rose and, despite our many disagreements, their tenaciously critical voices have always been a source of inspiration: I hope they will forgive me for having trespassed on their territory and analyzed it according to a different logic. And, last but not least, thanks to Paul Rabinow for his friendship, his fierce intellectual stimulation, and for our spirited disputes that have helped sharpen my thought in innumerable ways.

The idea for this book arose from a paper for a seminar on Displacement of Politics held at Santa Margherita, Genoa, June 24–26, 1999 organized by Sakari Hanninen and Allesandro dal Lago and published as "The politics of life itself," *Theory, Culture and Society*, 2001, 18 (6): 1–30 (an earlier version was published in Italian translation in *Aut Aut*, 2000, 298: 35–62). Versions have also been given at the University of Bath, the University of Stockholm, Goldsmiths College, University of London, University of East London, at the University of Copenhagen in May 2000, and at the conference on The Ethos of Welfare, University of Helsinki, September 2000. Thanks are due to those who invited me—Barry Barnes, Kenneth Hultqvist (whose untimely death is a great sadness), Mike Rustin, Barry Richards, Lene Otto, Lene Koch, and Ilpo Helen, and to those whose comments have helped me improve my argument. I would particularly like to thank all my Danish friends, especially the Health, Humanities and Culture Group and Uffe Juul Jensen, whose intellectual and gastronomic inspiration have contributed to my thought in many ways. Material from these papers is interwoven into chapters 1 and 2 of the present text. These chapters, and the introduction, also draw upon a paper entitled "The politics of life in the twenty-first century," given at the University of California, Irvine and at the University of California, Berkeley in 2003, and I would like to thank Inderpal Grewal and her colleagues for the invitation to Irvine, and Laurence Cohen and Aihwa Ong for the invitation to Berkeley.

Chapter 3 is developed from a paper entitled "Emergent forms of life" presented at the first Blankensee Conference, Emergent Forms of Life: Towards an Anthropology of Life Sciences, December 11–13, 2003. Thanks to Stefan Beck and Michi Necht for inviting me, and for allowing me to use the title of the conference in my argument.

Chapters 4 and 5 are based heavily on papers jointly written with Carlos Novas. I have edited the papers to make clear which aspects arise directly from his work, and to identify his specific contributions, but these chapters were jointly conceived and written, and I thank Carlos for his generosity in allowing me to present them in this form. Chapter 4 is derived from C. Novas and N. Rose, "Genetic risk and the birth of the somatic individual," *Economy and Society* Special Issue on configurations of risk, 2000, 29 (4): 484–513. Chapter 5 is derived from a paper prepared for a workshop in Prague in April 2001, under the title *Oikos* and *Anthropos*: Rationality, Technology, Infrastructure, organized by Aihwa Ong, Stephen Collier, and Paul Rabinow, and subsequently published as N. Rose and C. Novas, "Biological Citizenship," in Aihwa Ong and Stephen Collier, eds., *Global Assemblages: Technology, Politics and Ethics as Anthropological Problems* (Oxford: Blackwell, 2005), pp. 439–463.

Chapter 6 arose from an international symposium on Race in the Age of Genomic Medicine, organized by the BIOS center at the LSE in May 2003, and supported by the Wellcome Trust. Although I would not expect many of the participants to agree with my analysis, I thank those who came from many countries to this intensive two days of discussion, and made possible a debate that went beyond the rather stereotyped responses to these issues that characterize much discussion in the United States. An early version of this paper was given at the Department of African American Studies at Yale University in April 2004, and I would like to thank Paul Gilroy and Alondra Nelson for their invitation and hospitality, and for such a generous hearing of my arguments.

Chapter 7 has been given in many forms over the years: its first version was prepared for the History of the Present Research Group in 1998, and subsequently given to the Department of Social Science at the University of Loughborough in May 1999. Thanks to Mariam Fraser for inviting me, and to all those who have helped me revise the argument in the course of its various incarnations.

Chapter 8 in a much earlier version, was published as "The biology of culpability: pathological identities in a biological culture," *Theoretical Criminology*, 2000, 4 (1): 5–34: thanks to my good friend Pat O'Malley for suggesting that I write it, and to the editors and referees of that journal for comments.

I also draw upon other published and unpublished papers: "Normality and pathology in a biological age," *Outlines*, 2001, 1: 19–34 (*Outlines* is a Nordic interdisciplinary social science journal, published in English); "Biopolitics in the twenty first century—notes for a research agenda," *Distinktion*, 2001, 3: 25–44, and "The birth of the neurochemical self," prepared for the seminar series on "Social Sciences, Psychiatry, Biology," CESAMES (the Centre for Research on Mental Health and Psychotropics), University of Paris May 5, 2002; The politics of bioethics today, Biomedicalization, social conflicts and the new politics of bioethics, Vienna, October 2002; Biopower today (with Paul Rabinow); Vital Politics: Health, Medicine and Bioeconomics into the Twenty First Century, London School of Economics, September 5–7, 2003; Governing Risky Individuals in a Biological Age, Opening Keynote Address to Annual Conference of Forensic Section of Royal College of Psychiatry, Southampton, February 2004; Governing the will, presented in Munich and Oxford, February 2004, and Princeton, April 2004.

Thanks also to Mary Murrell of Princeton University Press for her enthusiasm for the idea that became this book, to Fred Appel who guided it so supportively through the review process, and to anonymous reviewers whose criticisms forced me to clarify my arguments.

As will be evident from these acknowledgements (which are incomplete—I apologize to those who I have not mentioned by name), my work is highly dependent on the research of many others and has only been possible because of the richness and productivity of the community of scholars who think, as I do, that something very significant—politically, ethically—is taking shape in the course of recent developments in the sciences and politics of life. I make no claims for originality in this book, but if I have managed to distill some concepts that, for a short time and in some small way, assist others' work, it will have served its purpose.

Nikolas Rose
London

Acronyms

ADHD — Attention Deficit Hyperactivity Disorder

AFM — Association Française contre les Myopathies

CAG — A "triplet" consisting of a sequence of three of the four nucleotides that make up DNA: **C** (cytosine), **A** (adenine), and **G** (guanine). Such triplets sometimes occur in repeats—CAGCAGCAG etc.—and some diseases such as Huntington's Disease occur when there are an unusually large number of such repeats. The so-called "genetic code" carried on molecules of DNA consists of tri-nucleotide units made up of combinations of four bases: **A** (adenine), **T** (thymine), **C** (cytosine), and **G** (guanine)

CAN — Cure Autism Now

CHADD — Children and Adults with Attention-Deficit/Hyperactivity Disorder

CNS — Central Nervous System

CSF — Cerebro Spinal Fluid

DAN! — Defeat Autism Now!

ECT — Electro Convulsive Therapy

EEG — Electroencephalogram

ELSI — Ethical, Legal and Social Implications (an acronym associated in particular with the allocation of a small propor-

	tion of the funds from the Human Genome project to the study of these implications of genetic sequencing)
FMRI	Functional Magnetic Resonance Imaging
GAD	Generalized Anxiety Disorder
GAIC	Genetics and Insurance Committee
HAP MAP	Haplotype Mapping. The International HapMap Project is a partnership of scientists and funding agencies from Canada, China, Japan, Nigeria, the United Kingdom, and the United States to develop a public resource that will help researchers find genes associated with human disease and response to pharmaceuticals using a technique known as Haplotype Mapping (see http://www.hapmap.org/)
HD	Huntington's Disease
IP	Intellectual Property
IVF	In Vitro Fertilization
LSE	London School of Economics and Political Science
MIND	The current name of the U.K.'s National Association for Mental Health
NAAR	National Alliance for Autism Research
NAMI	National Alliance on Mental Illness
NGRI	Not Guilty by Reason of Insanity
OECD	Organization for Economic Cooperation and Development
PDA	Personal Digital Assistant (a pocket or handheld computing device)
PET	Positron Emission Tomography
PGD	Pre-implantation Genetic Diagnosis
PMDD	Pre Menstrual Dysphoric Disorder
PXE	Pseudoxanthoma Elasticum, an inherited disorder that affects connective tissue in some parts of the body
RCT	Randomized Controlled Trial
SANE	Schizophrenia, A National Emergency (a U.K.-based charity working in the area of mental health)

SARS	Severe Acute Respiratory Syndrome
SNP	Single Nucleotide Polymorphism
SNRI	Serotonin-Norepinephrine Reuptake Inhibitor
SPECT	Single Photon Emission Computed Tomography
SSRI	Selective Serotonin Reuptake Inhibitor (sometimes expanded as Serotonin Selective Reuptake Inhibitor)
TMAP	Texas Medication Algorithm Project
USAID	United States Agency for International Development
WHO	World Health Organization

The Politics of Life Itself

Introduction

At the close of the twentieth century, many predicted that "we" were entering a "biotech century," an age of marvelous yet troubling new medical possibilities.[1] Some believed that the sequencing of the human genome would inaugurate an age of genetic manipulation with marvelous, perhaps terrifying consequences. Linking genomics with developments in reproductive technology, such as pre-implantation genetic diagnosis and cloning, they imagined a world of engineered people, with qualities and capacities fabricated on demand. Others believed that a new generation of psychopharmaceuticals would soon enable us to design our moods, emotions, desires and intelligence at will. Still others dreamed of the conquest of mortality, and a world in which humans had extended their lifespan indefinitely. Many of the biomedical techniques that were cited were already familiar: genetic screening, reproductive technologies, organ transplants, genetic modification of organisms, and the new generation of psychiatric drugs whose usual exemplar is Prozac. Others were said to be "just around the corner": genetic engineering, xenotransplantation, personalized medicine tailored to each individual's genotype as coded on a tiny chip, and the fabrication or regeneration of organs in vitro or using stem cells that could be differentiated into any kind of tissue.

These prospects have generated hopes and fears, expectation and trepidation, celebration and condemnation. While some invest great hope in the prospects of novel and effective cures for all sorts of diseases and afflictions, others warn of the dangers of treating human life as infinitely malleable, especially where the creation and use of human embryos in fertility treatment or research is concerned. Many politicians, universities, corporations, and private investors hope these biomedical advances will generate valuable intellectual property and drive a new and highly lucrative bioeconomy, but others believe that basic science is being suborned in the service of profit and that less glamorous factors affecting the health and illness of the majority are neglected in the search for therapies for the few that will advance careers and generate profits. Pharmaceutical companies have been singled out for particular criticism, accused of selling many new drugs at inflated prices and with false promises, ignoring potentially dangerous side effects, and medicalizing nondisease conditions such as baldness or lack of libido to create new markets in the ruthless pursuit of shareholder value. In many countries, biomedical developments that involve genetics have been particularly controversial, raising the specter of genetic discrimination and eugenics, notably where embryo selection is contemplated to avert hereditary conditions, but also in research that might identify the genetic bases of diseases, and even in pharmacogenomics that seeks the genetic variations giving rise to individual differences in pharmaceutical responses.

Politicians, regulators, theologians, philosophers, and others have been entangled in these debates. Governments have enacted laws to limit some of these developments, especially those relating to genetic selection in human reproduction. Many have set up committees and commissions to address the seemingly inescapable demand such possibilities seem to have engendered—that "a line must be drawn" between the permitted, the regulated, and the forbidden. Certain pressure groups campaign for restrictions to be overthrown to allow the research that might bring hope to their loved ones. Others campaign for restrictions to be tightened, in particular to protect the "sanctity of life" of the ovum from fertilization or even before. Some hope to settle these debates by appeal to a transcendental religious morality or an equally transcendental human ontology. For others, the issues are social, consequential, and situational—What kinds of societies do we want? What kinds of consequences might these developments have? Who should have the power to make decisions in each of the troubling situations where a choice has to be made about the selection of an embryo, the conduct of an experiment, the licensing of a drug, the termination of a life? A whole profession of bioethics—and a developing field of "neuroethics"—has sprung into existence to arbitrate these issues. Some suggest that we are entering a posthuman future, a prospect greeted by some "transhumanists" with rather poignant yearning, and by other with distress and consternation. Many intellectuals have been drawn into this debate: Francis Fukuyama, Leon Kass, and Jürgen Habermas are but the best known of those who have sought to establish normative limits, arguing that, however we might regard them in relation to nonhuman living organisms, such interventions on human beings are violations of our human nature—for them, human dignity, identity, and perhaps the fate of humanism itself depend upon the inviolability of human nature itself. We tamper with our "nature" at enormous risk, a risk, ultimately, to the human soul (Fukuyama 2002, Habermas 2003, Kass 2002, President's Council on Bioethics (U.S.) and Kass 2003).

This book is neither a set of speculations about the future nor a bioethical meditation on the present. Indeed such speculations and meditations are part of what I try to analyze. They themselves—their visions of the future, their fears and hopes, their evaluations and judgments—are elements in an emergent form of life.[2] The politics of this form of life, this "vital politics," is the focus of this book. Of course, politics has long been concerned with the vital lives of those who are governed. At the risk of simplification, one could say that the vital politics of the eighteenth and nineteenth centuries was a politics of health—of rates of birth and death, of diseases and epidemics, of the policing of water, sewage, foodstuffs, graveyards, and of the vitality of those agglomerated in towns and cities. Across the first half of the twentieth century this concern with the health of the population and its quality became infused with a particular understanding of the inheritance of a biological constitution and the consequences of differential reproduction of different subpopulations; this seemed to oblige politicians in so many countries to try to manage the quality of the population, often coercively and sometimes murderously, in the name of the future of the race. But the vital politics of our own century looks rather different. It is neither delimited by the poles of illness and health, nor focused on eliminating pathology to protect the destiny of the nation. Rather, it is concerned with our growing capacities to control, manage, engineer, reshape, and modulate the very vital capacities of human beings as living creatures. It is, I suggest, a politics of "life itself."[3]

While many of the themes in this contemporary politics of life are familiar, others are novel. Some of this novelty lies in more general shifts in rationalities and technologies of government, notably the transformations in the provision of security, welfare, and health associated with challenges to the social state in Europe and Australasia, and the rise of new "advanced liberal" governmental technologies (Barry et al. 1996, Rose 1989, 1996a, Rose and Miller 1992). In particular, these have involved a reorganization of the powers of the state, with the devolution of many responsibilities for the management of human health and reproduction that, across the twentieth century, had been the responsibility of the formal apparatus of government: devolving these to quasi-autonomous regulatory bodies—bioethics commissions, for example; to private corporations—like private fertility clinics and biotechnology companies selling products such as genetic tests directly to consumers; and to professional groups—such as medical associations—regulated "at a distance" by the powerful mechanisms of audits, standards, benchmarks, and budgets. These modifications in rationalities and technologies of government have also involved an increasing emphasis on the responsibility of individuals to manage their own affairs, to secure their own security with a prudential eye on the future. Nowhere have these been more telling than in the field of health, where patients are increasingly urged to become active and responsible consumers of medical services and products ranging from pharmaceuticals to reproductive technologies and genetic tests (Rose 1992, 1999). This complex of marketization, autonomization, and responsibilization gives a particular character to the contemporary politics of life in advanced liberal democracies.

Over and above these shifts, perhaps, the novelty of contemporary biopolitics arises from the perception that we have experienced a "step-change," a qualitative increase in our capacities to engineer our vitality, our development, our metabolism, our organs, and our brains. This step-change entails a change in scale. The biomedical knowledges and techniques that are currently taking shape have many differences, but they do have one thing in common. It is now at the molecular level that human life is understood, at the molecular level that its processes can be anatomized, and at the molecular level that life can now be engineered. At this level, it seems, there is nothing mystical or incomprehensible about our vitality—anything and everything appears, in principle, to be intelligible, and hence to be open to calculated interventions in the service of our desires about the kinds of people we want ourselves and our children to be. Hence the contestations that are taking place around each of these issues, from stem cells to smart drugs, are themselves shaped, in part, by the opportunities and threats that such a molecular vision of life seems to open up. As human beings come to experience themselves in new ways as biological creatures, as biological selves, their vital existence becomes a focus of government, the target of novel forms of authority and expertise, a highly cathected field for knowledge, an expanding territory for bioeconomic exploitation, an organizing principle of ethics, and the stake in a molecular vital politics.

A Cartography of the Present

To analyze the present, and the potential futures it may prefigure, is always a risky exercise. In analyzing contemporary vital politics, I do not think we can proceed simply by applying the now familiar tropes of genealogy and "histories of the present." Such genealogies seek to destabilize a present that has forgotten its contingency, a moment that, thinking itself timeless, has forgotten the time-bound questions that gave rise to its beliefs and practices. In making these contingencies thinkable, in tracing the heterogeneous pathways that led to the apparent solidity of the present, in historicizing those aspects of our lives that appear to be outside history, in showing the role of thought in making up our present, such genealogies sought to make that present open to reshaping. But today, to destabilize our present does not seem such a radical move. Popular science, media representations, pundits, and futurologists all portray our own moment in history as one of maximal turbulence, on the cusp of an epochal change, on a verge between the security of a past now fading and the insecurity of a future we can only dimly discern. In the face of this view of our present as a moment when all is in flux, it seems to me that we need to emphasize continuities as much as change, and to attempt a more modest cartography of our present. Such a cartography would not so much seek to destabilize the present by pointing to its contingency, but to destabilize the future by recognizing its openness. That is to say, in demonstrating that no single future is written in our present, it might fortify our abilities, in part through thought itself, to intervene in that present, and so to shape something of the future that we might inhabit.

To undertake such a cartography of the present, a map showing the range of paths not yet taken that may lead to different potential futures, it is important to recognize that we do not stand at some unprecedented moment in the unfolding of a single history. Rather, we live in the middle of multiple histories. As with our own present, our future will emerge from the intersection of a number of contingent pathways that, as they intertwine, might create something new. This, I suspect, will be no radical transformation, no shift into a world "after nature" or a "posthuman future." Perhaps it will not even constitute an "event." But I think, in all manner of small ways, most of which will soon be routinized and taken for granted, things will not be quite the same again. This book, then, is a preliminary cartography of an emergent form of life, and a draft of a history of the potential futures it embodies.

Mutations

The space of contemporary biopolitics has not been formed by any single event. The reshaping of medical and political perception and practice has come about through interconnections among changes along a number of dimensions. Without claiming exhaustivity, I delineate five pathways where I think significant mutations are occurring.

First, *molecularization*: The "style of thought" of contemporary biomedicine envisages life at the molecular level, as a set of intelligible vital mechanisms among molecular entities that can be identified, isolated, manipulated, mobilized, recombined, in new practices of intervention, which are no longer constrained by the apparent normativity of a natural vital order.

Second, *optimization.* Contemporary technologies of life are no longer constrained, if they ever were, by the poles of health and illness. These poles remain, but in addition, many interventions seek to act in the present in order to secure the best possible future for those who are their subjects. Hence, of course, these technologies embody disputed visions of what, in individual and or collective human life, may indeed be an optimal state.

Third, *subjectification.* We are seeing the emergence of new ideas of what human beings are, what they should do, and what they can hope for. Novel conceptions of "biological citizenship" have taken shape that recode the duties, rights, and expectations of human beings in relation to their sickness, and also to their life itself, reorganize the relations between individuals and their biomedical authorities, and reshape the ways in which human beings relate to themselves as "somatic individuals." This is linked to the rise of what I term a "somatic ethics"—ethics not in the sense of moral principles, but rather as the values for the conduct of a life—that accords a central place to corporeal, bodily existence.

Fourth, *somatic expertise.* These developments are giving rise to new ways of governing human conduct, and the rise of multiple subprofessions that claim expertise and exercise their diverse powers in the management of particular aspects of our somatic existence—geneticists specializing in particular classes of disorder working in alliance with groups of patients and families, specialists in reproductive medicine with their public or private clinics and devoted clientele, stem cell therapists whose work becomes known across the world via the Internet and who become the focus of pilgrimages of hope for cures for everything from spinal cord injuries to Alzheimer's disease. Around these experts of the soma cluster a whole variety of new pastoral experts—genetic counselors are perhaps the best exemplars—whose role is to advise and guide, to care and support, individuals and families as they negotiate their way through the personal, medical, and ethical dilemmas that they face. And, perhaps most remarkable has been the rise of a novel expertise of "bioethics" claiming the capacity to evaluate and adjudicate on these activities, which has been enrolled in the government and legitimation of biomedical practices from bench to clinic and marketplace.

Fifth, *economies of vitality.* Energized by the search for biovalue, novel links have formed between truth and capitalization, the demands for shareholder value and the human value invested in the hope for cure and optimality. A new economic space has been delineated—the bioeconomy—and a new form of capital—biocapital. Old actors such as pharmaceutical corporations have been transformed in their relation with science on the one hand and stock markets on the other. New actors such as biotech start-ups and spin-outs have taken shape, often seeking to stress their corporate social responsibility and combining in various ways with the forms of citizenship and expertise. Life itself has been made amenable to these new economic relations, as vitality is decomposed into a series of distinct and discrete objects—that can be isolated, delimited, stored, accumulated, mobilized, and exchanged, accorded a discrete value, traded across time, space, species, contexts, enterprises—in the service of many distinct objectives. In the process, a novel geopolitical field has taken shape, and biopolitics has become inextricably intertwined with bioeconomics.

I am wary of epochal claims, and it is necessary to recognize that none of these mutations marks a fundamental break with the past: each exhibits continuity alongside change. Yet, I suggest, from the point of view of the present, a threshold has been crossed. Something is emerging in the configuration formed by the intertwining of these five lines of mutation, and this "something" is of importance for those, like myself, who try to write the history of possible futures. This is why I suggest that we are inhabiting an emergent form of life.

■ ■ ■

In chapter 1 I examine these five mutations in more detail, describe their key characteristics, and set out my own view of their significance. In chapter 2 I focus on the ways in which these mutations are bound up with changing conceptions of life and changing forms of politics, and argue that, in the interrelations between these changes, in which neither politics nor life means quite the same as before, a new politics of life has taken shape. In chapter 3 I focus in particular on the implications of the shift away from biological and genetic determinism and develop my claim that the new world of vital risk and vital susceptibilities, demanding action in the vital present in the name of vital futures to come, is generating an emergent form of life. Each of the subsequent chapters explores one particular facet of the biopolitics of this emergent form of life in depth. In chapter 4 I focus on changing ideas of genetic risk and genetic prudence, describe the entanglements of genomic knowledge and expertise with particular regimes of the self, and examine the emergence of novel forms of genetic responsibility. In chapter 5 I develop these arguments in relation to changes in biological citizenship and examine some of the forms such biological citizenship currently takes. In chapter 6 I consider the implications of the mutations I have identified for transformations in the idea of race and ethnicity in the face of genomic medicine. In chapter 7 I examine the rise of new neurochemical conceptions of the self and its pathologies, and the associated rise on novel technologies of the neurochemical self. In chapter 8 I describe the implications of these new developments in molecular biology, neuroscience, behavioral genomics, and psychopharmacology for crime control and the criminal justice system.

■ ■ ■

The argument advanced through these chapters does not share the pessimism of most sociological critics, who suggest that we are seeing the rise of a new biological and genetic determinism. Instead I argue that we are seeing the emergence of a novel somatic ethics, imposing obligations yet imbued with hope, oriented to the future yet demanding action in the present. On the one hand, our vitality has been opened up as never before for economic exploitation and the extraction of biovalue, in a new bioeconomics that alters our very conception of ourselves in the same moment that it enables us to intervene upon ourselves in new ways. On the other hand, our somatic, corporeal neurochemical individuality has become opened up to choice, prudence, and responsibility, to experimentation, to contestation, and so to a politics of life itself. I thus conclude the book with a brief afterword, which turns directly to the question of ethics, drawing a distinction between the ethical ruminations of bioethicists and neuroethicists and a different sense of ethics, one that is embodied in the judgments individuals make of their actual and potential choices, decisions, and actions as they negotiate their way through the practices of contemporary biomedicine. I suggest that the apparatus of bioethics has achieved the salience that it has, in contemporary biopolitics, because of the problems of governing biomedicine in an age of choice and self maximization in which the body and its capacities have become central to technologies of selfhood. As Max Weber found an elective affinity between the Protestant ethic and the spirit of early capitalism, generating the forms of life that made foresight, prudence, calculation, and accumulation not just legitimate but potential indicators of salvation (Weber 1930), so there is an elective affinity between contemporary somatic ethics and the spirit of biocapitalism. Somatic ethics, that is to say, accords a particular moral virtue to the search for profit through the management of life. Yet, at the same time, it opens those who are seen to damage health in the name of profit to the most moralistic of condemnations. As biopolitics becomes entangled with bioeconomics, as biocapital becomes open to ethical evaluation, and as ethopolitics becomes central to our way of life, new spaces are emerging for the politics of life in the twenty-first century.

Chapter 1
Biopolitics in the Twenty-First Century

"Life is our life's work."
—Mission statement of the
pharmaceutical company, Pfizer Inc.

How might we approach the analysis of the biopolitics of the twenty-first century? I suggest that we might usefully approach this question along five lines, where significant mutations can be identified: molecularization, optimization, subjectification, expertise, bioeconomics. These will be explored in more detail in the chapters that follow. In this chapter I will introduce each of these themes and say a little about its significance. But first, a few words are in order about medicine itself.

Medicine—Then and Now

Michel Foucault's *Birth of the Clinic* (1973) remains a path-breaking analysis of the ways in which illness and medicine came to be spatialized upon the individual body. The book teaches a methodological lesson: the epistemological, ontological, and technical reshaping of medical perception at the start of the nineteenth century came about through the interconnections of changes along a series of dimensions, some of which seem, at first sight, rather distant from medicine. They include changes in the laws and practices of assistance, shifts in the organization of medical professions and medical pedagogy, new forms of record-keeping in hospitals allowing the production of new types of statistics of morbidity and mortality, pathological anatomy and post mortem dissection of those who died in hospital, and so forth. The mutation Foucault charted still defines a key dimension of our sense of health and illness: "the body itself" remains the focus of the clinical gaze. This is true even when that disease is problematized and addressed in terms of spatial and social associations, as in the early twenty-first century concerns about Severe Acute Respiratory Syndrome (SARS) and Avian influenza. Even when disease is situated in a field of attitudes, habits, and behaviors, as in attention to dangerous sexual or dietary practices, it is the body itself that becomes ill. Nonetheless, Foucault's book, first published in 1963, was written at the end of the "golden age" of clinical medicine. While the 1960s did not mark the "death of the clinic,"[1] the medical assemblage that took shape in the last quarter of the twentieth century was already very different from the clinical medicine born in the early nineteenth century.

The dynamics of these medical changes involved cumulative modifications along multiple dimensions over at least half a century. Many have described this new medical territory (e.g. Armstrong 1983, 1995, Arney and Bergen 1984, Clarke et al. 2003, Starr 1982).[2] Medical jurisdiction extended beyond accidents, illness, and disease, to the management of chronic illness and death, the administration of reproduction, the assessment and government of "risk," and the maintenance and optimization of the healthy body. The maintenance of the healthy body became central to the self management of many individuals and families, employing practices ranging from dietetics and exercise, through the consumption of proprietary medicines and health supplements, to self diagnosis and treatment.[3] Yet at the very same time as the scope of medical authority was extended in this way, this extension was challenged from many sides. Critics argued that we were experiencing a medicalization of social problems, witnessing aggressive medical imperialism based on unrealistic claims about the therapeutic powers of doctors, and that medics were intruding into moral and political matters that were properly not their concern. Simultaneously, social movements from feminism to disability rights advocates challenged the paternalistic power that doctors exercised over their patients and their lives. And, at the same time, there were attempts to "empower" the recipients of medical care, in ways that varied from country to country but included an increasing emphasis on "active citizenship," the rise of cultures of litigation and compensation, the transformation of patients into "consumers," and the growing availability of medical information on the Internet from a multiplicity of sources, which patients can access in order to shape their demands of their medics and to evaluate or challenge their opinions.

"Medicine" itself has also been transformed. It has become technomedicine, highly dependent on sophisticated diagnostic and therapeutic equipment.[4] It has been fractured by a complex division of labor among specialists. Doctors have lost the monopoly of the diagnostic gaze and of the therapeutic calculation: the clinical judgment of the practicing physician is hemmed in and constrained by the demands of evidence-based medicine and the requirements for the use of standardized, corporately framed diagnostic and prescribing procedures. The practice of medicine in most advanced industrialized countries has been colonized by, and reshaped by, the requirements of public or private insurance, their criteria for reimbursement, and in general their treatment of health and illness as merely another field for calculations of corporate profitability. In another sense, perhaps even more fundamental, medicine been reshaped by its intense capitalization. Basic and applied biological research—whether conducted in biotech companies or in universities—has become bound up with the generation of intellectual property, and illness and health have become major fields for corporate activity and the generation of shareholder value. In these processes human vitality has been opened up, at the molecular level, for technical innovation, economic exploitation, and for highly competitive forms of bioeconomics. This technologization and capitalization of medicine gives a particular form to the contested field of vital politics in the twenty first century. And this field is itself being reconfigured by a profound "molecularization" of styles of biomedical thought, judgment, and intervention.

Molecular Biopolitics

At one level, no doubt, most people—even those living within the remit of advanced technological biomedicine—still imagine their bodies at the "molar" level, at the scale of limbs, organs, tissues, flows of blood, hormones, and so forth.[5] This is the visible, tangible body, as pictured in the cinema or on the TV screen, in advertisements for health and beauty products, and the like. It is this molar body that we act upon and seek to perfect through diet, exercise, tattooing, and cosmetic surgery. And, indeed, this was the body—the body as a systemic whole—that was the focus of clinical medicine, as it took shape over the nineteenth century, revealed to the gaze of the physician after death in the post mortem dissection, visualized in the anatomical atlas, accessed in life through any number of devices, starting with the stethoscope, that would augment the clinical gaze and allow it to peer into the organs and systems of the living body.[6] Today, however, biomedicine visualizes life at another level—the molecular level. The clinical gaze has been supplemented, if not supplanted, by this molecular gaze, which is itself enmeshed in a "molecular" style of thought about life itself. As even a cursory reading of contemporary biomedical research shows, life is now understood, and acted upon, at the molecular level, in terms of the functional properties of coding sequences of nucleotide bases and their variations, the molecular mechanisms that regulate expression and transcription, the link between the functional properties of proteins and their molecular topography, the formation of particular intracellular elements—ion channels, enzyme activities, transporter genes, membrane potentials—with their particular mechanical and biological properties.

Ludwik Fleck's notion of a "style of thought" helps us understand what has happened here (Fleck 1979, Hacking 1992a, Rose 2000a). A style of thought is a particular way of thinking, seeing, and practicing. It involves formulating statements that are only possible and intelligible within that way of thinking. Elements—terms, concepts, assertions, references, relations—are organized into configurations of a certain form that count as arguments and explanations. Phenomena are classified and sorted according to criteria of significance. Certain things are designated as evidence and gathered and used in certain ways. Subjects are chosen and recruited. Model systems are imagined and assembled. Machines are invented and later commodified to make measures and inscriptions such as graphs, charts, and tables. All this is linked up within complex practical arrangements such as experiments and clinical trials. A style of thought also involves membership of a "thought community" in a discipline or subdisicpline, and an intimate knowledge of its relations of power and status. And, of course, a style of thought in an area of science also embodies a way of identifying difficulties, questioning arguments, identifying explanatory failures—a mode of criticism, of error seeking and error correction.

A style of thought is not just about a certain form of explanation, about what *it is* to explain, it is also about what *there is* to explain. That is to say, it shapes and establishes the very object of explanation, the set of problems, issues, phenomena that an explanation is attempting to account for. The brain, for the contemporary sciences of the brain, is not what it was in the 1950s; the cell, in cellular biology, is not what it was in the 1960s; "the gene"—if it still makes sense to call it that—is not what it was before genomes were sequenced, and so on. The new style of thought that has taken shape in the life sciences has so modified each of its objects that they appear in a new way, with new properties, and new relations and distinctions with other objects.

A style of thought is not merely a new discourse. The molecular knowledge of life that has taken shape since the 1960s has been linked to all sorts of highly sophisticated techniques of experimentation that have intervened upon life at this molecular level—not after the event but in the very process of discovery itself—as, for example, in the techniques of gene cutting and splicing, the polymerase chain reaction for creating multiple copies of precise segments of DNA outside living systems, the customized fabrication of DNA sequences to order, the manufacture of organisms with or without specific gene sequences. The laboratory has become a kind of factory for the creation of new forms of molecular life. And in doing so, it is fabricating a new way of understanding life itself.

Of course, many diagnoses and treatments of patients remain molar, conducted in terms of pathologies of organs or systems. But the mutation brought about by the molecularization of vitality is significant nonetheless. When a new infectious disease is encountered, for example, the immediate response is to seek the molecular structure of the causative agent. Thus, in the 2003 outbreak of SARS, the virus involved was genotyped within weeks of the first World Health Organization (WHO) alarm about the spread of the condition—even though the health strategy enacted was as molar as it could be, taking the form of quarantine, travel restrictions, and the policing of space that has been familiar since the medicine of the plague. In the pharmaceutical industry and therapeutic research more generally, it is at the molecular level that therapeutic agents are selected, manipulated, trialed, and developed, and in molecular terms that their modes of action are explained. A whole variety of healing practices from herbal cures to psychoanalysis seek a novel molecular legitimation for their apparently mysterious modes of action. The pharmaceutical industry explores traditional healing practices in the quest for molecular knowledge that can be extracted, developed, patented, and commodified. At a slightly slower pace, an expanding research program seeks the molecular bases of current clinical diagnoses, and is beginning to reshape those diagnoses on this basis. On the one hand, many phenotypically distinct conditions now appear to be related at the molecular level—the level of biochemical mechanisms and of genetic variations. For example, the discovery of the genetic basis of PXE (Pseudoxanthoma Elasticum) on chromosome 16, an inherited disorder in which elastic tissue in the body becomes mineralized, may have implications for the molecular and genetic basis of hypertension and cardiovascular disease—since the mineralization of the midsize arteries in PXE mimics the general ageing of the arteries—as well as macular degeneration.[7] On the other hand, disorders previously classed together, such as monopolar depression, are beginning to be fragmented into subgroups, in part as a result of investigations of the molecular bases of the variation in responses of patients diagnosed with depression to the new generation of molecularly crafted antidepressants that claim selecting to target specific sites in neurotransmission believed to be involved in different forms of depression.

Visualization techniques were crucial here (Cartwright 1995a, 1995b). In part it is by means of new technologies of visualization that life has been made amenable to thought at the molecular level, as a set of intelligible vital mechanisms among molecular entities (Rose 2001). By the start of the twenty-first century, in addition to the X Rays and medical films developed during the first half of the twentieth century, there were a multitude of screening devices that rendered that interior organic body visible: mammograms, ultrasound, fetal images and, for the brain, EEG traces, PET, SPECT, fMRI scans, and many more (Kevles 1997). Such visualization techniques increasingly operate through digital simulation. Some reconstruct an apparent mimetic realism at the molecular level though the use of algorithms that manipulate digital information—this is the case in fMRI scanning (Beaulieu 2000). Others visualize life in terms of manipulable strings of information, as in the DNA sequences we still, just about, refer to as genes (Kay 2000, provides numerous illustrations of different ways in which "genes" have been visually represented; see also Keller 2000). But visualization alone was not enough. Molecular genomics has depended upon the invention of a whole range of technologies for decomposing, anatomizing, manipulating, amplifying, and reproducing vitality at this molecular level—from DNA binding dyes that made large chromosomal structures visible under the microscope, through restriction enzymes that cut DNA at specific base sequences, electrophoresis on gels to separate DNA fragments by length, radioactive markers that bind to particular base sequences, the construction of "clones" or fragments of DNA that can be copied in large numbers and collected in clone libraries in order to provide enough DNA for accurate analysis, and the polymerizing chain reaction for producing large amounts of short stretches of DNA.[8] Combined together, in extraordinary subtle ways, these techniques opened "the gene" to knowledge and technique at the molecular level.[9]

Once vitality is anatomized at this level, intervention is no longer constrained by the normativity of a given vital order. No doubt the fragmentation of the body into transferable tissues which could, often with difficulty, be freed from their marks of origin and re-utilized in other bodies, began with blood and blood products.[10] Organs began to be mobilized, initially with difficulty, later to become potent and controversial objects of commodification.[11] The elements of reproduction—eggs, sperm, and later embryos—also became separable from any particular body, mobilized around circuits of laboratories, clinics, and other bodies. But now tissues, cells, and DNA fragments can be rendered visible, isolated, decomposed, stabilized, stored in "biobanks," commoditized, transported between laboratories and factories re-engineered by molecular manipulation, their properties transformed, their ties to a particular individual living organism, type, or species suppressed or removed.[12] Molecularization strips tissues, proteins, molecules, and drugs of their specific affinities—to a disease, to an organ, to an individual, to a species—and enables them to be regarded, in many respects, as manipulable and transferable elements or units, which can be delocalized—moved from place to place, from organism to organism, from disease to disease, from person to person. Whether it is the transfer of genes along with their properties—luminescence, salt tolerance—from one species to another, the transfer of treatments from one disease to another, or the transfer of tissues, blood plasma, kidneys, stem cells, molecularization is conferring a new mobility on the elements of life, enabling them to enter new circuits—organic, interpersonal, geographical, and financial. Mobilization of vitality is itself not new, nor is combination, one needs only to think of the very long history of plant collections and breeding. And molecularization is not sufficient on its own; as we shall see, many other factors must be added—notably standardization, regulation, and even ethics—to make up circuits of vitality. But what is crucial, for present purposes, is that "molecular biopolitics" now concerns all the ways in which such molecular elements of life may be mobilized, controlled, and accorded properties and combined into processes that previously did not exist. At this molecular level, that is to say, life itself has become open to politics (see also the discussion in Franklin 2000).

Technologies of Optimization

Perhaps it is not too much to suggest that an epistemological change is occurring. The "biology" that came into existence in the nineteenth century was a biology of "depth." It tried to discover the underlying organic laws that lay behind and determined the functioning of closed living systems. But contemporary biology operates, at least in part, in a "flattened" field of open circuits. This seems counterintuitive, I know—surely to refer to "genes" is to attribute the real basis of our human nature to phenomena at the deepest of all depths? And yet, I think that the truth discourse of contemporary genomics no longer sees genes as the hidden entities that determine us.

Consider, for example, "systems biology." Making use of the information generated by the Human Genome Program, by emerging proteomics, and by the advanced informatics and computational models that have been developed, biologists, engineers, mathematicians, and experts in computation work together to model the interactions between basic components of biological systems such as DNA sequences and proteins. They generate mathematical models of their interactions, thus identifying functional properties that emerge from the connections and interactions among the components of that system at the level of the organism or ecosystem, and generating predictions as to future states. Systems biology entails reverse-engineering genetic and metabolic data, developing computer simulations of gene regulation and metabolism, generating predictions and hypotheses within these computational simulations, comparing these with data from experiments in animal model systems, and so forth. Synthetic biology, which actually seeks to engineer organisms from such models is but a variant here, although a rather striking one. In the contemporary molecular biology that systems biology exemplifies, the search is not for simplifying underlying laws but precisely the reverse: for simulations of dynamic, complex, open systems, combining heterogeneous elements, to predict future vital states and hence to enable intervention into those vital systems to reshape those futures. And, as I have already indicated, in the interventions that proliferate in this flattened world, almost any vital element can, in principle, be freed from its ties to cell, organ, organism, or species, set free to circulate and to be combined with any other, provided certain conditions are met. An epistemological change, then, and perhaps also an ontological change is in process.

Indifferent, perhaps, to this epistemological and ontological radicalism, contemporary biomedicine is enthusiastically engaged with the biological re-engineering of vitality. Sarah Franklin draws upon the phrase used by Ian Wilmut, one of the creators of Dolly the sheep, to characterize this engagement: we have entered the age of "biological control." "This means that we can no longer assume that the biological 'itself' will impose limits on human ambitions. As a result, humans must accept much greater responsibility toward the realm of the biological, which has, in a sense, become a wholly contingent condition" (Franklin 2003: 100). Contemporary medical technologies do not seek merely to cure diseases once they have manifested themselves, but to control the vital processes of the body and mind. They are, I suggest, technologies of optimization.

What is a technology? Conventionally, we often think of technologies as equipment or techniques: diagnostic technologies such as brain imaging or genetic testing using fast throughput sequencers; therapeutic technologies such as new methods of drug delivery; surgical technologies such as those for the replacement of organs, joints, or the reconstruction of bone degeneration, and so forth. But, for me, a technology is more than this. It is an assemblage of social and human relations within which equipment and techniques are only one element: "Technology, here, refers to any assembly structured by a practical rationality governed by a more or less conscious goal . . . hybrid assemblages of knowledges, instruments, persons, systems of judgment, buildings and spaces, underpinned at the programmatic level by certain presuppositions and assumptions about human beings" (cf. Rose 1996b: 26, Brown and Webster 2004). Thus, as many have argued, new reproductive technologies entail much more than the craft skills of doctors using new instruments and techniques. They engender certain ways of thinking about reproduction, for the subject and for the expert, certain routines and rituals, techniques of testing and practices of visualization, modes of advice giving and the like (Franklin 1997, Rapp 1999, Strathern 1992). Organ transplantation is not merely a triumph of surgical techniques but requires new sets of social relations bringing together donors and recipients across time and space, entailing and generating new ideas about end of life, new senses of ownership of the body and rights to a cure, as well as the complex financial and institutional relations that make the procedure possible (Lock 2002, Scheper-Hughes 2000, 2003a, 2003b). These new biotechnologies, then, must be understood as hybrid assemblages oriented toward the goal of optimization.

These are not merely medical technologies or technologies of health, they are technologies of life. Previously it seemed that life inhered in the inescapable natural workings of the vital processes themselves. All medicine was able to hope for was to arrest the abnormality, to re-establish the natural vital norm and the normativity of the body that sustained it. But these norms no longer seem so inescapable, these normativities appear open to alteration. Once one has witnessed the effects of psychiatric drugs in reconfiguring the thresholds, norms, volatilities of the affects, of cognition, of the will, it is difficult to imagine a self that is not open to modification in this way. Once one has seen the norms of female reproduction reshaped by assisted conception, the nature and limits of procreation and the space of hopes and fears around it are irrevocably changed. Once one has seen the norms of female aging reshaped by hormone replacement therapy, or the norms of aging male sexuality reshaped by Viagra, the "normal" process of growing old seems only one possibility in a field of choices, at least for those in the wealthy West. As Hannah Landecker has put it, "the usual formula, 'biotechnology changes what it is to be human,' should have an interim step included in it in order to understand this process of change in any detail: 'biotechnology changes what it is to be biological.' "[13] The old lines between treatment, correction, and enhancement can no longer be sustained. The ways in which they are to be redrawn shapes the new territory of molecular biopolitics.

These new technologies, then, do not just seek to cure organic damage or disease, nor to enhance health, as in dietary and fitness regimens, but change what it is to be a biological organism, by making it possible to refigure—or to hope to refigure—vital processes themselves in order to maximize their functioning and enhance their outcomes. Their key feature is their forward vision: these technologies of life seek to reshape the vital future by action in the vital present. Two dimensions are of particular interest to me: those of susceptibility and enhancement.

Susceptibility indexes the problems raised by attempts to identify and treat persons in the present in relation to ills that they are predicted to suffer in the future. In the early days of the Human Genome Project, when the term "geneticization" came into fashion among critics, it was often thought that the sequencing of the human genome would establish such a single "normal" sequence—a composite or "consensus genome." It was suggested that this sequence would serve as a norm of health against which all discrepancies would be judged as morbid abnormalities. Many predicted a new form of molecular surveillance that would categorize individuals as healthy or pathological on the basis of the sequences of bases on their genome, and would divide them up and administer their lives in the light of this implacable biological truth (see, for example, Flower and Heath 1993). But the sequencing of the human genome did not produce a single "normal" sequence. Not only were there far fewer sequences coding for protein than had been anticipated but there were millions of loci on the genome where individuals differed from one another by as little as a single base in the chains of As, Cs, Gs, and Ts that make up "the genetic code" (an A is substituted by a C, for example).[14] Every sequence identified as a "gene" now appeared marked by such Single Nucleotide Polymorphisms, or SNPs. While there are some relatively rare "single gene disorders" and others that are related to such anomalies as extended repeats of particular sequences of bases—as in the extended CAG repeats that underlie the development of Huntington's Disease—for common complex disorders such as heart disease, diabetes, and cancer, it is at the SNP level, rather than "the gene for" a disorder, that the search is on for the genomic variations that might increase susceptibility to disease. Hence the focus is on the development of genetic tests, for children and adults, for embryos, and even for unfertilized ova, that will identify genomic variations at the SNP level, often in combinations at multiple loci on many chromosomes, that are thought to increase the probability of a disease developing. And, once identified, it is hoped that remedial action will be possible, with options ranging from selective implantation of embryos though gene therapy and preventive drug therapy, to lifestyle changes.

In one sense, the contemporary focus on susceptibility is merely an extension of two other modes of thought that have a long history—that of predisposition and that of risk. A predisposition, since at least the eighteenth century, was an inherited taint or flaw that would, in the right (or wrong) circumstances, manifest itself in illness or pathology. The idea of inherited predispositions came to the fore, of course, in the second half of the nineteenth century, as all manner of problems of social pathology and danger became understood in terms of degeneracy (Chamberlin and Gilman 1985, Pick 1989, Rose 1985). For some, degeneracy was a result of the damaging effect of urban existence on town dwellers, of the migration from the country into the towns, the weakening of the constitution of the immigrants, the weakly offspring they gave birth to, and the further deterioration of those offspring as they were afflicted by all manner of pathologies from tuberculosis to prostitution and frank insanity: this was spiraling the constitution of the population downwards generation by generation. For others, this process was exacerbated by the promiscuous breeding of degenerates, aided by misguided charity, resulting surely in the deterioration of the national stock. Taken up in the eugenics movement, the idea of degeneracy would be central to the biopolitics of the first half of the twentieth century.

Contemporary concerns with genetic susceptibility thus rework much older beliefs that weaknesses were inherited as predispositions—that might lurk unseen until triggered by external events ranging from excessive drinking to accidents or age—and might be averted by adopting a careful and moderate way of life. They also rework some well-established technologies of risk assessment, risk prediction, and risk management. Emerging out of epidemiological research on the prevalence of disorders and diseases among different sectors of the population—differentiated by age, gender, race, family history, weight, diet, use of alcohol and cigarettes, and the like—the use of risk scales to assess the likelihood of an individual developing a disorder is now familiar. While such allocation to a risk category, however, is usually probabilistic and factorial—that is to say, not arising out of an identification of a clear etiological pathway to the disease—the dream of the contemporary diagnostics of susceptibility is molecular precision, based in the identification of the precise genomic variations whose products—a low activity enzyme, a defective transporter—form part of the pathway of the disease itself. But, like risk thinking, the idea of susceptibility brings potential futures into the present and tries to make them the subject of calculation and the object of remedial intervention. This generates the sense that some, perhaps all, persons, though existentially healthy are actually asymptomatically or pre-symptomatically ill. Technologies of life not only seek to reveal these invisible pathologies, but intervene upon them in order to optimize the life chances of the individual. Hence new forms of life are taking shape in the age of susceptibility, along with new individual and collective subjectifications of those "at risk," and, of course, new extensions of the powers of expertise potentially to all who are now understood as "pre-patients."

Enhancement, like susceptibility, is future oriented. Almost any capacity of the human body or soul—strength, endurance, attention, intelligence and the lifespan itself—seems potentially open to improvement by technological intervention. Of course, humans, at almost any place and time one cares to investigate, have tried to improve their bodily selves—using prayer, meditation, diet, spells, physical and spiritual exercises, and much more to increase their health, fertility, sporting prowess, longevity, acuity, and almost everything else. And, in all these places and times, there have been experts of bodily improvement with their own potions and systems, as well as lay beliefs about the life-enhancing powers of particular activities, foods, thoughts, and the like. What is new, then, is neither the will to enhancement, nor enhancement itself. In part, I suspect, the feeling of novelty and disquiet arises from the sense that we are moving, in the words of Adele Clark and her colleagues, "from normalization to customization" (Clarke et al. 2003: 181–82). Previously expert medical interventions were utilized in order to cure pathologies, to rectify generally accepted deviations from desirable functioning or to promote biopolitical strategies through lifestyle modification. Now recipients of these interventions are consumers, making access choices on the basis of desires that can appear trivial, narcissistic, or irrational, shaped not by medical necessity but by the market and consumer culture. In part, also, the feeling of disquiet about contemporary enhancement technologies arises from the belief that they have become more powerful, precise, targeted, and successful—powerful because they are grounded in a scientific understanding of bodily mechanisms (for helpful discussions, see Elliott 2003, and Parens 1998). The artificially enhanced body is no longer a cyborg—a fusion of human and artifact—as in the augmentation of bodily powers from spectacles and hearing aids, through the use of medical devices from saline drips and colostomy bags to heart pacemakers. Such "cyborgism" finds its contemporary apotheosis in the bizarre experiments of Steve Mann of the University of Toronto with his wearable computers,[15] or Kevin Warwick of the University of Reading with his implants that try to make neural signals communicate directly with computers or devices such as wheelchairs and artificial limbs.[16] Unlike these uses of robotics and computing, which seem to make the human being *less* biological, the new molecular enhancement technologies do not attempt to hybridize the body with mechanical equipment but to transform it at the organic level, to reshape vitality from the inside: in the process the human becomes, not less biological, but *all the more* biological.[17]

Perhaps most worrying to critics is the belief, or prospect, that unlike previous practices of self-improvement, which required exercise of the will, training over long periods, hardships and endurance, these new enhancement techniques can be acquired without much exertion. Critics fear what Sarah Franklin has termed "design on demand"—the promise of improvement of almost any aspect of human vitality on request at private clinics for those who can pay, purchased on the Internet for a few dollars, or consumed in the form of a pill without effort. For some it is the reshaping of the body that is particularly worrying—not just cosmetic surgery but limb lengthening, gender reassignment, facial remodeling of children with Down's syndrome, and the like (see Frank 2004 for a provocative discussion). For others it is intervention in sexuality and reproduction that is particularly troubling—from Viagra for extending the sexual capacities of aging men to the use of reproductive technologies allowing postmenopausal women to have children. Reflecting on such developments, Ian Hacking has suggested that they lead us to reassess criticisms of Cartesian dualism.[18] Medical developments in replacement body parts—hips, corneas, hearts, kidneys, and the prospect of face transplants[19]—and new surgical techniques that enable a person to consciously observe doctors re-engineering their organs in real time on the operating room monitor, reinforce the idea of an analogue body, with interchangeable body parts, distinct from the mind. We are, he suggests, becoming Cartesian—our body is indeed as it was already envisaged by Descartes. In fact, as Hacking implies, the body long ceased to be a natural given. Hence the politics of "enough" (McKibben 2003), which hopes to call a halt to such augmentation and transformation, is both historically naïve and ethically wistful, yearning for a past that exists only in the imagination.

Hacking initially took the metaphor of digital minds to think through analogous interventions to augment human mental capacities. But, as he recognizes, this metaphor is already outdated. The computer model of mental processes, with its focus on a strangely abstracted idea of cognitive processes, has been rendered redundant by our capacity to observe and anatomize the living brain. But if not digital minds, what then? What if our minds too become bodily, fleshly things, to be anatomized, dissected, re-engineered? Suppose we can identify and re-engineer the neural pathways and enzyme activities responsible for variations in human impulses and our capacity to control them—what then for ideas of free will and criminal responsibility?[20] Suppose we can reshape our moods, emotions, and desires at will, without effort other than through the consumption of a pill—the myth of Prozac and cosmetic psychopharmacology. For some, this strikes at the heart of their sense of what it is to be human (President's Council on Bioethics 2003).[21] Suppose we can enhance our cognition, as in the suggestion that drugs apparently successful in mitigating early memory loss in Alzheimer's disease may be about to open the way to a whole slew of pharmaceuticals to improve memory, intelligence, concentration, and the like? Without waiting for the development of such drugs as HT-0712 ("Viagra for the brain"),[22] companies selling all manner of nutritional supplements are marketing products that promise such results across the Internet. As those in education worry about the ethical implications of some students with the resources and knowledge being able to improve their cognition by taking drugs prior to examinations, those in the sporting field worry about the use of enhancement technologies—from drugs to potential genetic manipulation—to boost performance artificially. A new world of inequity appears to beckon and a new subdiscipline is born—"neuroethics." When ethics becomes neuronal, it implies that our technologies of subjectivity have also become neuronal; that is to say, it appears that in some significant respects we have become "neurochemical selves."[23] Around this new sense of ourselves, and with the apparent development of new capacities to intervene upon the mind through manipulating the brain, a new biopolitics—neuropolitics—has taken shape.

Subjectification and Ethopolitics

Across the twentieth century, the responsibilities of states in Europe and North America, and to some extent elsewhere, expanded from the collective measures to ensure health that were widely adopted in the nineteenth century—pure water, sewers, food quality, and so forth—to the active encouragement of healthy regimes in the home and interventions into the rearing of children. While both the wealthy and the poor had always engaged in a range of practices to maintain health, now the maintenance and promotion of personal, childhood, and familial health—regimen, personal hygiene, healthy child-reading, the identification and treatment of illness—became central to forms of self-management that authorities sought to inculcate into citizens and hence to their own hopes, fears, and anxieties. Over this period, those who were citizens of the advanced industrial societies of the West became committed to these norms of health and hygiene disseminated through the practices of state, medical and philanthropic authorities, as matters of their own self-maintenance and self-formation (see, for one of many historical accounts, Valverde 1991).

By the second half of the twentieth century, health had become one of the key ethical values in such societies. A plethora of medical and philanthropic organizations engaged in campaigns of health education and health promotion, and made demands on political authorities in the name of health. Additionally, actual or potential patients and their families and advocates, now became key actors in the economics, politics, and ethics of health. Encouraged by health educators to take an active interest in their own health, and "activated" by the new cultures of active citizenship, many refused to remain "patients," merely passive recipients of medical expertise. They became consumers actively choosing and using medicine, biosciences, pharmaceuticals, and "alternative medicine" in order to maximize and enhance their own vitality. They demand information from their doctors, expect successful therapies, and are liable to complain or even go through legal channels if they are disappointed. Health, understood as an imperative, for the self and for others, to maximize the vital forces and potentialities of the living body, has become a key element in contemporary ethical regimes.

Thus, while medicine has long had a role in shaping subjectivities, there are some significant features that distinguish contemporary forms of biomedical subjectification from their predecessors. Paul Rabinow was one of the first to recognize this phenomenon—he coined the term "biosociality" to characterize the new forms of collective identification that are taking shape in the age of genomics (Rabinow 1996a). His research had led him to identify new types of group and individual identities and practices arising out of the new techniques of genetic diagnosis and monitoring of risks and susceptibilities. Such groups meet to share experiences, lobby for funding research into "their" disease, and change their relations to their children, their environment, and their forms of life in the plight of genetic knowledge. He also foresaw the ways in which they would develop novel kinds of relations with medical specialists, clinics, laboratories, and with medical knowledge, surrounded by "a heavy panoply of pastoral keepers to help them experience, share, intervene, and "understand" their fate" (1996a: 102). His analysis arose from his observations, in France in the first half of the 1990s of the mobilization of patients, relatives, and others affected by the dystrophies—organized in the form of a nongovernmental organization—the *Association Française contre les Myopathies* (AFM) (1999). Faced with a situation in which their disease appeared to be neglected, some of those affected families turned away from the older "charitable" model of support and advice to patients and families, and toward the search for treatment and cure. They collaborated with genomic researchers in providing blood samples for DNA analysis, with the hope of localizing and mapping the genes responsible for the disorder. They supported French efforts to map the human genome, and raised funds through a television appeal to set up a genomic research laboratory—Généthon. Rayna Rapp and her colleagues drew on these ideas in their research on families whose children were affected by genetic diseases, and who adopted new tactics of patient activism, in particular, tactics of lobbying politicians to support genetic causes and cures—they term this "genetic citizenship" (Heath et al. 2004).

Writing with Carlos Novas, I have identified similar developments in relation to diseases as widely divergent as bipolar affective disorder and Huntington's Disease, and have suggested analysis in terms of "biological citizenship" (Rose and Novas 2004).[24] We used this wider term because we wanted to highlight the ways that citizenship has been shaped by conceptions of the specific vital characteristics of human beings, and has been the target of medical practices since at least the eighteenth century in the West. We need only to consider the history of race thinking to realize that the characteristics of actual, desirable, impossible citizens have at least in part been understood and acted on in terms of their biology, their vital organic character and characteristics as members of an ethnos, a race, a nation, or a civilization. In the processes of nation-building in European states and their colonies, from at least the mid-nineteenth century, ideas and practices of citizenship involved ways in which citizens should conduct themselves in relation to their health and reproduction. And for the biopolitics of the first half of the twentieth century—whether in its eugenic or its welfarist forms—the body of the citizen, the individual citizen and the collective citizen body of the people, the nation or the Volk, was a prime value.

Biopolitics, here, was not exhausted by sterilization, euthanasia, and the death camps. Many "citizenship projects" were organized in the name of health. In the education of German citizens in the Third Reich, in eugenic education campaigns in the United States, Britain, and many European countries, making social citizens involved instructing those citizens in the care of their bodies—from school meals to toothbrush use, inculcation of the habits of cleanliness and domesticity, especially in women and mothers, state regulation of the purity of food, interventions into the workplace in the name of health and safety, instructing those contemplating marriage and procreation on the choice of marriage partners, family allowances, and much else. The citizen here was not merely a passive recipient of social rights, but was also obliged to tend to his or her own body and, for a woman, those of her spouse and offspring While the state would engage in measures for preserving and managing the collective health of the population, whether this be in seeking to shape reproduction or trying to eliminate toxins, individuals themselves must exercise biological prudence, for their own sake, that of their families, that of their own lineage, and that of their nation as a whole.

Biological notions of citizenship have also been bound up with projects "from below," such as the feminist campaigns for the legalization of contraception in the first half of the nineteenth century. More recently, citizenship claims have been embodied in campaigns such as those at Bhopal, where those affected by biological damage battle for compensation. Adriana Petryna has explored these issues as they took shape in the Ukraine after the disastrous meltdown of the nuclear reactor at Chernobyl (Petryna 2002). She shows how, in this instance, citizens demanded that their rights to compensation were recognized, and claimed a redistribution of political resources on account of their damaged biology. While biological citizenship, here, was enacted by means of demands upon state authorities, such citizenship can take many forms; indeed it takes its character from the more general modes of citizenship in particular governmental regimes. Thus the forms of biosociality identified by Rabinow are shaped by more general practices of citizenship and subjectivity in the governmental regimes that I have termed "advanced liberal." Patient support groups, and the multitude of genetic support groups that have sprung up in Britain, Europe, and North America—from those organized around a single chromosome, such "The Chromosome 18 Registry and Research Society" founded by Jannine Cody, the mother of a child with 18 q minus syndrome in 1990, to multiorganization alliances such as the Genetic Alliance, an international coalition comprised of millions of individuals with genetic conditions and more than 600 advocacy, research, and health care organizations that represent their interests—partake of the ethic of active citizenship that has taken shape in advanced liberal democracies.[25] This is an ethic in which the maximization of lifestyle, potential, health, and quality of life has become almost obligatory, and where negative judgments are directed toward those who will not, for whatever reason, adopt an active, informed, positive, and prudent relation to the future. Perhaps it is inevitable that many contemporary biological citizens should now feel that they have acquired rights to the treatment of their sickness and disabilities and that others—politicians, health authorities, doctors—should be held accountable and be required to recompense or compensate them for their conditions. Indeed it is not rare to see groups of such active and aggrieved citizens contest with one another as to the priority and righteousness of their own particular "states of injury" (Brown 1995, Rose 1999).

Biomedicine, throughout the twentieth century and into our own, has thus not simply changed our relation to health and illness but has modified the things we think we might hope for and the objectives we aspire to. That is to say, it has helped make us the kinds of people we have become. Social theorists have recently focused on historical transformations in the self, often analyzing these in terms of increasing individualization and reflexivity (e.g., Beck et al. 1994). My focus is related but different. I make no claims about changes in human personality or psychology—this would require a very different type of investigation. My analysis concerns not what human beings are, but what they think they are: the kinds of human beings they take themselves to be (Rose 1985, 1989, 1996b). And, I suggest, we are increasingly coming to relate to ourselves as "somatic" individuals, that is to say, as beings whose individuality is, in part at least, grounded within our fleshly, corporeal existence, and who experience, articulate, judge, and act upon ourselves in part in the language of biomedicine. From official discourses of health promotion through narratives of the experience of disease and suffering in the mass media, to popular discourses on dieting and exercise, we see an increasing stress on personal reconstruction through acting on the body in the name of a fitness that is simultaneously corporeal and psychological. Exercise, diet, vitamins, tattoos, body piercing, drugs, cosmetic surgery, gender reassignment, organ transplantation: the corporeal existence and vitality of the self has become the privileged site of experiments with the self.

This somatization of ethics extends to the mind. Over the first sixty years or so of the twentieth century, human beings came to understand themselves as inhabited by a deep interior psychological space, and to evaluate themselves and act upon themselves in terms of this belief (Rose 1989). But over the past half century, that deep space has begun to flatten out, to be displaced by a direct mapping of personhood, and its ills, upon the body or brain, which then becomes the principle target for ethical work. In the twentieth century, we came to ground our ethical practices in an understanding of ourselves as creatures inhabited by an inner world, the font of all our desires and the place where we might discover the secret source of all our troubles. But these relations to ourselves are being transformed in the new games of truth that we are caught up in. New sciences of brain and behavior forge direct links between what we do—how we conduct ourselves—and what we are. These games of truth work at a molecular level, the level of neurons, receptor sites, neurotransmitters, and the precise sequences of base pairs at particular locations in what we now think of as the human genome. These molecular phenomena, rendered visible and transformed into the determinants of our moods, desires, personalities, and pathologies, become the target of new pharmaceutical techniques. And these techniques do not merely promise coping, nor even cure, but correction and enhancement of the kinds of persons we are or want to be.[26] Here too, in relation to our moods, desires, cognitive capacities, and affects, it is in corporeal terms that our truth and destiny is imagined: our corporeality, now at the molecular level, is the target of our judgments and of the techniques that we use to improve ourselves.

Thus we can see that in advanced liberal democracies, where individuals are enjoined to think of themselves as actively shaping their life course through acts of choice in the name of a better future, "biology" will not easily be accepted as fate or responded to with impotence. Of course, an ethic organized around the ideals of health and life produces anxiety, fear, even dread at what one's biological future, or that of those one cares for, might hold. But while this may engender despair or fortitude, it frequently also generates a moral economy in which ignorance, resignation, and hopelessness in the face of the future is deprecated. At least in part, fears and anxieties about morbidity and mortality are being reframed within an ethos of hope, anticipation, and expectation (Brown 1998, Franklin 1997, Novas 2001). And this moral economy of hope is also an economy in the more traditional sense, for the hope for the innovation that will treat or cure stimulates the circuits of investment. Hence the ethos of hope links together many different actors—of actual or potential sufferers for a cure, of scientists and researchers for a breakthrough that will make their name and advance their career, of doctors and health care professionals for a therapy that will help treat their patients, of biotech companies for a product that will generate profit, of governments for industrial and commercial developments that will generate employment and stimulate economic activity and international competitiveness.

I think this economy of hope is one dimension of a wider shift in what I have termed "ethopolitics" (Rose 1999). By ethopolitics I refer to attempts to shape the conduct of human beings by acting upon their sentiments, beliefs, and values—in short, by acting on ethics. In the politics of our present, notably in the revival of communitarian themes, the ethos of human existence—the sentiments, moral nature, or guiding beliefs of persons, groups, or institutions—has come to provide the "medium" within which the self-government of the autonomous individual can be connected up with the imperatives of good government. If "discipline" individualizes and normalizes, and "biopolitics" collectivizes and socializes, "ethopolitics" concerns itself with the self-techniques by which human beings should judge and act upon themselves to make themselves better than they are. While ethopolitical concerns range from those of life-style to community, they coalesce around a kind of vitalism, disputes over the value accorded to life itself: "quality of life," "the right to life" or "the right to choose," euthanasia, gene therapy, human cloning, and the like. This biological ethopolitics—the politics of how we should conduct ourselves appropriately in relation to ourselves, and in our responsibilities for the future—forms the milieu within which novel forms of authority are taking shape.

Experts of Life Itself

These developments in the biomedical government of somatic selves are not primarily mobilized by politicians, or by the kinds of professionals that were invented over the twentieth century to make liberal freedom possible—social workers, therapists, personnel managers, and all those others who claimed to understand how we should live better lives. Biopolitics today depends upon meticulous work in the laboratory in the creation of new phenomena, the massive computing power of the apparatus that seeks to link medical histories and family genealogies with genomic sequences, the marketing powers of the pharmaceutical companies, the regulatory strategies of research ethics, drug licensing bodies committees and bioethics commissions, and, of course, the search for the profits and shareholder value that such truths promise. It is here, in the practices of contemporary biopower, that novel forms of authority are to be found.

In part, those authorities are the clinicians, whose expertise, as I have already indicated, reaches far beyond the diagnosis and treatment of diseases. Medics have long had a role that extends way beyond illness and cure. We have only to think of the forms of "medical police" that took shape in the eighteenth century, to realize that doctors have been central to social investigation, town planning, hygienic reforms, the management of dead bodies, the regulation of foodstuffs, and much more (Foucault 1999, Roberton 1812, Rosen 1958). Further, since at least the nineteenth century, doctors have played key roles in the criminal justice system, in insurance strategies and technologies and, throughout the twentieth century, in the organization and management of working life. Medicine, that is to say, has been central to the development of the arts of government; not only the arts of governing others, but also the arts of governing oneself. For at the very moment when health and illness became amenable to a positive knowledge and to explanations and interventions in terms of the biology of the organic living body, medics took up their role as experts of lifestyle (cf. Rose 1994: 69–70). As the quest for health has become central to the telos of living for so many human beings in advanced liberal democracies, people have come to experience themselves and their lives in fundamentally biomedical terms, and with the best of intentions on all sides have become bound to the ministrations and adjudications of medical expertise, and/or those paramedical alternative and complementary forms of expertise that have partaken of much the same logic.

But the somatic experts involved are no longer simply medical, and their advice and interventions on life itself extend rather widely. There are nurses, midwives, health visitors. There are the multiple kinds of therapists, not just psychological therapists but speech therapists, occupational therapists, art therapists, physiotherapists, and a host of others. There are nutritionists, dieticians, health promotion experts, remedial gymnasts, experts on exercise and fitness, and multiple advisers on shaping a form of life in the name of health. And there are the counselors—addiction counselors, sex counselors, family and relationship counselors, mental health counselors, educational counselors and, of course, genetic, family planning, fertility, and reproduction counselors. Of most interest to me here are the new kinds of "pastoral powers" that are emerging in the context of what Margaret Lock has termed "premonitory" knowledge—that is to say, the kind of knowledge deployed by genetic counselors, but which might well be extended to encompass predictive and future-oriented information based upon neuronal evidence such as brain scans that may indicate risk of future disease or, as some are suggesting, undesirable behavioral traits such as impulsivity (Lock 2005). The sites of such pastoral power are likely to proliferate in the new age of susceptibility and presymptomatic diagnoses, as premonitory knowledge with variable levels of certainly, emerges in relation to more and more "threats to health." This is not the kind of pastoralism where a shepherd knows and directs the souls of confused or indecisive sheep. It entails a dynamic set of relations between the effects of those who council and those of the counseled. These new pastors of the soma espouse the ethical principles of informed consent, autonomy, voluntary action, and choice and nondirectiveness.[27] In an age of biological prudence, where individuals, especially women, are obliged to take responsibility for their own medical futures and those of their families and children, these ethical principles are inevitably translated into microtechnologies for the management of communication and information that are inescapably normative and directional. These blur the boundaries of coercion and consent. They transform the subjectivities of those who are counseled, offering them new languages to describe their predicament, new criteria to calculate its possibilities and perils, and entangling the ethics of the different parties involved. It is in this sense of managing the present in terms of an uncertain medical future, and in the face of technological medicine and pastoral expertise, that I have suggested, following Rayna Rapp, that all of us will soon follow those "ethical pioneers"—AIDS activists and women experiencing new reproductive technologies—in developing a new pragmatic ethics of vitality and its management (cf. Rapp 1999).

But somatic expertise is not just proliferating in the "application" of biomedical knowledge, it is central to the very truth discourses of biology. In the new molecular style of thought that characterizes the life sciences, the distance between fundamental science and the clinic is bridged by all sorts of mediations. As Ludwik Fleck showed us, each style of thought has its own "thought collective," and this is certainly true for contemporary molecular thought in biomedicine (Fleck 1979). From the stem cell experts to the molecular gerontologists, from the neuroscientists to the technologists of cloning, new specialists of the soma have emerged, each with their own apparatus of associations, meetings, journals, esoteric languages, star performers, and myths. Each of these is surrounded by, augmented by, a flock of popularizers, science writers, and journalists. While often disowned by the researchers themselves, they play a key translational and mediational role in forming the associations—made up of politicians, lay people, patient groups, research councils, and venture capitalists and investors—on which such expertise depends.

Surrounding these somatic experts is another branch of expertise—bioethics.[28] Bioethics has mutated from a sub-branch of philosophy to a burgeoning body of professional expertise. Ethics was once inscribed within medical personages, imbued by long training and experience at the bedside, and supported by a code of conduct and enforced, where required, by professional bodies themselves. For medical researchers, for the fifty years following World War II and the debate on ethics in the wake of the Nazi doctors and the revelation of other medical experiments, the ethics of research was ensured by a set of principles and overseen by research ethics committees.[29] But now—from national bioethical committees, local Institutional Review Boards, to a whole apparatus of bioethical approved patient information and consent forms for any medical procedure or piece of biomedical research—we have witnessed a bioethical encirclement of biomedical science and clinical practice. Similarly, we can observe a bioethical reshaping of the self-representations of commercial actors in the biotech sectors, especially those involved in pharmaceuticals or genetic services for patients. In a market driven by the search for shareholder value, where consumption of medical and pharmaceutical products is itself shaped by brand images and brand loyalty, where confidence in products is crucial, and where there are spirals of unrealistic hope and manipulated distrust, corporations engage bioethicists on their advisor boards and use a whole variety of techniques to represent themselves as ethical and responsible actors.[30] What generates the insatiable demand for bioethics in the political and regulatory apparatus of advanced liberal societies?[31]

One can certainly regard the expansion of bioethics, and its imbrication within regulatory strategies, as one answer to a kind of "legitimation crisis" experienced by genetic and other biotechnologies in advanced liberal democracies (Salter and Jones 2002, 2005). Further, as biotech companies seek to commodify products—DNA sequences, tissues, stem cells, organs—it is clear that ethics has a crucial function in market creation. Products that do not come with appropriate ethical guarantees, notably assurances as to the "informed consent" of donors, will not find it easy to travel around the circuits of biocapital. It is also clear that the routinization of ethical concerns in the bureaucratic procedures of research governance can serve to insulate researchers rather than to constrain them, and that the now almost inescapable inclusion of ELSI[32] considerations in calls for grants and in successful proposals may, perhaps accidentally, serve to assuage critical voices. Similarly, in those jurisdictions where bioethicists work in clinical settings, it is clear that they can function to shield medical authorities, hospital managers, clinicians, and others from the consequences of contested and controversial decisions, such as those relating to the termination of life support to a putatively brain-dead individual.

We need, therefore, to open this peculiar persuasion of bioethics to critical investigation. What forms of expertise does bioethics claim, or is it ascribed to support its authority? And what determines the issues that "become" bioethical? While bioethicists return again and again to highly individualized issues such as autonomy, confidentiality, and rights and protections in high tech medicine, they seldom address the ethical issues raised by the mundane, routine, global depredations of illness and premature death (Berlinguer 2004). Why should informed consent in reproductive technology be "bioethical" and the rising rate of female infertility not? Why should the "dignity" of the person at the end of life be a bioethical issue, but not the massive "letting die" of millions of children under five years of age each year from preventable causes? The mere presence of illness, death, medical technology, and professional decision making do not in themselves necessitate bioethics. So what is it about the biopolitics of life itself, as it has taken shape in certain types of societies, that provides the spaces within which bioethical authority seems to be required and simultaneously circumscribes the issues to which such ethical concerns appear relevant (Rose 2002)?

Bioeconomics: The Capitalization of Vitality

Conducted at a molecular level, biology and medicine require long periods of investment, the purchase of expensive equipment, the maintenance of well-staffed laboratories, a multiplication of clinical trials, financial commitments for measures required to meet regulatory hurdles—in short, allocation of funds on a large scale over many years before achieving a return. Increasingly such investment comes from venture capital provided to private corporations, who also seek to raise funds on the stock market, and it is subject to all the exigencies of capitalization, such as the obligations of profit and the demands of shareholder value.[33] These biotech companies do not merely "apply" or "market" scientific discoveries: the laboratory and the factory are intrinsically interlinked—the pharmaceutical industry has been central to research on neurochemistry, the biotech industry to research on cloning, genetech firms to the sequencing of the human genome.[34] Thus we need to adopt a "path dependent" perspective on biomedical truth. Some critics, especially of the pharmaceutical industry, suggest that such dependency distorts, that biotech companies determine what they will or will not consider true in order to meet their own commercial interests. My own view is slightly different. Where funds are required to generate potential truth in biomedicine, and where the allocation of such funds depends inescapably upon a calculation of financial return, commercial investment shapes the very direction, organization, problem space, and solution effects of biomedicine and the basic biology that supports it. This is less a matter of the manufacture and marketing of falsehoods than of the production and configuring of truths. The reshaping of human beings is thus occurring within a new political economy of life whose characteristics and consequences we have yet to map: MedImmune Inc. ("Dedicated to advances in biotechnology, and devoted to patient support"), Gene Logic Inc. ("Discovery happens here"), (Celera Genomics ("Discovery can't wait"), DeCode Genetics ("Decoding the language of life"), Genentec ("In business for life"). Biopolitics becomes bioeconomics.[35]

Catherine Waldby initially proposed the term "biovalue" to characterize the ways that the bodies and tissues derived from the dead are redeployed for the preservation and enhancement of the health and vitality of the living (Waldby 2000). More generally, we can use the term to refer to the plethora of ways in which vitality itself has become a potential source of value: biovalue as the value to be extracted from the vital properties of living processes (Novas and Rose 2000, Waldby 2002). Indeed a rather similar conception is explicitly proposed by the Organisation for Economic Co-operation and Development, for example in its "Proposal for a Major Project on the Bioeconomy in 2030," which aims to "construct scenarios 'to image' the bioeconomy in the future landscape" in order to draft a policy agenda for governments in respect to this sector. They define "the bioeconomy" as that part of economic activities "which captures the latent value in biological processes and renewable bioresources to produce improved health and sustainable growth and development" (Organisation for Economic Co-operation and Development 2004). As Sarah Franklin has noted, Edward Yoxen pointed to the significance of the economic exploitation of biology as long ago as 1981, suggesting that the informational language for analyzing nature, which started to take shape in the 1920s, allowed for the possibility of a technological capitalization of life; this was "not simply a way of using living things that can be traced back to the Neolithic origins of fermentation and agriculture. As a technology controlled by capital, it is a specific mode of appropriation of living nature—literally capitalizing life" (Yoxen 1981: 112, cited in Franklin 2000: 190). We should, no doubt, question the motives that Yoxen attributes to capital. Further, I shall argue later that the information metaphor of life is only one mode in which it can be made adequate to capitalization. However, the point remains: as the OECD report remarks, bioeconomic circuits of exchange have as their organizing principle the capturing of the latent value in biological processes, a value that is simultaneously that of human health and that of economic growth.

Once more, we need to be cautious about overstating the novelty of these developments. Humans put the vital properties of the natural world in service for themselves from their inception, with the domestication of animals and plants. They turned these properties into technologies, when they harnessed the milk producing capacities of cows, and the silk producing capacities of the silkworm to the generation of biovalue: capturing, domesticating disciplining, instrumentalizing, the vital capacities of living creatures.[36] In a sense, then, contemporary projects to embody human desires and aspirations within living entities—organisms, organs, cells, molecules—in order to extract a surplus, be it food, health, or capital, can be traced to these early events. Yet something has changed. The very emergence of the term itself—bioeconomics—brings into existence a new space for thought and action. As Peter Miller and I have argued elsewhere, the government of an "economy" becomes possible only through discursive mechanisms that represent the domain to be governed as an intelligible field with its limits, characteristics whose component parts are linked together in some more or less systematic manner (Miller and Rose 1990). For the bioeconomy to emerge as a space to be mapped, managed, and understood, it needs to be conceptualized as a set of processes and relations that are amenable to knowledge, that can be known and theorized, that can become the field or target of programs that seek to evaluate and increase the power of nations or corporations by acting within and upon that economy. And the bioeconomy has indeed emerged as a governable, and governed, space.

In part, this is exemplified by the routinization of the term "biocapital" itself: this term is an active agent in the constitution of the bioeconomy. Thus the third annual conference of BioCapital Europe was held in Amsterdam in March 2005—an event for pharma and biotech companies across Europe, from 4AZA Bioscience in Belgium to U3 Pharma from Germany, sponsored by PriceWaterhouseCoopers, Bird and Bird, and Ernst and Young, among others.[37] In Australia, around the same time, the state of Queensland established a AU$100m biocapital fund to establish globally enduring bio-businesses. In May 2005, BioSpace, a leading on-line information source for the biotech and pharma industry, published the fifth edition of *BioCapital*, which showcases a variety of biopharmaceutical companies located within the mid-Atlantic region, including AstraZeneca, Celera, Gene Logic, and Wyeth, and includes an interactive BioCapital Hotbed map that also highlights research institutes, nonprofit organizations, and universities within the area.[38] Also, the term "biocapital" is used in the title of numerous investment and consultancy organizations worldwide. Marxists and post-Marxists may disagree about whether "biocapital-

ism" is a novel "mode of production," but the existence and significance of biocapital, as a way of thinking and acting, cannot be disputed.

Today, a plethora of documents and statistics map out the emerging bioeconomy, some with the aim of making it amenable to calculation and exploitation, and others seeking to open it to a variety of programs of regulation and government. These mapping projects incorporate a long tradition of statisticalization of health, disease, and medicine, and the documentation of health care system costs. The numbers that proliferate concerning biotechnology—rates of investment, numbers of companies, rates of return on capital, numbers of products brought to market, divided by sector, country, region, charted over years to show growth or decline—constitute the bioeconomy through the ways in which they inscribe it in a docile form amenable to thought, discussion, analysis, diagnosis, and deliberation (Rose 1991). Let us glimpse the world they make thinkable. We can start with the statistics of the health care sector. By the start of the twenty-first century, spending on health constituted a very significant sector of GDP in advanced industrial countries and one that was growing year by year. In the United States in 2002, health care spending at US$1.6 trillion was double that in 1972, and grew by 9.3 percent over the previous year, which was the sixth year of successive growth. Levit, Smith, and their colleagues, analyzing the "rebound" of spending on health in the United States for the readers of *Health Affairs*, tell us that health care spending now accounts for nearly 15 percent of the GDP, driven by rising costs for hospitalization, physician services, home health care, and especially prescription drugs (Levit et al. 2004). A key area, here, is the market for pharmaceuticals. IMS Health—"the one global source for pharmaceutical market intelligence, providing critical information, analysis and services that drive decisions and shape strategies . . ."[39]—estimates that the retail pharmacy market for the twelve months prior to May 2004 in the United States was US$167.9 billion, up 10 percent over the previous year; in the United Kingdom it was valued at US$14.2 billion, up 11 percent, with even higher rates of growth in Latin America.[40] And pharmaceuticals are just one element in the constitutive relations between truth, health, and capitalization that form the contemporary bioeconomy.

Projects to govern the bioeconomy in almost every geographical region are characterized by novel alliances between political authorities and promissory capitalism.[41] An apparently virtuous connection between health and wealth mobilizes the large budgets for research and development invested by national governments and private foundations, the dealings of the commercial health care and health management industries, the operation of the pharmaceutical and biotechnology companies, the flows of venture and shareholder capital. This is especially the case where a new theme has come to dominate political reason concerning the government of the economy—the theme of the "knowledge economy."[42] For example, speaking at the European Bioscience Conference in Lisbon in November 2000, U.K. Prime Minister Tony Blair said, "biotechnology is the next wave of the knowledge economy and I want Britain to become its European hub."[43] The hope he expresses for a virtuous alliance of state, science, and commerce in the pursuit of health and wealth is one that is shared by many other political authorities. These were most famously and controversially illustrated in the political support—in Iceland, Sweden, and a number of other countries—for private companies to be licensed to undertake genetic sequencing of populations, and to allow them to combine this with publicly held genealogical data and medical records, in the hope that they would identify the genomic bases of common complex disorders. In the case of deCode in Iceland, which was heavily criticized by social scientists from other countries, these hopes were not fulfilled, at least in the short term (Palsson and Rabinow 1999, Rose 2003).[44] UmanGenomics in Sweden sought to use bioethical shields to insulate itself from some of the criticisms; it too found that its business model was not viable (Abbott 1999, Høyer 2002, 2003, Nilsson and Rose 1999, Rosell 1991). This did not dissuade a number of other countries from pursuing such public-private partnerships, notably those from former "strong state" traditions emerging from Soviet domination, such as Lithuania and Estonia, where comprehensive medical and genealogical records, together with relatively stable populations and some unusually prevalent medical conditions seemed to provide a favorable basis for enterprises that might generate employment, boost industry, and promote both public and shareholder value.[45] Genetic stock was now a marketable commodity, although one that has, to date, not generated the value for which some hoped.

The hope invested in the bioeconomy has energized official investigations, enquiries, and reports in many countries. Thus in 2003 the U.K. House of Commons Trade and Industry Committee Report on Biotechnology identified biotechnology, especially biomedical biotechnology, as a key economic driver and estimated that, in 2002, the U.K. biotechnology industry had a market capitalization of £6.3 billion, accounting for 42 percent of the total market capitalization of European biotechnology (pharmaceutical biotechnology is the dominant branch).[46] In 2003, Ernst and Young reported that the U.S. biotech sector was a US$33.6 billion industry, with a total of 1,466 companies, 318 which are public (Ernst & Young 2003b). They also reported that "In Australia . . . total revenues among publicly traded companies increased 38 per cent from $666 million in 2001 to $920 million in 2002. The number of . . . people employed in the industry jumped 24 percent from 5,201 to 6,464." "The Japanese government anticipates the nation's biotech work force will surge to 1 million by 2010, an enormous increase over the estimated 70,000 today. Government officials plan to double their investment in biotechnology in the next five years" (Ernst & Young 2003a). This is not simply another case of predatory Western capitalism plundering the resources of the poor. A report of a U.K. government mission to India in 2003 was headed with a quote from then Indian Prime Minister Atal Behari Vajpayee: "Biotechnology is a frontier science with a high promise for the welfare of humanity": at that time there were 160 biotechnology companies in India with combined revenues of US$150 million, driven by developments in the health care sector; the industry was expected to grow to US$4.5 billion by 2010, and to generate a million or more jobs. Singapore's revenues from biomedical manufacturing are projected to reach US$7 billion by 2005. In China, world number three in terms of overall R&D spend by 2003, the government spent about $180 million building a biotech industry from 1996 to 2002, a figure predicted to triple over the following three years. Despite or because of its one child policy, China has an active sector of reproductive medicine, and IVF and PGD is widespread. China is also a world leader in research on stem cells, with its own set of lines, and is already involved in clinical trials. The Stem Cell Research Centre in South Korea has guaranteed government funding of US$7.5 million for the next ten years. In Asia such developments are underpinned by long-term government funding and investment in infrastructure. And political investment to support the development of the biotechnology sector in each country and region was driven, at least in part, by fears of the consequences of loosing out in this intense international competition.

By the start of the twenty-first century, the value of the biomedical biotechnology complex—biotech companies (working on everything from therapeutic stem cells to DNA paternity testing), pharmaceutical companies, manufacturers of machinery, equipment, reagents, and much more—was immense. Some critics delighted in suggesting that the bioeconomy was a "bubble" economy and that the bubble was already beginning to burst (Ho et al. 2003). But the market information providers, reporting on the situation in 2005 (for those who can afford to pay for their reports), do not support this view. Consider, for example, Ernst and Young's "Global Biotechnology Report 2005," titled, like its predecessors, *Beyond Borders* (Ernst & Young 2005): Biotech is moving "beyond borders" they suggest because it is "rapidly evolving, restructuring and *recombining*. . . . With biotech spreading across the globe, and strong progress in Asia . . . the answers to challenges are being found at the global level, as obstacles in one region are overcome by leveraging strengths and capabilities in another part of the globe" (1). Pointing in particular to the improvement of regulatory and IP regimes in China and India, to the Biopolis vision in Singapore, and to the fact that "from Malaysia to Michigan, governments are developing strategic plans with ambitious goals for biotech," they point out that "the global industry raised a whopping $21.2 billion in 2004" for early stage development, yet even this was not enough to meet the challenge of finding early stage capital.[47] While the "global biotechnology industry's revenues grew by 17 percent in 2004, to $54.6 billion," and it raised $21.2 billion in capital from private equity investors and others in the capital market, it was still making net losses of $5.3 billion, and many companies seeking to raise funds from IPOs (initial public offerings) did not obtain the valuations they sought and suffered falls in share prices. Times may be "challenging," as the report often remarks, not least from developments in regulation and legislation in many regions, for example, in the United States, the debates over the ethics of stem cell research, and the tendency of key policymakers to "scrutinize research agreements between academic medical centers, clinicians and biotech/pharmaceutical companies" and to question "potential conflicts of interest" (35). In Europe, after "enduring some life-threatening storms and refocusing their resources in recent years," capital markets are recovering and the biotech industry is "turning a corner" and focusing on bringing products to market, despite continuing concerns about the burden of regulation especially in relation to drug safety. The Asian biotech sector "continues to grow aggressively" and "biotech companies in the region increased their top line-revenues by 36 percent in 2004" although they too face "challenges" as investment from Western companies is hampered by worries over IP protection, and governments and nonbiotech industrial conglomerates have to provide the capital that, in the West, would be raised in other ways (67). Yet the promissory allure of biocapital remains strong.

Indeed, notwithstanding these "challenges," national and local politicians in countries across the globe continue to foster the growth of a biotech sector and seek to find a niche in this global bioeconomy. South Africa's Cape Cluster strategy, for example, highlights the "niche drivers" of market opportunities and political will in relation to five key factors: "South Africa's unique richness in biodiversity; Prevalence of undeserved unique disease creating local demand (HIV, malaria, tuberculosis); Unique genetic populations, both isolated immigrant and diverse African; Strong clinical environment (South Africa was the site of the first heart transplant); Low cost of research and development (R&D) and first-world intellectual property management" (13). Like the OECD, which planned to spend two million euros on its future-oriented scenario, governments set up foresight and horizon-scanning exercises to map the future potential of this biotechnological industrial revolution in the medical management of health, disease, and life and formulated strategies at international, national, and local levels—research funding, technology transfer, support for start-up and spin out firms, tax breaks for research and development, low regulatory hurdles—to encourage the development of this sector of the economy.

The circuits traced out by these contemporary economies of vitality are thus conceptual, commercial, ethical, and spatial. These spaces range from the atomic, the molecular, the cellular, the organic, the spaces of practices (laboratories, clinics, consulting rooms, factories), of cities and their economies (Shanghai, Mumbai, the Cape), of nations and their regulatory frameworks and economic strategies, and of the virtual spaces of the Internet that ensure the immediate availability at any point in the world of the totality of data on the genome. The circuits are mobilized by a variety of relations. Major pharmaceutical companies based in North America or Europe trial their experimental drugs in Africa, Asia, Eastern Europe, and Latin America, the results flowing back to base, feeding into the production of profitable new products for the market in the developed world and playing their part in the generation of shareholder value.[48] Biosocial communities comprised of those affected by diseases thought to have a genetic component often ask their members across the world to donate blood and tissues, store them in tissue banks, and make them available for biomedical research (Corrigan and Tutton 2004, Taussig 2005). Geneticists themselves travel the world to collect tissue samples from families with diseases for genomic analysis wherever they may live.[49] Researchers from Europe or the United States, often employed by biotech companies, travel to remote areas, extract tissues from their "isolated" populations, and transport them back to Europe or the United States for genomic analysis and, potentially, for the identification of markers for disease susceptibility that might produce patentable inventions.[50] The production of the exploitable knowledge of vitality today thus involves multiple transnational circuits to mobilize and associate material artifacts, tissues, cell lines, reagents, DNA sequences, techniques, researchers, funding, production, and marketing.

Circuits of vitality are not themselves new—think, for example, of the longstanding practices of "ethnobotanical" collections of seeds and plants, or of the exchange of biological material and model organisms such as fruit flies, which were central to modern genetics (Balick and Cox 1996, Kohler 1994). But today, a kind of "dis-embedding" has occurred: vitality has been decomposed into a series of distinct and discrete objects, that can be stabilized, frozen, banked, stored, accumulated, exchanged, traded across time, across space, across organs and species, across diverse contexts and enterprises, in the service of bioeconomic objectives. For some this capitalization of human vitality is profoundly troubling. Inevitably, it raises questions about the borders of life, and those troubling entities—notably embryos and stem cells—whose position on the binaries of life / nonlife and human / nonhuman is subject to dispute.[51] Leaving this issue to one side for the present, many have been highly critical of the development of a market in human tissue. Dorothy Nelkin, in one of the first detailed analyses of this market, argued that biotechnology firms reduce and decontextualize the body, stripping it of it cultural meanings and personal associations, reducing it to a utilitarian object, as indicated by the way in which the language of bioscience has become "permeated with the commercial language of supply and demand. Body parts are *extracted* like a mineral, *harvested* like a crop, or *mined* like a resource. Tissue is *procured*—a term more commonly used for land, goods, and prostitutes (Andrews and Nelkin 2001: 5). It is not clear, however, whether this critique concerns the extent to which such practices were legitimated by informed consent and conformed to the wishes and beliefs of the patients or subjects involved, whether what is objectionable is the intrusion of commerce into the otherwise apparently benign world of medicine, whether it was that the benefits flowed to private capital rather than to the individuals involved or the community at large, or whether the objection was to the very fact of commodification of elements of human vitality.[52]

What is clear, however, is that the classical distinction made in moral philosophy between that which is not human—ownable, tradeable, commodifiable—and that which is human—not legitimate material for such commodification—can no longer do the work that is required to resolve this issue: that distinction is itself what is at stake in the politics of the contemporary bioeconomy. The tensions between the intensifying somatic ethics in the West, with the centrality it accords to the management of one's own health and body to contemporary self-fashioning, and the inequities and injustices of the local and global economic, technological, and biomedical infrastructure required to support such a somatic ethic, seems to me to be a constitutive feature of contemporary biopolitics.[53]

Beyond Sociocritique: A Politics of Life Itself

As I have suggested, most sociological commentary on these developments in biomedicine has been highly critical or at least deeply suspicious. I take a different view. I suggest that while many of these changes are shaped by processes that are open to criticism—the relentless search of biocorporations for profits and shareholder value, the quest of scientists for funds and career advancement, the attraction to many doctors of "heroic" medicine rather than the mundane work of treatment and prevention, the rise of a secular morality in which life and health are the only ends thought worth pursuing, and so forth—in the process we are seeing the emergence of an innovative new ethics of biological citizenship and genetic responsibility. Our somatic, corporeal, neurochemical individuality now becomes a field of choice, prudence, and responsibility. It is opened up to experimentation and to contestation. Life is not imagined as an unalterable fixed endowment. Biology is no longer destiny. Vitality is understood as inhering in precise, describable technical relations between molecules capable of "reverse engineering" and in principle of "re-engineering." Judgments are no longer organized in terms of a clear binary of normality and pathology. It is no longer possible to sustain the line of differentiation between interventions targeting susceptibility to illness or frailty on the one hand, and interventions aimed at the enhancement of capacities on the other.

There are many practices where identification of high risk and biological incorrigibility can switch the affected individual, or potential individual, onto the circuits of exclusion. But the dream—of doctors, geneticists, biotech companies, and many "afflicted individuals" and their families—is of presymptomatic diagnosis followed by technical intervention at the biological level to repair or even improve the suboptimal organism. The political vocation of the life sciences today is tied to the belief that in most, maybe all cases, if not now then in the future, the biologically risky or at risk individual, once identified and assessed, may be treated or transformed by medical intervention at the molecular level. In this regime, each session of genetic counseling, each act of amniocentesis, each prescription of an antidepressant is predicated on the possibility, at least, of a judgment about the relative and comparative quality of life of differently composed human beings and of different ways of being human. As biomedical technique has extended choice to the very fabric of vital existence, we are faced with the inescapable task of deliberating about the worth of different human lives—with controversies over such decisions, with conflicts over who should make such decisions and who cannot. This new politics is not one in which authorities claim, or are given, the power to make such judgments in the name of the quality of the population or the health of the gene pool. On the one hand, new forms of pastoral power are taking shape in and around our genetics and our biology. In this pastoral power, questions about the value of life itself infuse the everyday judgments, vocabularies, techniques, and actions of all those professionals of vitality—doctors, genetic counselors, research scientists, and drug companies among them—and entangle them all in ethics and ethopolitics. On the other hand, the politics of life itself poses these questions to each of us—in our own lives, in those of our families, and in the new associations that link us to others with whom we share aspects of our biological identity. Our very biological life itself has entered the domain of decision and choice; these questions of judgment have become inescapable. We have entered the age of vital politics, of somatic ethics, and of biological responsibility.

Chapter 2

Politics and Life

Politics and life—these two words convey something that is neither self-evident nor unchanging. That is to say, while the words may remain the same, their meaning and function, in lay and professional discourses, has varied greatly over time, as have the practices associated with each. The histories of these two terms are intertwined, as are the practices associated with each. It is relevant for my argument in this book, therefore, to examine them in more detail. Such an examination would, if properly conducted, require a book of its own. Even focusing on the parts of that history relevant to Euro-America, we should probably start with the Greeks. But my focus is more limited: it is to clarify what is specific about these terms in contemporary biopolitics. Where better to begin than with "the meaning of life."

Life[1]

Georges Canguilhem remarks that "interpreted in a certain way, contemporary biology is, somehow, a philosophy of life" (Canguilhem 1994: 319). In her illuminating discussion of "life itself," Sarah Franklin points out that the birth of the modern idea of life is closely linked to the rise of the life sciences: "founded on notions of evolutionary change, the underlying connectedness of all living things, and a biogenetic mechanisms of heredity through which life reproduces itself" (Franklin 1995). But, as she argues, life cannot simply be taken as a pre-existing phenomenon—our understanding of life has been transformed many times since the word "biology" was first proposed for a novel science of life in 1802.[2] An understanding of the birth and transformation of the idea of life itself, that is to say, requires an exercise in historical epistemology, along the lines developed in George Canguilhem's essay of 1966, "Le Concept et la Vie" (Canguilhem 1966, reprinted in Canguilhem 1968, see Franklin 2000). Canguilhem's pupil, Michel Foucault, in a book published in the same year, summarized his own more radical view of the epistemological mutation that occurred in the nineteenth century, arguing that, for the eighteenth century, biology did not exist: "the pattern of knowledge that has been familiar to us for a hundred and fifty years is not valid for a previous period . . . if biology was unknown there was a very simple reason for it: that life itself did not exist. All that existed was living beings, which were viewed through a grid of knowledge constituted by *natural history*" (Foucault 1970: 127–28) also quoted by Franklin 1995).[3]

Of course, in one sense Foucault's claim of a rupture is overstated: Canguilhem has given us detailed analyses of the vitalists of the eighteenth century, with their emphasis on the intimate relations between "man and nature," their rejection of mechanism, and their conceptions of the vital forces and powers that animate every living body.[4] But Foucault was here focusing on the ways in which the natural history of what he termed "the Classical Age" exemplified a certain epistemological structure, in which to know was to classify, to locate that which was to be known in a table or grid. To know a plant or an animal was to place it in a taxonomy, identifying it by allocating it to its proper genus and species on the basis of its observable characteristics. This configuration, he suggests, begins to mutate at the close of the eighteenth century (he points in particular to the work of Pallas and Lamarck) with the installation of a fundamental division of nature into two kingdoms—organic and inorganic: "'There are only two kingdoms in nature' wrote Vicq' d'Azyr in 1786, 'one enjoys life and the other is deprived of it' . . . The organic becomes the living, and the living is that which produces, grows and reproduces; the inorganic is the non-living, that which neither develops nor reproduces; it lies at the frontiers of life, the inert, the unfruitful—death" (Foucault 1970: 232). And it is that death, inescapably lodged within life, to which life is opposed, which it struggles against, and which comes to define its vitality. When a depth opens up beneath the taxonomic table, the depth of vitality within each entity that can belong to the organic world, only then did biology, in the form that it would exist for one hundred fifty years, become possible.[5] That biology, and the vitalism with which it was sometimes intertwined, was defined by an epistemology of depth, a style of thought that understood the visible characteristics of plants, animals, and humans in terms of the underlying organic laws that determined them, and the functions that they served in preserving life and averting death.

Foucault's argument was, no doubt, shaped in part by his study of medicine that predated these claims. In *Birth of the Clinic*, published in France in 1963, a study of the transformations in French medical thought and practice in the period roughly from 1780 to 1830, Foucault also diagnoses a shift from a medicine of surfaces and classifications to a medicine of depth, of organs and functions (Foucault 1973). Over and beyond the controversies and variations that he traces, one sees a shift in which the medicine of species, focused on the classification of diseases, gave way to a medicine that focused on the individual organic body of the sick person: disease "is no longer a pathological species inserting itself into the body wherever possible; it is the body itself that has become ill" (136). The clinical gaze, trained at the bedside, shaped by anatomical atlases and the experience of pathological anatomy, must read beneath the appearance, distribution, and progression of visible symptoms to the vital living interior of the patient, to identify the underlying pathology that gave rise to them and is the key to their intelligibility.

Foucault does not claim that the whole of clinical medicine followed the vitalism of Bichat or accepted his doctrine that life was the totality of functions that resist death. There were many controversies over the physiological medicine of F.J.V. Broussais, who defended Bichat's legacy when he argued that "disease was not an ontological "other" but the result of altered functions—too much or too little of regular processes" (Porter 1997: 312). However, as Roy Porter puts it, in his magisterial history of Western medicine: "If the patho-anatomists were consumed with disease and death, Broussais' physiology opened windows onto the laws of life—and incidentally influenced the positivist philosophy of Auguste Comte (1798–1857), who analyzed society in physiological terms as an organism" (314). The forms of biological and medical thought that took shape from this point onwards, and the forms of social thought that were modeled on them, were framed within this notion of the organic and functional unity of the living body and the constant threat of disruption that might lead to a cessation of those vital functions. In that sense, resisting death is both the key to an understanding of vitality and pathology, and the definition of life itself. "Since 1816, the doctor's eye has been able to confront a sick organism. The historical and concrete a priori of the modern medical gaze was finally constituted" (Foucault 1973: 192).

The body that twentieth-century medicine inherited from the nineteenth century was visualized by such a clinical gaze, as it appeared in the hospital, on the dissection table, and was inscribed in the anatomical atlas. The body was a vital living system, or a system of systems—it was an organically unified whole. The skin enclosed a "natural" volume of functionally interconnected organs, tissues, functions, controls, feedbacks, reflexes, rhythms, circulations, and so forth. This unified clinical body was located within a social body made up of extracorporeal systems—of environment, of culture—also conceptualized in terms of large-scale flows—of air, water, sewage, germs, contagion, familial influences, moral climates, and the like. The eugenic strategies that I discuss later in this chapter took their character from this way of linking the individual and the social body. The contemporary biomedical body differs from this eugenic body in many ways. Most notably, it is conceived on a different scale. In the 1930s, biology came to visualize life in terms of phenomena at the submicroscopic region—between 10^{-6} and 10^{-7} cm. (Kay 1993: 5). Life, that is to say, was molecularized (see the papers collected in Chadarevian and Kamminga 1998). This molecularization was not merely a matter of the framing of explanations at the molecular level. Nor was it simply a matter of the use of artifacts fabricated at the molecular level. It was a reorganization of the gaze of the life sciences: their institutions, procedures, instruments, spaces of operation, and forms of capitalization.

In 1966, reflecting on the implications of Watson and Crick's famous article of 1964 on the double helical structure of DNA, Canguilhem suggested that one of the many revolutionary consequences would be a redefinition of life: life as meaning (Canguilhem 1966). Contemporary biology, he argued, in changing the scale at which it studies the phenomena of life such as the regulation of function, has "dropped the vocabulary of classical mechanics, physics and chemistry . . . in favor of the vocabulary of linguistics and communication theory. Messages, information, programs, codes, instructions, decoding: these are the new concepts of the life sciences. . . . The science of life no longer resembles a portrait of life . . . and it no longer resembles architecture or mechanics. . . . But it does resemble grammar, semantics and the theory of syntax. If we are to understand life, its message must be decoded before it can be read" (translation quoted from Canguilhem 1994: 316–17). The informational epistemology, which began to take shape in the 1950s, was undoubtedly fundamental to almost half a century of genomics (see the account given in Kay 2000). Combined with a conception of life derived from evolutionary theory—that life is that which can reproduce itself and be subject to natural selection—this stress on the essential continuity and commonality of all living entities undoubtedly formed the framework for much thinking in molecular biology for some four decades.[6] It would be foolish to deny the revolutionary consequences of informational models of inheritance and development for the Human Genome Project and the multitude of endeavors to identify the genomic bases of development, cell biology, metabolism, aging, diseases, and much more. It would be similarly foolish to deny the hold that this new information-based genomics has on governments, foundations sponsoring research, venture capitalists, corporate investment and strategy, the pharmaceutical industry, and the stock market. And perhaps most important, it would be foolish to deny the hold that this idea of genes as the "blueprint for life" coding for our human-ness, intertwined with older ideas of heredity and the gene as the unit of inheritance, has had on popular imagination. Yet perhaps two developments might make us pause before accepting, wholesale, the suggestion that "life as information" has replaced "life as organic unity" as the new biological epistemology.

THE RISE AND FALL OF "THE GENE"

Developments in the life sciences themselves undercut the claim that information is the new epistemology of biology. As is well known, across the twentieth century, the idea of the gene came to designate both the unit of inheritance and the unit of development. The gene was first hypothetical, then given a real location in the cell nucleus, then identified with segments of deoxyribonucleic acid on strands that could be stained and made visible under the microscope and hence termed "chromosomes." Each species had a characteristic number and configuration of these chromosomes, and also, it was presumed, of the genes strung along them; these genes became conceived as the elements that determined the development of an organism, and which, during the process of reproduction, were responsible for passing characteristics, both normal and pathological, to offspring. Critics argued vociferously against the reductionism that was often a part of such arguments, claiming that simple genetic determinism was naïve, that the search for the "gene for" diseases ignored complex developmental, social, and environmental processes by which phenotypes were generated; that single gene disorders were a minority and that determinism could not account for common complex disorders, even less for conditions whose very definitions were in dispute—intelligence, schizophrenia. Such critics included some biologists, notably developmental and behavioral biologists and those who were themselves politically radical (Bateson and Martin 1999, Rose 1998b). But the "gene for" paradigm seemed hard to dislodge.

Over the last four decades of the twentieth century, this idea of the gene was indeed reframed within an informational metaphor, most commonly expressed as a matter of reading: "reading the book of life." This image was embodied in the proposals and assumptions underlying the Human Genome Project (Kay 2000). The set of genes along the human chromosomes were spoken of as the digital instructions for making a human being, and it was estimated that the human genome contained between

100,000 and 300,000 genes, with most, for example Walter Gilbert, settling for a figure of around 100,000 (Gilbert 1993). Even though many, including Gilbert, warned against simple genetic determinism, and recognized that individual genes could encode many different functions with different implications in different diseases, there was still an assumption that each disease might be linked to one or a small number of mutations and that the gene, rather than the protein, the cell, or the organism, was the key unit for analysis.

But this "vision of the grail," as Gilbert termed it, proved to be as ephemeral as most such visions. When the first two papers on the sequence of the human genome were published in 2001, one from the International Human Genome Sequencing Consortium, the other from the team established by Celera Genomics, the estimated number of sequences coding for protein was much less—each team estimated it lay from between 30,000 and 40,000 (Lander et al. 2001, Venter et al. 2001). Some molecular biologists already realized that genomics was moving beyond the "gene for" paradigm: Sydney Brenner (Brenner 2000) noted for example, that the genome of the fruit fly Drosophila Melanogaster was estimated to contain only 14,200 coding sequences, when the genome of the much simpler nematode worm Caenorhabditis elegans had been estimated to contain about 20,000 (Adams et al. 2000, Anon 1998). Given that an average jet plane contains some 200,000 unique parts, even at 30,000 coding sequences—one third more "genes" than a nematode worm and only twice the number for the fruit fly—it was evident that the human genome could not be the "digital parts list" for making a human being. Indeed the 2001 paper by Venter et al. had argued that the reductionist approach had to be abandoned in favor of models of complexity: "networks that exist at various levels and at different connectivities, and at different states of sensitivity to perturbation" (Venter et al. 2001: 1347). And by 2004 a reanalysis of the gene-containing portion of the human genome, identifying significant shortcomings in the shotgun sequencing method Venter had used in Celera's sequencing program, had reduced the number of protein coding sequences still further, to between 20,000 and 25,000 (Collins et al. 2004). Within the style of thought of contemporary genomics, it is accepted that one coding sequence can be involved in the synthesis of several different proteins, and that one protein can entail the interaction of several distinct coding sequences from different regions of the genome. Hence the focus shifted from the gene to processes of regulation, expression, and transcription (transcriptomics), from the gene to those small variations at the level of a single nucleotide termed Single Nucleotide Polymorphisms (SNPs), and indeed from the gene to the cell and the process for the creation of proteins (proteomics).

In his refutation of the Lamarckian doctrine of the inheritance of acquired characteristics, August Weismann demonstrated that somatic changes in the lifetime of an organism did not result in the modification of the reproductive cells (Weismann 1893 [1982]). While Weismann's barrier, for now, remains unchallenged, its reformulation by James Watson as the "central dogma" of genetics—that there is a one-way path from DNA via RNA to protein—can no longer be sustained. Not only does RNA code for DNA, in some circumstances, by means of the enzyme reverse transcriptase, but there are also multiple pathways back from somatic tissues to the DNA sequences during development and in the normal course of cellular metabolism. Further, it is now recognized that the very large proportion of the DNA sequence that does not code for protein, previously labeled "junk DNA," is crucial to the regulatory apparatus of the cell, much of it apparently coding for RNA that shapes the conditions under which particular protein coding sequences are "turned on" and hence expressed.[7] The discovery of multiple types of noncoding RNA—of "noncoding RNA genes and the modern RNA world," to quote the title of Sean Eddy's influential review—has demonstrated that there is much involved, both in transmission across generations and in development of any specific organism, that goes beyond the informational model of DNA as "the book of life" or the "code of codes" (Eddy 2001).[8] Hence the new interest in epigenesis, understood not merely as the mechanisms regulating gene expression that produce phenotypic effects from gene activity during differentiation and development, but also as heritable characteristics, some acquired over the life of an organism, that are not part of nuclear DNA sequences, for example those involving proteins or chemical groups attached to the DNA that modify its function.[9] With some hesitation, but with increasing frequency, researchers are coming to accept that DNA sequences alone do not comprise the master plan of organic existence. Not only might factors other than base sequences themselves be involved in inheritance, but within the development and life of an organism, DNA sequences gain their functional significance only within highly differentiated cellular networks in which RNA-based regulatory mechanisms play a crucial role.[10] We can, then, agree with Evelyn Fox Keller, who has suggested that, as the twentieth century ended, so did "the century of the gene" (Keller 2000). A genetic style of thought is giving way to a postgenomic emphasis on complexities, interactions, developmental sequences, and cascades of regulation interacting back and forth at various points in the metabolic pathways that lead to the synthesis of enzymes and proteins. And in the process informational epistemologies seem to have reached their limit; they can no longer capture what researchers do as they represent and intervene in the vital complexities that constitute life at a molecular level.

Life itself, that is to say, resists its reframing in terms of information, and its properties have refused to be summarized or summoned up in the logics of language, or reduced to a reading out of the "book of life."[11] This is not to deny the significance of information processing to contemporary visions of life itself. Undoubtedly the capacity to represent genetic sequences in terms of the different possible combinations of a small set of elements, and the extraordinary processing capacity of computers to analyze and sequence DNA fragments, made it possible, by the end of the twentieth century, to sequence genomes much faster than had been thought even a decade earlier. Coupled with the delocalizing and interconnecting powers of the Internet, which made such sequences available in real time to any researcher across the world with access to the World Wide Web, these developments transformed molecular biology. And given the extraordinary capacity of even quite simple computers to model events at the level of gene sequences, transcription, protein synthesis, and protein folding, and to make these visually available to researchers, computation opened up the molecular and submolecular world of DNA and RNA sequences, amino acids, proteins, and cell biology to the new molecular gaze of biomedicine. Genomics, transcriptomics, proteomics, and much more depend, for their possibility, on the powers and capacities of computers, and their programmers, to model complex systems. Systems biology depends for its capacities on the integration of informatics, computer simulation, engineering, and experiments in vitro and in vivo. But while life itself may be modeled, simulated, experimented with, and rendered intelligible in silicon, while it can even be augmented by linking living molecules and tissues with processes in silica, we would be mistaken to think that for this reason molecular biology now understands life itself as information. When "synthetic biologists" try to create life, it is not a task in silicon they set themselves, though computational power is no doubt crucial in many respects: the solution, if it is to be found, must be found in the life-liness of the created entities themselves.

Between Life and Nonlife

This leads us to the second reason to pause before accepting the claim that the epistemology of contemporary sciences of life is informational. This concerns the problematic boundaries of "life" itself. Our very sense of what is or is not life, living, or alive is often exactly what is at stake in the politics of the present. A host of entities inhabit a transitional zone where their life-liness is precisely in question. The focus of organizations terming themselves "pro-life" directs our attention to one set of those troubling liminal entities at stake in new reproductive technologies—

sperm, unfertilized and fertilized ova, blastocysts, embryos at various stages of development within and outside the womb, in particular those considered by some to be spare or surplus, together with what Sarah Franklin has termed "the enormous population of frozen embryos that have become official legal entities, with protected status under the law of most countries" (Franklin 2004: 74). To these one might add those entities that previously did not exist, that are generated at the intersection of reproductive technologies and stem cell technologies—stem cells and stem cell lines, the embryonic stem cell clusters, containing cells from all three germ layers, termed "embryoid bodies," which can be induced to differentiate into various types of tissue or to express particular enzymes.[12] Like frozen embryos, these stem cells and their progeny have entered into public debate and their life-liness, and their rights and protections, the legitimacy of the practices that bring them into existence, and their potential fate, have become the subject of intense political debate whose form differs from country to country (perhaps most starkly highlighted by comparisons between the United Kingdom, Germany, Israel, and the United States) (Gottweis 2002b, Lenoir 2000). Then again, one can think of the issues raised in the field of organ transplantation, where human beings can be once, twice, or even thrice dead: here, once more, the lines between life and death are geographically variable and subject to intense political, bioethical, and biomedical debate, as well as intense capitalization (Lock 2002, Scheper-Hughes 2000, 2003b).

These entities, and the troubles they provoke, are not matters of genes or information: what is at issue is vitality at the level of the organism, where the very meaning and limits of life itself are subject to political contestation. Perhaps one can even speak of a new vitalism. The Oxford English Dictionary defines vitalism as "The doctrine or theory that the origin and phenomena of life are due to or produced by a vital principle, as distinct from a purely chemical or physical force" (Haigh 1984, Joad 1928, Wheeler 1939). No doubt this kind of naturalistic vitalism now seems untenable to most. However some contemporary social theorists, taking their cue from Gilles Deleuze, and inspired by his readings of Bergson, are arguing for a novel kind of postmodern materialist vitalism (Braidotti 1994, 2002, Burwick and Douglass 1992, Deleuze 1988a, Marks 1998). While I share many of these views, I take a rather different tack. My aim is not so much to call for a new philosophy of life, but rather to explore the philosophy of life that is embodied in the ways of thinking and acting espoused by the participants in this politics of life itself. What beliefs do they themselves hold or presuppose about the special qualities of living things? What forms of differentiation of life and nonlife do they enact, either explicitly or embedded in their practices? What differences exist in the obligations that they accept toward entities at one side or the other of that divide? What, then, are the politics of our contemporary philosophies of life?

Politics

Critics tend to argue that, like previous appeals to biological nature, contemporary developments in biomedicine, especially those involving genetics, will embody a conception of human beings that *individualizes* human worth, *essentializes* variations in human capacities, *reduces* social phenomena to the aggregate of individual actions, and *discriminates against, constrains, or excludes* those found biologically abnormal or defective. Thus they have warned of the dangers of a new determinism—"geneticism"—based on a false but seductive mystique of the power of the genes, and entailing all sorts of discrimination on genetic grounds (Dreyfuss and Nelkin 1992, Lippman 1992, Nelkin and Lindee 1995, Nelkin and Tancredi 1989). While most accept that this would not take the form of a state-imposed masterplan, they speak of the rise of a new individualized eugenics, a eugenics "by the back door" fueled by a potent mixture of the commercial aspirations of the biotech companies and the desires of parents, especially wealthy parents, for a perfect child in an age of manipulated consumerism and reproductive choice (Duster 1990, Hubbard and Ward 1999, Rifkin 1998). From such a perspective, the rhetoric that celebrates the potential of biomedicine and biotechnology to improve health, welfare, and quality of life denies the collective and environmental determinants of variations in human health and life chances, individualizes and fatalizes these variations, and obscures the threat that new biological practices of control will discriminate against those who are considered biologically inferior, and will coerce, restrict, and even eliminate those whose biological propensities are believed—by doctors, parents, or perhaps even by political authorities—to be defective.

Undoubtedly these possibilities exist. Indeed, as we shall see, every act of choice opened up by the new biomedicine does indeed involve a judgment of value in a field of probabilities shaped by hopes. Even the use of a pharmaceutical therapy implies that one kind of life is better than another, that one might hope for a life free of pain or anxiety disorders, and that there is a possibility, based on probabilistic evidence from clinical trials and medical experience, that a particular drug might help. Indeed I write this sentence on the day in July 2005 when the French biotech company IntegraGen announced that it will soon launch the first genetic test diagnostic of autism. A judgment of the differential value of different forms of life is engaged whenever preimplantation genetic diagnosis is used in the hope that parents can avoid having a child that carries a mutated genetic sequence that might lead it to a life with a crippling or terminal illness or a troubled subjectivity.[13] This field of choices, judgments, values, and hopes about life itself is the territory of our new vital politics. But history is not repeating itself. Even as we try to choose one kind of life over another, to give birth to a child free of illness or disorder rather than one afflicted, we are not on the verge of a new eugenics, or even a revived biological determinism.

In the new field of biopolitics, where interventions are scaled at the molecular level, biology is not destiny but opportunity—to discover the biological basis of an illness, of infertility, of an adverse drug reaction in a cascade of coding sequences, protein syntheses, and enzyme reactions is not to resign oneself to fate but to open oneself to hope. The nonimplantation of a potentially afflicted embryo is not to condemn a defective or inferior person to death; it embodies the hope that biological information may enable potential parents to maximize the chance that they will have a child who will lead a fulfilled life. For one who lives, to identify the biological bases of an affliction is, in principle at least, to render that condition open to intervention, transformation, and rectification at the molecular level—to "reverse engineer" the condition and then to rectify the anomaly or compensate for the missing elements. To put it crudely, biology has become imbued with dreams of technological reformation.

And, perhaps more fundamentally, as we have seen, biology is not destiny because the new molecular knowledges of the human condition are not deterministic, even though they do still have some simpleminded determinists among their proponents. The kinds of explanations generated in genomics, proteomics, transcriptomics, and cell biology are not simple, linear, and direct causal chains. To use a much abused term, they are "complex." While causal chains can be traced, between a coding sequence and a protein for example, the actual cellular mechanisms involved in the creation of that protein in vivo depend on interactions between multiple events at different levels, involving the regulation of gene expression by factors in the cellular environment. And the mode of action of that protein, once synthesized, involves a nexus of activations and terminations, cascades, feedback loops, regulatory mechanisms, epigenetic processes, interactions with other pathways, and much more. The complexity of such cellular mechanisms, their operation in time (hence directionality, interaction, feedback) and in space (hence movements, circuits, passage across membranes and between cells, activation of secondary systems) ensures that the relations here, even at the cellular level, are stochastic, open and not closed, and hence probabilistic. At levels above the cell, the knowledges produced by the new biology are also probabilistic rather than deterministic. Even in the very simplest and starkest example, of a condition long believed to be caused by a mutation in a single gene—

Huntington's Disease—which has been localized to the short arm of Chromosome 4, and identified as an expanded CAG repeat—the test that enumerates the numbers of repeats does not tell the affected individual or their family when they will become ill, or how rapidly the condition will progress, let alone when they will die. The genetic bases of common complex disorders are certain to involve interactions of many coding sequences, at many different loci, some of which are protective and others of which increase risk, shaped by development, biography, environment, and much more. The new life sciences, of which genomics is only one aspect, thus open up a space of uncertainty, not certainty. While the calculation of risk often seems to promise a technical way of resolving uncertainty, risk calculations offer no clear-cut algorithm for the decisions of doctors or their actual or potential patients. Contemporary biopolitics thus operates in practices of uncertainty and possibility: We need new critical tools to analyze it.

Biopolitics

In the first volume of *The History of Sexuality*, Michel Foucault remarks that "For millennia, man remained what he was for Aristotle: a living animal with the additional capacity for a political existence; modern man is an animal whose politics place his existence as a living being in question" (Foucault 1978).[14] He develops these arguments in a chapter entitled "Right of Death and Power over Life": For a long time, he claims, one of the privileges of sovereign power was the right to decide life and death—the right of a ruler to seize things, time, bodies, ultimately the life of subjects. But, he suggests, since the classical age, deduction has become merely one element in a range of mechanisms working to generate, incite, reinforce, control, monitor, optimize, and organize the forces of individual and collective life. While external wars are bloodier than ever, and regimes inflict holocausts upon their own populations, he did not consider these wars to be waged in the name of the sovereign, but in the name of the existence of everyone: "entire populations are mobilized for the purpose of the wholesale slaughter in the name of life necessity . . . It is as managers of life and survival, of bodies and the race, that so many regimes have been able to wage so many wars, causing so many men to be killed" (137).

From the eighteenth century, at least in Europe, Foucault argued, political power was no longer exercised solely through the stark choice of allowing life or giving death. Political authorities, in alliance with many others, took on the task of the management of life in the name of the well-being of the population as a vital order and of each of its living subjects.

These new tasks of management emerged around a range of specific problems, for example those of illness, sanitary conditions in the towns, problems of security in old age, and of accidents. In the course of struggling to understand and intervene upon these problems, the obligations of political and other authorities extended to the vital processes of their subjects—not just the power to let them live or make them die, but a multitude of attempts to manage their life, to turn their individual and collective lives into information and knowledge, and to intervene on them. From this moment on, politics would have to address the vital processes of human existence: the size and quality of the population; reproduction and human sexuality; conjugal, parental, and familial relations; health and disease; birth and death.

In volume 1 of *The History of Sexuality*, Foucault proposed a now familiar bipolar diagram of biopower, or power over life. One pole of biopower focuses on an anatamo-politics of the human body, seeking to maximize its forces and integrate it into efficient systems. A second pole is one of regulatory controls, a biopolitics of the population, focusing on the species body, the body imbued with the mechanisms of life: birth, morbidity, mortality, longevity. He claims that this bipolar technology, emerging in the seventeenth century, seeks "to invest life through and through" (1978: 139). He argued that by the nineteenth century these two poles were conjoined within a series of "great technologies of power" of which sexuality was only one. Now new kinds of political struggle could emerge, in which "life as a political object" was turned back against the controls exercised over it, in the name of claims to a "right" to life, to one's body, to health, to the satisfaction of one's needs (145). The distinction between discipline and regulation—between strategies seeking the management of individual bodies and those focused on the collective body of the population—blurs, as different authorities seek to act upon the one through action upon the other: to act upon the health of the population as a whole, for example, by inculcating hygiene into the bodily habits of each individual within their domesticated households, and to act upon the habits of each individual by reshaping the urban spaces and milieu within which they were to conduct their lives.

Foucault initially linked the expansion of politics to the management of life with the expansion of the ambitions and obligations of the state. But of course, in the eighteenth and nineteenth centuries, such intellectual and political strategies were pioneered outside the state—by philanthropists, medical reformers, religious organizations, among others. And he recognized that "the great overall regulations that proliferated throughout the nineteenth century . . . are also found at the sub-State level, in a whole series of sub-State institutes such as medical institutions, welfare funds, insurance, and so on" (Foucault 2002). Indeed it became clear that the field of biopolitics, even in the high point of the social state, was not the implementation by the state of a single strategy of regulation. Biopolitics took many forms, from the management of cities, space, and sociality in the name of minimization of disease, to attempts to maximize the quality of the race through the administration of birth and death. It was been inextricably bound up with the rise of the life sciences, the human sciences, clinical medicine. It has given birth to techniques, technologies, experts, and apparatuses for the care and administration of the life of each and of all, from town planning to health services. It was a fragmented field of contested truths, heterogeneous and often conflicting authorities, diverse practices of individual and collective subjectification, competing ways of thinking and acting, and divergent opinions about what were the most important, and most appropriate, objectives for authoritative action. But, nonetheless, the birth of biopolitics gave a kind of "vitalist" character to the existence of individuals as political subjects.

Biopower is more a perspective than a concept: it brings into view a whole range of more or less rationalized attempts by different authorities to intervene upon the vital characteristics of human existence—human beings, individually and collectively, as living creatures who are born, mature, inhabit a body that can be trained and augmented, and then sicken and die. Given the intrinsic connections between the management of populations and their characteristics, and the government of bodies and their conducts, I will use the term "biopolitics" to refer to the specific strategies brought into view from this perspective, strategies involving contestations over the ways in which human vitality, morbidity, and mortality should be problematized, over the desirable level and form of the interventions required, over the knowledge, regimes of authority, and practices of intervention that are desirable, legitimate, and efficacious.

Eugenics

The image of biological politics that is stamped in the minds of most contemporary critical intellectuals is that of the eugenics of the first half of the twentieth century. Eugenics, as is well known, comprised a whole set of strategies, which had in common the presupposition that it was desirable, legitimate, and indeed necessary to secure the future welfare of the nation by acting upon the differential rates of reproduction of specific portions of the population, so as to encourage the best to procreate and to limit the procreation of those thought to be of lower, inferior, defective, or diseased stock (of the many general accounts of eugenics, the best is Kevles 1985). As Lene Koch has pointed out, references to eugenics in many contemporary evaluations of biomedicine have a rhetorical function.[15] Some invoke eugenics to *distinguish* the present from the past: thus contemporary molecular geneticists usually argue that their discipline, in common with the rest of medicine, has decisively rejected eugenics in favor of individualized, voluntary, informed, ethical, and preventive medicine organized around the pursuit of health. Others invoke eugenics to *link* the present with the past, suggesting that despite its differences, contemporary biomedicine, in combination with genetics, still judges human life and worth, insofar as it intervenes upon the chances of life in order to eliminate differences coded as defects. But if we are to be clear about the specificity of the present, beyond rhetoric, it is important to be precise about the nature of eugenics as a biopolitical strategy. Francis Galton coined the term "eugenics," defining it thus in 1883: "Eugenics is the science of improving stock, which is by no means confined to judicious mating, but which, especially in the case of man, takes cognisance of all influences that tend in however remote a degree to give the more suitable races or strains of blood a better chance of prevailing speedily over the less suitable than they otherwise would have had" (Galton 1883: 25n). As was clear in his writings, as they developed from the 1860s to the 1890s, his objective was to demonstrate that the standard of the human race could be improved by deliberate political action that encouraged "beneficial" evolutionary developments and discouraged those that would tend to reduce the quality of the race.[16]

Crucial to note here is the emphasis on population, conceived of as an interbreeding unity with certain characteristics: as Galton put it, "the science of heredity is concerned with Fraternities and large Populations rather than with individuals and must treat them as units" (Galton 1889: 35). Indeed, in this respect Galton was entirely Darwinian, for as François Jacob put it, "The most radical transformation of the biological attitude wrought by Darwin was to focus attention, not on the individual organisms, but on large populations" (Jacob 1974: 174). For Galton and his followers, the theory of evolution was based on the laws of large numbers and worked on populations; what degenerated, or could be improved, was the population as a whole. While one must perforce act on the population by regulating the rates of reproduction of the individuals in different sectors, the aim of that action on individual reproduction is the improvement of the quality of the population itself. And "population," here, designated a relatively enclosed human collectivity, territorialized upon the nation. While races were, of course, not coterminous with nations, nonetheless if eugenics was to be taken seriously it would be the nation, as the unit of politics both in terms of external contests and internal governance, that would be required to act. Further, for this type of thought a population contains individuals that can be ranked according to their quality, where quality is an overall summation of their evolutionary fitness, and one that is manifested in both physical and mental characteristics and hence can be indexed, say, by a single measure such as that of intelligence. For Galton, the key was "energy": "In any scheme of eugenics, energy is the most important quality to favor; it is . . . the basis of living action, and it is eminently transmissible" (Galton 1883: 27). He shared this way of thinking with many others in the second half of the nineteenth century; all held a conception of degeneracy based on the idea that individuals inherited a constitution of a certain quality whose strength or weakness was manifested in a whole range of physical and mental characteristics: energy, intelligence. And, as became central to eugenic arguments, if it was the case that those of weakest constitution bred faster than those of the strongest—and many felt that there were good reasons to believe that this was so—the consequence for the quality of the population as a whole, and hence for the vitality of the nation, was obvious.[17]

It was within this way of thinking that eugenicists urged politicians and all those in power to improve the body politic, and to relieve it of the economic and social burdens of disease and degeneracy in the future, by acting upon the reproductive decisions and capacities of individuals in the present. Thus eugenic responsibilities fell upon states, or statesmen, to act both negatively—to prevent the excessive breeding of those of worst quality, or the dilution of the quality of the population by an influx of those less fit from outside through immigration—and positively—to encourage those who were fittest to reproduce for the good of the nation as a whole. Initially promoted by intellectuals and various pressure groups, by the end of the nineteenth century eugenic ideas were taken up by states in many countries, in the context of a recoding of politics as a struggle between nations thematized as races. As Foucault argues, this was at the point where the State had to look like, function and present itself as the guarantor of the integrity and purity of the race, and had to defend itself against the race or races that were infiltrating it, introducing harmful elements into its body, and which therefore had to be driven out for both political and biological reasons (Foucault 2002). "Population," "quality," "territory," "nation," "race"—these were the terms that were fused in the discourse of eugenics and the terms that gave it its specific, and ultimately murderous, character.

Many contemporary commentators suggest that this eugenic wish to control the biological makeup of the population underpins biological politics today. For both Giorgio Agamben and Zygmunt Bauman a thanatopolitics of population purification lies at the very heart of modernity (Agamben 1998, Bauman 1989). They think that this is immanent within the very ethos of biopolitics: to manage the health of the "body politic" inescapably requires the control and elimination of "foreign bodies." The National Socialism of the German Third Reich certainly exemplifies this link between the administration of life and the administration of death. For example, in a book of 1936 entitled *Rassenhygiene als Wissenschaft und Staatsaufgabe*, Ottmar von Verschuer asserts that politics should mean "giving form to the life of the people" (von Verschuer, 1936: 8, quoted from Agamben 1998: 148). Life itself, the vital reality of a people, must become the overriding criterion that should guide the exercise of political authority. This requires a nationally organized and politically directed program to improve the quality of the national stock and eliminate taints or weaknesses that might threaten it. Its tactics range from propaganda and education stressing the burden on the race imposed by those with defective constitutions, to eugenic legislation on marriage, the assessment of reproductive worth by experts working on behalf of the state, and the prevention of those members of defective or inferior subpopulations from reproducing through sterilization or extermination. Once each life has a value that may be calculated, and some lives have less value than others, such a politics has the obligation to exercise this judgment in the name of the race or the nation. All the eugenic projects of selective reproduction, sterilization, and incarceration follow.

In *Homo Sacer*, Giorgio Agamben finds this diagram of biopower at the very heart of the present (Agamben 1998). He argues that death—the threat of death, the fear of death, the example of death, the calculated exercise of death—is not merely a premodern negative out of which a positive, productive biopolitics emerges. On the contrary, he suggests, thanatopolitics—a politics of death—is actually the first principle of biopolitics. At the very moment when political sovereignty was established over a territory, power was linked to the living bodies of its subjects, if only because this is what enables the sovereign alone to make legitimate political use of their death. The fundamental biopolitical structure of modernity retains this link with sovereignty: it is dependent on the belief that life itself is subject to a judgment of worth, a judgment that can be made by oneself (suicide) but also by others (doctors, relatives) but is ultimately guaranteed by a sovereign authority (the state). Thus, he claims, thanatopolitics operated in the name of sovereign control over the bodies and lives of each and of all is the inescapable other side of the positive biopolitics of the life of a population. Indeed he asserts that "the camp" is the "biopolitical paradigm of the modern": when the collective body of the people becomes the principal resource for politics, the purging of defective individuals becomes an essential part of the care of life.

It is true, as Agamben points out, that the thanatopolitical rationalities and strategies of modern (twentieth century) geopolitics were not throwbacks to a past political configuration. But I think he is wrong in suggesting that the camp is the diagram of the biopolitics of the present, and that exclusion and elimination are the hidden truth or ultimate guarantee of contemporary biopolitics. Life may, today more than ever, be subject to judgments of value, but those judgments are not made by a state managing the population en masse. Contemporary biopolitics in advanced liberal polities does not take the living body of the race and its vital components as resources whose fitness is to be maximized in a competitive struggle between states. And while the lives, illnesses, and troubles of many may be ignored or marginalized in contemporary political economies of vitality, to let die is not to make die—no "sovereign" wills or plans the sickness and death of our fellow citizens. While we can agree with Agamben that today, life itself is both beyond value and subject to recurrent judgments of value, the troubles of our own times are not reactivations of the past. The four terms that delineated eugenics—population, quality, territory, and nation—do not characterize the molecular biopolitics of the present.

However, Agamben and Bauman are undoubtedly correct in one respect. Over the course of the twentieth century there was no such clear distinction between preventive medicine and eugenics, between the pursuit of health and the elimination of unfitness, between consent and compulsion. Even under National Socialism—which was, as Foucault points out "a paroxysmal development . . . a coincidence between a generalized biopower and a dictatorship that was at once absolute and retransmitted throughout the entire social body" (Foucault 2002: 260)—biopower was a complex mix of the politics of life and the politics of death. As Robert Proctor has shown, it entailed not merely the exercise of state power but strategies for governing life developed by many other authorities. Nazi doctors and health activists, not acting under the direction of a sovereign state, waged war on tobacco, sought to curb exposure to asbestos, worried about the over use of medication and X Rays, stressed the importance of a diet free from petrochemical dies and preservatives, campaigned for whole-grain bread and foods high in vitamins and fiber, and many were vegetarians (Proctor 1999).

More generally, at the start of the twentieth century, two great state-sponsored biopolitical strategies were taking shape across the nation-states of Europe and North America and in many of their colonies. The former sought to maximize the fitness of the population by adding an individualized attention to the habits of subjects to the earlier hygienic concern with securing the external conditions of health through town planning, sewage systems, and the like. Taking its cue from the moralizing interventions into the habits of the poor that had proliferated in the second half of the nineteenth century, this "neohygienist" program sought to instill habits conducive to physical and moral health into each individual via the machinery of the domesticated home and school, linked with such disciplinary and tutelary measures as medical inspection of school children, health visitors, clinics, and so forth (Donzelot 1979, Rose 1985). Health here formed a kind of transactional zone between political concerns for the fitness of the nation and personal techniques for the care of self.

Eugenics was the second great biopolitical strategy. This is not the place for a detailed examination of eugenics arguments, organizations, and practices, but some further analysis is relevant if we are to be clearer about the specificity of eugenics as a political rationality and its links to an array of technologies of government. As we have seen, eugenics sought to maximize the fitness of the population, but it privileged one site—that of reproduction. Eugenic programs thus used a combination of incentives and compulsion to modulate the wish or ability of individuals in certain categories to procreate: those judged to have hereditary disease, to be deranged, feebleminded, or physically defective, those who were deemed habitually or incorrigibly immoral or antisocial, especially alcoholics and those guilty of sexual crimes. "Positive" eugenic measures ranged from exhortation to family allowances. "Negative" eugenic measures ranged from abortion, to segregation, to more or less involuntary sterilization, to "euthanasia." Eugenics was not disreputable or marginal: it defined one dimension of mainstream thinking about the responsibilities of politicians, professionals, scientists, and individuals in the modern world. Scientific and popular books written in these terms proliferated, eugenic organizations and societies were established in many countries, programs of eugenic research were initiated, national and international congresses, conferences, and symposia were held, and doctors, social workers, and many other professionals were trained in eugenic ways of thinking and encouraged to carry out their practices in eugenically informed ways (Anon. 1919, Bibby 1939; for just a few examples from the English language materials only, see Bishop 1909, Carmalt and Connecticut State Medical Society 1909, Carr-Saunders 1926, Castle 1916, Crookshank 1924, Dampier and Whetham 1909, Darwin 1928, Galton and Eugenics Education Society (Great Britain) 1909, Lidbetter 1933, Pearson 1911, 1912, Saleeby 1914).

There were many disputes between those allied with the eugenic movement and those who argued for preventive medicine and public health education. But the operational relations between these strategies were more complex: eugenic themes were present in much politics of public health, and public health and preventive medicine seemed to many to be compatible with eugenic thought. The belief in the necessity of rational planning of the quality of the population underpinned early immigration controls in the United States (Walker and Dewy 1899, Walker 1924). Attempts to act on reproduction were widespread, ranging from popular advice on the choice of marriage partners, through the development of children's allowances and welfare benefits for mothers, to segregation and sterilization of those thought to be physically or morally unfit. Infused with a more or less virulent racism, eugenic policies of forced or coerced sterilization of those considered threats to the quality of the population—notably inhabitants of mental hospitals, the "feeble-minded," and those deemed incorrigibly immoral or antisocial—spread across the United States, Europe, to Latin America and beyond.[18] In the 1920s and 1930s, such sterilization laws were passed in many states in the United States, in Switzerland, Denmark, Finland, Germany, Norway, Estonia, Iceland, Mexico (Vera Cruz), Cuba, Czechoslovakia, Yugoslavia, Lithuania, Latvia, Hungary, and Turkey. While a number of intellectuals toyed with the idea of "the lethal chamber" for those of lower races who threatened to undermine the quality of the population, this lethal solution was enacted only in Nazi Germany.

The murderous form that eugenics took in Germany derived from a fusion of eugenic ideas with other images of race, with debates about euthanasia for those whose lives were thought "not worthy of life," and with longstanding concerns about racial hygiene, and the fear of disease, notably typhus, penetrating into the body politic from degenerate social groups, especially those from the East. Hugh Raffles, in his study of Nazi insectification, shows how bacteriology, parasitology and entomology intertwined with concerns about German population purity: initial moves to safeguard the national homeland by concentrating those thought to be carriers of disease in camps and delousing them, transmuted into strategies to eradicate disease by transforming the people themselves, notably the Jews, into the parasitic lice, and hence to "delouse" the German *Volk* itself.[19] Raffles quotes Heinrich Himmler, in April 1943: "Antisemitism is exactly the same as delousing. Getting rid of lice is not a question of ideology. It is a question of cleanliness."[20]

But such insectification did not characterize all eugenic discourses. Even in Germany cleanliness was only one way in which the elimination of those of weaker stock was framed. Another common theme was economic. Robert Proctor's superb book on medicine under the Nazis includes examples of illustrations from *Volk und Rasse* showing the different annual costs to the Prussian government of educating normal (125 Reichsmarks), slow (573 RM), educably mentally ill (950 RM), and blind or deaf school children (1,500 RM); another poster shows an image of a healthy German working man carrying two deformed individuals on his shoulders and bears the slogan, "You Are Sharing the Load! A Genetically Ill Individual Costs Approximately 50,000 Reichsmarks by the Age of Sixty" (Proctor 1988: 182). Indeed, the modification of public attitudes and individual judgments by education and counseling was central to eugenics. It should not surprise us then that even in Nazi Germany, at least as far as mental pathologies were concerned, sterilization and euthanasia were not always imposed by a coercive State apparatus. In the context of the eugenic domination of the medical and scientific literature, many German doctors took their own decisions on eugenic grounds; in the context of a widespread campaign of propaganda and public education, parents often requested eugenic measures for their own children (Burleigh 1994).

Sterilization for eugenic reasons continued into the postwar period in a number of democratic nations.[21] Many scientists and policymakers saw nationally planned and state directed eugenic practices as quite compatible with rationalities of welfare. Sweden is the exemplar.[22] Here, from 1935 up until 1975, under a nationally organized program carried out by liberal and humanistic doctors, a total of 62,000 people—many who were merely thought to be antisocial or sexually promiscuous—were prevented from reproducing by sterilization, much of which was involuntary or coerced. While the initial targets had been informed by concerns about racial purity, the program continued into the postwar years, where its targets were largely women who were thought to be antisocial, sexually active, and without good judgment. It was publicly rationalized on the basis that the size of families of those with a history of antisocial behavior must be controlled. This was a kind of pastoral eugenics operating unproblematically within the context of a developing paternalistic welfare state. It was modeled on the responsibilities of a good shepherd—who must, of course, be prepared to take sometimes harsh decisions in order to reduce the burden that weak or sickly sheep would otherwise place on the flock as a whole (cf. Foucault 2001). Eugenic arguments, informed by a wish to improve the quality, fitness, and health of the national population, also underpinned pronatalist policies such as family allowances and prenuptial examinations in France, and the key actors saw no contradiction between eugenic and hygienic strategies to combat "social evils" (Carol 1995, Drouard 1999). And, at least up until the 1950s in Britain and the United States, eugenic considerations infused reproductive advice to prospective parents in the new profession of genetic counseling: such advice was considered especially important for those subjects with a family history of inherited defects or disease who lacked the moral capacity to appreciate the implications and hence control their reproduction.[23]

State-organized or state-supported tactics for modification of reproductive decisions and capacities in the name of population health thus played a part in the medical and biological politics of many liberal democratic societies throughout the twentieth century in the name of preventive medicine and public health. However great the moral and political distance between euthanasia, compulsory sterilization, and genetic counseling, it is clearly inadequate to mark the specificity of contemporary biopolitics by counterposing positive to negative policies, voluntary to compulsory measures, coercion, and persuasion. The specificity of the biopolitics of the first half of the twentieth century lies elsewhere: in the links established between population, quality, territory, nation, and race. It involved more than the idea that, other things being equal, healthy individuals were more desirable than those who were unhealthy. Health was understood in terms of quality—of the individual and of the race—and quality was understood in a quasi-evolutionary manner, as fitness. The problem space of eugenics was framed in terms of the political importance attached to the fitness of the national population considered en masse, as it competed with other national populations. Population fitness was liable to threats from within and without, and national governments had the obligation to guard against these threats and to take measures to enhance that fitness through policies formulated by, and enacted through, the apparatus of the state.

Eugenics was grounded in the truth discourse of the biology of the first four decades of the twentieth century, not just in Germany, but also in Britain, France, the Nordic countries, and the United States.[24] But that truth has been transformed. In the aftermath of World War II, much intellectual and political work went into disconnecting the links between concerns about genetic health of individuals and concerns about the quality of the population en masse. Genetics was to transform itself into a liberal discipline. Geneticists reflected upon their own past, and reshaped the problem space of genetics in terms of the search for the roots of disease. The norm of individual health replaced that of the quality of the population. Geneticists would cease to think in terms of broad social categories. They would try to penetrate beneath the misleading appearances of pathology and normality to the underlying determinants, the genes and their modes of functioning at a molecular level (Paul 1998a, 1998b) This molecularization is central to the thought style of contemporary biology, and hence to the practices and politics of contemporary biomedicine.

In addition, the political rationalities of our present are no longer inspired by the dream of the taking charge of the lives of each in the name of the destiny of all (see my argument in Rose 1999). The ideal of an omnicompetent social state that would shape, coordinate, and manage the affairs of all sectors of society has fallen into disrepute. The idea of "society" as a single, if heterogeneous, domain with a national culture, a national population, a national destiny, coextensive with a national territory and the powers of a national political government has entered a crisis. The idea of a "national culture" has given way to that of "cultures," national identity to a complex array of identity politics, "community" to communities. In this new configuration, the political meaning and salience of health and disease have changed.

Of course, programmers of preventive medicine, health promotion, and health education still take, as their object, "the nation's health." Today, however, the rationale for political interest in the health of the population is no longer framed in terms of the consequences of population unfitness as an organic whole for the struggle between nations. Instead it is posed in economic terms, the costs of ill health in terms of days lost from work or rising insurance contributions, or moral terms, the imperative to reduce inequalities in health. While international comparisons are undoubtedly still significant, their contemporary political function is no longer that of marking the potential vulnerability of a polity in geopolitical rivalry; rather, they serve as public indices of the extent to which nations have instituted successful policies for the governance of health. National health indicators, here, do not measure the fitness of a population as a whole; they function as aggregates of the health status of individual citizens and families. Further, in fostering health, the ideal relation of state to people is no longer that aspired to in the "social" state. The contemporary state does not "nationalize" the corporeality of its subjects into a body politic on which it works en masse, in relation to the body politics of other states competing in similar terms. The state is no longer expected to resolve society's needs for health. The vitality of the species—the nation, the population, the race—is rarely the rationale and legitimation for compulsory interventions into the individual lives of those who are only its constituent elements. In this domain as in so many others, the images now are of the enabling state, the facilitating state, the state as animator.

On the one hand, the state retains the responsibility that it acquired in the eighteenth or nineteenth century (the precise timing varying across national contexts) to secure the general conditions for health: regulating the sale of foodstuffs, organizing pure water and sewage disposal, sometimes mandating the addition of health-promoting elements into the diet—vitamins, fluoride in water, and the like. On the other hand, within such a health-promoting habitat, the state tries to free itself of some of the responsibilities that it acquired throughout the twentieth century for securing individuals against the consequences of illness and accident. Thus we have seen an intensification and generalization of the health-promotion strategies developed in the twentieth century, coupled with the rise of a private health insurance industry, enhancing the obligations that individuals and families have for monitoring and managing their own health. Every citizen must now become an active partner in the drive for health, accepting their responsibility for securing their own well-being. Organizations and communities are also urged to take an active role in securing the health and well-being of their employees and members. This new "will to heath" is increasingly capitalized by enterprises ranging from the pharmaceutical companies to food retailers. And a whole range of pressure groups, campaigning organizations, and self-help groups have come to occupy the space of desires, anxieties, disappointments, and ailments between the will to health and the experience of its absence. Within this complex network of forces and images, the health-related aspirations and conduct of individuals are governed "at a distance," by shaping the ways they understand and enact their own freedom. Biopolitics today no longer operates in a problem space defined by population, quality, territory, and nation. Individual substitutes for population, quality is no longer evolutionary fitness but quality of life, the political territory of society gives way to the domesticated spaces of family and community, and responsibility now falls not on those who govern a nation in a field of international competition but on those who are responsible for a family and its members. It is not that we should now celebrate the march of progress. But it is in this new politics, the politics of what we have come to know as freedom, that the costs of the new biomedicine, and the choices it opens for us, are to be found.

Population

Considerations of the absence, termination, or prevention of life are not absent in contemporary biopolitics: we need only to think of contraception, abortion, preimplantation genetic diagnosis, debates about the right to die, and much more. And the different values attached to different forms of life are even more evident if we consider the vast geographical discrepancies in morbidity and mortality that existed as we entered the twenty-first century. At the end of the twentieth century some 12.2 million children under five years of age in less developed countries died every year—equal to the combined total population of Norway and Sweden—mostly from causes that could be prevented for a few cents (U.S.) per child. A person in Malawi had a life expectancy of thirty-nine years; in the most developed countries life expectancy was twice this at seventy-eight years. This is "letting die" on a massive and global scale. But it does not operate according to the biopolitical rationalities of the camp; it is not grounded in any political rationality that seeks to adjust the qualities of the population as a whole in the name of national political objectives. Perhaps the specificity of our present becomes clearer if we look at two relatively recent examples where population was the explicit object of politics.

LIMITS TO GROWTH: THE "POPULATION EXPLOSION"

First, consider the politics of population limitation that took shape in the 1970s as exemplified by the 1972 report from the Club of Rome entitled *Limits to Growth* (Meadows 1972). Using a model derived from system dynamics for its analysis, the report concluded that "if the present growth trends in world population, industrialization, pollution, food production, and resource depletion continue unchanged, the limits to growth on this planet will be reached sometime within the next one hundred years. The most probable result will be a rather sudden and uncontrollable decline in both population and industrial capacity." Fundamental to their prescription to avert this problem was birth control, to stabilize population by limiting family size to two children, especially in those countries where it currently greatly exceeded that, but even this path was no guarantee of success: "We end on a note of urgency. . . .We have repeatedly emphasized the importance of natural delays in the population-capital system of the world. These delays mean, for example, that if Mexico's birth rate gradually declined from its present value to an exact replacement value by the year 2000, the country's population would grow from 50 million to 130 million. . . .We cannot say with certainty how much longer mankind can postpone initiating deliberate control of its growth before it will have lost the chance for control."(Meadows 1972:182–3)

These dire warnings resonated with a raft of analogous concerns about the impact of population growth on economic wealth and the need for governments—especially those of less developed states—to introduce policies to curtail reproduction, especially among the poor, as a pre-requisite to modernization. These varied from the authoritarian—the sterilization campaigns in India are the best known—through those that gradually came to adopt principles of informed consent, to what was euphemistically termed "voluntary surgical contraception" in Mexico. They were based on demographic data and algorithms linking population growth to economic performance developed by geographers and mathematicians, embedded in educational programmers for development workers and others, proselytized by numerous private pressure groups and policy advisory bodies, and built into the policies of development agencies such as the "Office of Populations" of the "Bureau for Global Programs" of the United States Agency for International Development (USAID). The "population time bomb" became part of the common sense of public opinion in the West, and a major justification for aid from advanced industrial societies to poorer countries was that this would enable them to limit their population and hence the danger that their population growth posed.

By the start of the 1990s, Robey et al. (1992) reported that voluntary female sterilization was the most prevalent contraceptive method worldwide, used by over 138 million married women of reproductive age compared to 95 million in 1984. For many critics, despite the rhetoric of informed consent and choice they deployed, such campaigns were aimed at particular segments of the population considered problematic or undesirable, and amounted to a form of global eugenics.[25] However I think this is misleading. If we use the term "eugenics" to apply to any intervention on the reproduction, morbidity, and mortality of the population, it covers everything from contraception through abortion to public health, and its use becomes merely part of a general critical rhetoric. Limiting population size for economic reasons, however reprehensible the methods that may be used, is not the same as seeking to maximize racial fitness in the service of a biological struggle between nation-states.

THE QUALITY OF THE RACE: MANAGING THE CHINESE POPULATION

We can, however, find recent instances of explicitly eugenic policies. Perhaps the clearest example is that of China, at least up until the start of this century (I am drawing here on Dikötter 1998). It would be misleading to assume that biopolitical conceptions of inheritance, degeneration, population, and race in China in the late nineteenth and early twentieth century were radically different from those in Europe. Eugenic ideas were widely held by a number of professional groups in republican China in the 1920s, although, according to Dikötter they were not institutionalized to any significant extent, perhaps due to the Japanese invasion and its aftermath. After the Communist victory in 1949, eugenics was condemned for its class bias. However the concept of birth planning was not condemned, indeed it was enunciated by Mao himself in 1956–57, and in the 1960s it was seen as an integral part of the socialist project of social and economic planning (Greenhalgh 2005). However it was not until the 1970s that population control targets were set by China's central planners, and from the end of the 1970s a number of University Institutes were established specializing in the study of population policy. As Susan Greenhalgh has shown, this was part of a sustained project to render the Chinese population knowable and calculable through the amassing of population statistics from a wide range of sources (Greenhalgh 1986, 2003, 2005). The population was problematized in terms of its "abnormality" and "backwardness" on the one hand, and its consequences for Chinese modernization on the other, and the view that the size of the Chinese population was itself a threat to modernization underpinned a program of population planning, the setting of targets and the major birth control strategy, most notably the "one child policy" with its combination of legislation, propaganda aimed at transforming subjectivities to limit the will to reproduce, together with widespread sterilization.

Modernization did not merely require controlling the size of the Chinese population, but also its quality.[26] With the increasing significance of medical experts and population specialists involved in the one child policy, one of the goals of reproductive advice became that of improving the "quality" of births. As Dikötter puts it: "a variety of different issues, including birth control, sex education, infant health care and prenatal screening techniques, are articulated within a eugenic frame which is prominent in its denial of personal choice. . . . Eugenics provides an overarching rationale for a range of reproductive and demographic concerns which are constrained by a policy which prioritizes the needs of board collectivities of interests such as "the state," "the economy; or 'future generations" over the possible desires and choices of individuals" (Dikötter 1998: 123). This style of thinking was embodied in health manuals giving advice to pregnant women on matters ranging from diet to the risks of birth defects in women above a certain age, legal restrictions on early marriage, the need for premarital medical examination, and genetic counseling to discover concealed defects in a prospective partner that may influence the quality to offspring and much besides. There was great concern to eliminate those with "mental retardation" and about the dangers of inbreeding, which were thought to increase the chances of an encounter between deleterious genes.

The Marriage Laws of the People's Republic of China of 1950 and 1980 explicitly identified categories of people unfit for reproduction, especially those where both partners had a history of hereditary disease, or where both partners had a history of "mental disease." The focus of concern was often with preventing births of those with fairly major birth defects—whose lives were often described as not worth living—but there is some evidence that it spread to milder conditions such as harelip. Despite moves to integrate disabled people that were also developed during the 1980s, and the passage of a Law on the Protection of Disabled People in 1990 establishing equal rights in law, and prohibiting discrimination, some doctors remain of the view that neonates born with such disabilities should have their lives terminated.[27] Repeating themes common in European eugenics of the first half of the twentieth century, leading medical experts calculated the number of genetically defective people in the People's Republic and estimated their cost to the nation; the journal *Population and Eugenics* was particularly concerned with the consequences of differential reproduction, with the least gifted members of society breeding faster than the best; and there were a number of national exhibitions on eugenics and health aimed at eugenic education in order to improve the success of the state and the prosperity of the race.

Throughout the 1980s a number of University Institutes were established to study population quality, and the China Birth Defect Monitoring Centre was established at Chengdu, with responsibility for directing eugenics at a national level (Dikötter 1998: 168). There was research on scanning techniques, diagnosis of birth defects in fetuses, and techniques of abortion. In relation to mental disorders, local regulations were passed banning mentally retarded people from marrying unless they had previously been sterilized, and requiring any mentally retarded pregnant women to have their pregnancies terminated; some areas extended these regulations to cover the sterilization of any prospective marriage partner who suffered from a chronic mental disorder such as schizophrenia or manic depression. Those who refused were to be fined and reported to the authorities (173). These regulations were formalized in the 1994 Law on Maternal and Infant Health Care, which contained four articles (Articles 10, 16, 18, and 20) that aroused international concern, as they seemed to require all couples seeking marriage to have a premarital genetic checkup, physicians to give advice on the inadvisability of marriage where there was a history of genetic disease considered inappropriate for child-rearing, and marriage to be permitted in such cases only if both sides agreed to take long-term contraceptive measures or to be sterilized. Further, they seemed to require physicians to advise abortion if a fetus was thought to have a genetic disease (Dikötter 1998: 174).[28]

It is difficult to assess how effectively these measures were implemented. Some leading Chinese philosophers and bioethicists spoke out against them at the time.[29] There is also evidence that many in the Chinese population did not support withdrawal of medical treatment from disabled neonates. However, the elements of this political rationality of eugenics were clear.[30] Indeed the eugenic character of Chinese population politics was obvious enough to require an explicit rejection during the 2002 and 2003 debates and reforms of the Marriage Law—reforms that abolished compulsory medical inspection prior to marriage, and established a new State Family Planning and Population Commission.[31] Leading Chinese geneticists involved in the Chinese Human Genome Ethical, Legal and Social Implications Committee also recognized and repudiated the eugenic character of these policies when, in 2000, they issued a statement declaring that from that time on research concerning human genomics must focus on disease prevention and treatment, instead of serving eugenics.[32] And, in the first decade of the twenty-first century, in principle at least, the Euro-American model of informed consent became the official criterion governing all medical research and treatment in China, although in some cases it remains the family, rather than the individual, who is empowered to consent.

Flexible Eugenics?

These contrasting examples serve to highlight the differences between the politics of life in advanced liberal democracies and the politics of population in territories governed according to different rationalities. Nonetheless, many critics of contemporary reproductive technologies still rest their arguments on a critical application of the term "eugenics"—liberal eugenics, individualized eugenics, flexible eugenics—to describe that politics. Thus, for instance, Karen-Sue Taussig, Deborah Heath, and Rayna Rapp suggest that "we see the persistence of eugenic thinking in the United States today, where many people across a broad spectrum of social groups consider the genome to be the site at which the human future must or can be negotiated . . . an ever increasing number of actors and practices are conscripted into a world defined genetically, in which reductive determinism looms large"—although they accept that, others will contest such determinism or "develop interventions—molecular and otherwise—that they consider choice enhancing" (Taussig et al. 2003: 62). Eugenics, here as for many others who have made similar arguments in less sophisticated ways, refers to the combination of reference to genetics, a judgment of the desirability of different body types, and a form of action designed to act on this now flexible body so as to improve it, filtered through a culture of individualism and choice. These authors ground their analysis in empirical evidence as to the complicated ways in which parents, clinicians, researchers, counselors, and those affected from conditions ranging from epidermolysis bullosa to achondroplasia are negotiating the dilemmas that confront them. But there is nothing analogous to Nazi insectification here, or remotely similar to the view that those who are born afflicted by any of these conditions are of less worth than others, although many parents are worried by the prospect that their children may live lives blighted by pain and shortened by disease. Seldom, if ever, are the actions or judgments of any of the actors in these practices shaped by the arguments that the nation is somehow weakened geopolitically by the presence of "diseased stock" within the population. What we have here, then, is not eugenics but is shaped by forms of self-government imposed by the obligations of choice, the desire for self-fulfillment, and the wish of parents for the best lives for their children. Its logics and its costs deserve analysis on their own terms.

The styles of biological and biomedical thinking that inform ways of governing others and ourselves in the advanced liberal polities of the West are no longer those concerning the quality of the race and the survival of the fittest. As innumerable examples show, for instance the difficult and painful debate about the legitimacy of rationing socially funded or reimbursed health care, the experience of eugenics has left an indelible mark on the politics of individual and collective health, throwing suspicion on any forms of state management of vital forces that does not operate under the signs of autonomy, consent, and individual rights. Indeed, over the last half century, at least in Euro-America, the pathologies, malfunctions, deficiencies, or suboptimal capacities of individuals and groups have become key sources of biovalue, and competition between states and corporations in the circuits of biovalue takes a very different form from the national struggles for supremacy that underpinned in the rationalities of eugenics. It is for these reasons that I think we should conceptualize the economy of contemporary biopolitics as operating according to logics of vitality, not those of mortality. While biopower, today, certainly has its circuits of exclusion, letting die is not making die. This is not a politics of death, though death suffuses it and haunts it, nor even a politics of illness and health; it is a matter of the government of *life*.

Risk

Perhaps Agamben and Bauman are right in suggesting that the link between individual and collective in the first half of the twentieth century—both neohygienic and eugenic—took a characteristically "modern" form. If so, I would argue that within the political rationalities that I have termed "advanced liberal" the contemporary relation between the biological life of the individual and the well-being of the collective is posed somewhat differently (Rose 1996a). It is no longer a question of seeking to classify, identify, and eliminate or constrain those individuals bearing a defective constitution or to promote the reproduction of those whose biological characteristics are most desirable, in the name of the overall fitness of the population, nation, or race. Rather, it consists in a variety of strategies that try to identify, treat, manage, or administer those individuals, groups, or localities where risk is seen to be high. The binary distinctions of normal and pathological, which were central to earlier biopolitical analyses, are now organized within these strategies for the government of risk. Such strategies are organized at a number of levels. There are actuarial or epidemiological strategies that seek to reduce aggregate levels of risk across a population. There are strategies for the management of high risk groups. And, increasingly, there are strategies based on identification of, and preventive intervention for, risky individuals.

Risk here denotes a family of ways of thinking and acting that involve calculations about probable futures in the present followed by interventions into the present in order to control that potential future (on the rise of risk thinking, see Hacking 1990).[33] Mortality and morbidity were key sites for the development of conceptions of the future as calculable, predictable, and as dependent upon identifiable factors some of which were manageable. The politics of health in the nineteenth century was underpinned by the collection and tabulation of numerical information on populations and its analysis in terms of frequencies and probabilities. This was the rationale for hygienic strategies that tried to identify, manage, and reduce aggregate levels of morbidity and mortality by modifying the factors within a geographical area, a habitat, or a portion of the life course that were statistically associated with increased levels of morbidity or mortality. Thus, in England, Edwin Chadwick used statistical and probabilistic reasoning in his arguments for the reform of sewage systems, and William Farr compiled evidence on the geographical distribution of morbidity and mortality to inculpate water supplies in the spread of cholera (Osborne 1996). In the first half of the twentieth century, such ideas about the regularity and predictability of illness, accidents, and other misfortunes within a population were central to the birth of the very idea of society, and to the emergence in different countries of social insurance strategies for the spreading of the risks entailed in the very activity of living across an insured national population (Ewald 1986, 1991). Over the second half of the twentieth century, biopolitical concerns with the minimization of risks to health—control of environmental pollution, reduction of accidents, maintenance of bodily health, nurturing of children—became intrinsic not just to the organization of health and social services, but to expert decisions about town planning, building design, educational practice, the management of organizations, the marketing of food, the design of automobiles, and much more. That is to say, for over 150 years, risk thinking has been central to biopolitics.

Demands for collective measures of biopolitical risk management, far from lessening, are proliferating and globalizing.[34] And strategies aimed at reducing the probability of untoward events across a population—targeting risky practices and locales rather than risky or at risk individuals—are spreading to many other kinds of unwanted events, notably crime control (Feeley and Simon 1992, 1994). However, again starting in the nineteenth century, a second set of strategies has also operated in terms of a type of risk thinking. These strategies attempted to find factors that would enable the identification of high risk groups and hence permit authorities to intervene upon those falling within such groups in a preventive or prophylactic manner. The use of risk profiling to demarcate risk pools has a very long history, going back to the very beginnings of the insurance industry. Risk profiling, rather than acting "actuarially" or insurantially upon the population at large, uses probabilistic and epidemiological knowledge to identify factors associated with higher risks of particular forms of ill health, reproductive problems, or other forms of pathology, and then allocates individuals to risk pools using an algorithm made up of these factors. In the field of health, as risk profiling developed from the mid-twentieth century, risk profiles, scales, and indices which were used to identify those who fell within groups with a significantly higher than average risk, in order to intervene preventively upon them.

Most readers will be personally familiar with such procedures for the allocation of individuals to risk groups, on a genealogical basis, in terms of a family history of illness or pathology, and/or on a factorial basis, in terms of combinations of factors statistically linked to a condition. Men with high blood pressure are risk profiled in terms of age, weight, family history, smoking, and so forth, are allocated to a risk group using a scale based on epidemiological and clinical research, and if at high risk, may be advised to make changes to behavior, diet, or lifestyle, or preemptively placed on a drug regime intended to reduce the risk of the occurrence of such disorders. Pregnant women are risk profiled by their doctor or midwife, and if allocated to a high risk group for miscarriage, premature birth, or associated difficulties, are subject to enhanced surveillance by midwives and gynecologists (Weir 1996). The earliest reproductive advice based on genetics also operated in terms of the identification of those who were members of high risk groups on the basis of epidemiological factors, family history, and probabilities. So did the eugenic programmers of sterilization of "the feeble minded," psychiatric patients, and sexual offenders in the early twentieth century that I have already discussed. In addition, from the early twentieth century, in many countries, a number of individuals and organizations began to give reproductive advice, focusing on prospective marital partners or would-be parents who had a family history of certain types of disease or disability thought to be "hereditary." They were given advice concerning the risks and probabilities of their children carrying the condition and advised not to marry or not to have children, and offered a termination of pregnancy, even in jurisdictions where abortion was illegal on other grounds. Genetic advising in the first half of the twentieth century was an explicitly eugenic tool (Kevles 1985). In the years following World War II when a large number of heredity clinics were established in many countries, their major goal was to prevent birth defects and help couples make "eugenic" rather than "dysgenic" decisions (Fine 1993: 103). And in the closing decades of the twentieth century, antenatal screening, for example, testing samples of maternal serum for raised levels of proteins associated with certain genetic abnormalities, became widespread for pregnant women whose age or family history placed them in high risk groups, coupled with the option of termination (Rapp 1999). In the first decade of the twenty-first century, screening of embryos for major chromosomal defects and for single gene genetic disorders is offered routinely in preimplantation genetic diagnosis, and is under discussion for disorders with more complex and uncertain genetic basis, for example for the SNP level variations that increase the chances of development of certain types of breast cancer in adults. Once more, the suggestion of the critics is that this amounts to a "new eugenics." But, as I have already argued, I think that if the term "eugenic" is not to become an all purpose and analytically meaningless rhetorical device, we should retain it for those biopolitical strategies that partook of its original sense. Our contemporary biopolitics is no less problematic, no less entangled with relations of power and judgments of the differential worth of different forms of life, the nature of suffering, and our individual and collective obligations to the future, but it deserves to be analyzed on its own terms.

The New Pastorate

A few doctors and geneticists still argue that any responsible citizen who believes that they carry genetic defects should try to reduce the risk of transmission to future generations (Glass 1971, Robertson 1993): such statements lead some critics to diagnose an implicit eugenics in the very idea of genetic counseling.[35] But I think it is necessary to be more specific about the make-up of such contemporary logics of control.[36] Decision making in the biomedical context takes place within a set of power relations that we could term "pastoral." I have referred to Swedish eugenics as pastoral, in a sense close to that developed by Foucault—a form of collectivizing and individualizing power concerned with the welfare of the "flock" as a whole. But this contemporary pastoral power is not organized or administered by "the state," even if we use this term to encompass the whole complex of legislative provision, state-funded research organizations, and national committees of inquiry into the medical and ethical aspects of the new biomedicine. It takes place in a plural and contested field traversed by the codes pronounced by ethics committees and professional associations, by the empirical findings generated by researchers, the attitudes and criteria used by employers and insurers, the tests developed and promoted by psychologists and biotech companies, the advice offered by self-help organizations, and even, one might add, the critical perspectives contributed by religious organizations and sociological critics.

Crucially, this pastoral power does not concern itself with the flock as a whole. For at least three decades professionals involved in this work have explicitly rejected the view that they are, or should be, seeking to limit the reproductive capacities of those at risk of passing an inherited condition or disability to their offspring, let alone the suggestion that they are, or should be, concerned with the contribution of individual biological characteristics to future population quality, or even the future cost to the nation of caring for children with disabilities.[37] But nor does this pastoralism simply entail a priest-like shepherd knowing and mastering the soul of the individual troubled sheep. Perhaps one might best describe this form of pastoral power as *relational*. It works through the relation between the affects and ethics of the guider—the genetic counselors and allied experts of reproduction who operate as gatekeepers to tests and medical procedures—and the affects and ethics of the guided—the actual or potential parents who are making their reproductive decisions, and their networks of responsibility and obligation.

These new pastors of the soma espouse the ethical principles of informed consent, autonomy, voluntary action, and choice and non-directiveness.[38] But in the practices of this pastoral power, such ethical principles must be translated into a range of microtechnologies for the management of communication and information. These blur the boundaries of coercion and consent. They transform the subjectivities of those who are to give consent or refuse it, through discursive techniques that teach new ways of rendering aspects of oneself into thought and language, new ways of making oneself and one's actions amenable to judgment. And they reshape the telos of these encounters in specific ways, for example in terms of psychological notions of mental health, or in terms of the recent, but currently ubiquitous idea of "quality of life"—each now defined within, and measured by, any number of rating scales. Nonetheless this pastoral power differs from Christian pastorship, where the vectors of power flowed one way, requiring the submission of the sheep to the will of the shepherd, and the internalization of that absolute will in the form of self-examination and self-mastery. These counseling encounters entail intense bidirectional affective entanglements between all the parties to the encounter, and indeed generate multiple "virtual" entanglements with parties not present—distant relatives, absent siblings, potential offspring. In these entanglements, the ethical relations of all the subjects to themselves and to one another are at stake, including the experts themselves. The consultation acts as an intensifier of ethicality. It mobilizes affects of shame and guilt, and of the respective claims, scope, and limits of freedoms for the self and obligations to others. It activates the conflicts within the counselors between the ethics of care and the ethics of guidance. It requires the counselors to fold into themselves, in a way that is by no means trivial or transient, some of those anxious and fateful undecidabilities that possess those whom they counsel.

It is true that whatever the explicit agenda of nondirective genetic counseling, evidence suggests that those who were counseled about their membership of risk groups before the availability of the kinds of predictive genetic testing I discuss below, were less likely to have children (Carter et al. 1971). But the perils of the contemporary biopolitics of group risk are not a repetition of the past. Rather than a wholesale politics of population management, these are more mundane yet no less hazardous. There are technical problems, for example the validity and appropriateness of the factors used to calculate the risk profiles through which individuals are allocated to risk groups, their generalizability to others given national and cultural variations, the effects of changes since the time when the scales were constructed, and so forth. There are the problems of false positives and false negatives that are built into the very project of applying probabilistic reasoning to determine the treatment of individuals—these have become infamous in decision-making practices concerning compulsory treatment or detention of "risky" psychiatric patients, and those arising from advice to women with a family history of breast cancer who may be contemplating prophylactic mastectomy. There are the problems that flow from the fact that, once known to fall within a risk group, the individual may be treated—by others and by themselves—as if they were, now or in the future, certain to be affected in the severest fashion.

These problems have been much discussed in relation to discrimination in insurance and employment (Gostin 1991, Hubbard and Ward 1999, Nelkin and Tancredi 1989). They also arise when risk assessments are linked to preventive interventions. We can see this where individuals allocated to a high risk group for cancer or heart disease, despite being healthy, must nonetheless conduct their lives under the shadow of medical authority if they are to be "responsible." We can see it in the area of behavioral risk, in the projects for preventive screening and intervention into the lives of young people in the inner cities at risk of violent, aggressive, or offending behavior, which tend to justify preventive intervention into the lives of "the usual suspects," that may itself be the first step in the moral career of spoiled identity (Rose 2000a). In addition, there are the ethical problems that arise in the relation between experts and their clients when trust in numbers replaces other forms of trust—that is to say, when decisions as to action seem to arise automatically from judgments "black boxed" within an "objective" calculating device—whose authors are not available for debate and contestation (cf. Porter 1995, Rose 1998). And, as innumerable accounts of "the risk society" have pointed out, risk registers and databases have the potential for an expansion and refinement of strategies of control based on surveillance (Lyon 1994). Nonetheless, the politics of risk management is not the politics of population quality, not least because it operates in terms of a different style of thought about vitality itself and its political consequences. The varieties of biopolitics, that is to say, are not only characterized by different and heterogeneous political rationalities; each such rationality is bound up with particular ways of thinking about its object, individual and collective human vitality, and its political consequences—that is to say, a way of thinking about life.

Vital and Social Norms

In *The Normal and The Pathological*, Georges Canguilhem argued that there was a fundamental ontological and epistemological difference between vital and social norms (Canguilhem 1978). Vital norms, he suggested, arose from and manifested the normativity of life itself, of the organism as a living being, of its adaptability to its environment. Social norms, on the other hand, manifested only adaptation to a particular artificial order of society and its requirements for normativity, docility, productivity, harmony, and the like. Some suggest that the new politics of life has once more mistaken social norms for vital ones, for example that the molecular politics of psychiatry, with its claims to be able to manipulate and transform personality and emotion, has incorporated the social into the vital, and has taken key features of vitality itself—for example sadness—as errors open to correction in the name of a social, not a vital, norm of health. But, at least as far as human life is concerned, the social and the vital have never been so distinct. Our very ideas of what it is to be a normal human being have been made possible by historically specific institutional and technical developments, not least by biopolitics itself. Thus, new norms of human capacities and longevity were born in the sanitization and hygienization of urban existence in the nineteenth century; new norms of recovery and cure were born in the clinical hospitals; new ways of posing the relations between fate, life, and health were born in the development of epidemiological knowledges of the distributions of disease and their social correlations. Our contemporary vital norms are no less, but no more, shaped by their conditions of formation as those of previous generations. On the one hand our very personhood is increasingly being defined by others, and by ourselves, in terms of our contemporary understandings of the possibilities and limits of our corporeality. On the other hand, our somatic individuality has become opened up to choice, prudence, and responsibility, to experimentation, to contestation. This, then, is the problem space that defines the biopolitics of our contemporary emergent form of life.

Chapter 3

An Emergent Form of Life?

Many social theorists, bioethicists, and philosophers worry that biomedical knowledge is effacing the distinction between the natural and the artificial and, in doing so, is raising fundamental questions about human nature, free will, human dignity, and crucial moral values.[1] For example, Leon Kass and his colleagues, in the October 2003 report of the President's Council on Bioethics, entitled *Beyond Therapy: Biotechnology and the Pursuit of Happiness* (President's Council on Bioethics [U.S.] and Kass 2003), identify four areas in which medical biotechnology is now moving "beyond therapy" to pursue goals of augmentation or transformation of life: (a) better children—prenatal diagnosis, embryo selection, genetic engineering of embryos, and behavior modification with drugs especially in relation to ADHD, (b) superior performance in sport, (c) ageless bodies—the whole range of technologies designed to increase the healthy human lifespan, and (d) happy souls—memory alteration and, in particular, mood improvement through the SSRI family of drugs. We will not be surprised to hear that Kass—proponent of a morality based upon "the wisdom of repugnance" (Kass 1997)—and his colleagues, who include Francis Fukuyama, author of another text troubled by such developments (Fukuyama 2002), seek to prescribe some limits to the use of biotechnology in furthering these secular goals and desires of human beings.

Similar concerns are at play in Jürgen Habermas's reflections on *The Future of Human Nature* (Habermas 2003). Like many of those commenting on these issues from within Germany, developments in genetics, especially reproductive genetics, are a particular focus of his anxieties. Habermas reluctantly concedes that the use of new reproductive technologies for therapeutic goals can be morally justified—for example, to select against embryos on the basis of disease predictions. But he believes that the manipulation of the human genome in preimplantation genetic diagnosis, and the technologies that it foreshadows, would change morally fundamental distinctions between the grown and the made, between chance and choice. In so doing, they would instrumentalize what he terms the "prepersonal life" of the child to come, constraining the ethical freedom of the genetically modified person by undermining their capacity to see themselves as the undivided author of their own life. This would also undermine what, in the "species ethic" favored by Habermas, underpins the ethical self-understanding of the human species: the necessity for human beings to regard all other such beings as morally equivalent members of a single community in which each deserves equal respect. We can, it seems, only play our part in the moral universe of our species on the presupposition that "each of us carries the sole responsibility for giving ethical shape to his or her life"—a presupposition destroyed by such interventions as preimplantation genetic diagnosis, which, for Habermas, involves the "alien determination" of the life of a future person, which would have deeply harmful consequences for that individual him or herself, and for the moral structuring of our form of life (92, 81).

Such concerns from philosophers and social theorists are seldom based on an examination of the realities of contemporary biomedicine. Nor do they take much notice of the body of empirical evidence concerning how individuals—doctors, nurses, patients, prospective parents, children produced using novel reproductive technologies, and consumers of psychiatric medication—actually understand, experience, or ethically evaluate these issues, choices, and dilemmas in their everyday lives. Their dilemmas owe more to popular science predictions and speculations about even more breathtaking advances in genomic medicine, reproductive technology, neuroscience, and psychopharmacology "just around the corner" (McKibben 2003, Silver 1998, Stock 2003). Is it possible, the social theorists, bioethicists and philosophers ask themselves, for concerned individuals, religious organizations, bioethicists, national governments, or international conventions to get control of these developments and mitigate or avert the potential damage to humanity itself that lurks within this engineered future?

There have undoubtedly been great technological transformations in medicine over the past fifty years, and no doubt there will be more. However we must remember that the major advances in life expectancy and health in the wealthy West do not owe much to highly technologized medical intervention. Likewise, the ills that afflict most human beings now and in the foreseeable future require no high tech solutions—merely clean water, sufficient food, a living wage, and moderately competent politicians and bureaucrats—and they are unlikely to be significantly ameliorated by developments in biomedicine. Thus we cannot take this promissory biomedical culture at face value. On the contrary, we need to be alert to the ways in which these predictions of fundamental transformations—imminent, but somehow always just out of reach—function in the present. Contemporary bioscience and biotechnology, no doubt following a pattern familiar from other technologies, thrives on such expectations of epochal changes just around the corner; such claims generate publicity, inflate share prices, mobilize funding agencies, enhance careers and, no doubt, generate a sense of excitement and mission for those working in the field (Brown 2003). While it may be true that many of the phenomena of life—from reproduction to emotion—now seem to be understandable as mechanisms, we are a long way from being able to re-engineer them at will, although it often appears as if our current limits are merely technical and will be overcome. The cautionary note struck by some social scientists is welcome: there is little evidence at present to suggest that we are experiencing an epochal change in the therapeutic capacities of medicine. As Paul Nightingale and Paul Martin have argued, "biological knowledge derived in the laboratory is not easily translated into useful clinical practices" (Nightingale and Martin 2004: 567): many obstacles have to be overcome before advances in basic biological knowledge generate new medical technologies. No doubt, by the time they have been translated into the clinic, many of the medical possibilities on offer will seem as routine and non-contentious as in vitro fertilization appears today, a far cry from the febrile debates over "test tube babies" sparked by the birth of Louise Brown in July 1978. But the expectations themselves form a crucial part of the contemporary field within which these biomedical sciences and technologies are taking shape. As others have pointed out, such future orientation is a central feature of contemporary technological practices and the imaginations that they inhabit and sustain (Brown et al. 2000).

In one sense, as Bachelard taught us, the imagining of potential futures is intrinsic to all those practices we term science: sciences are not phenomenologies but phenomeno-technologies. They try to conjure up in reality, by technological means, what they have already conjured up in thought. Phenomena "must be selected, filtered, purified, shaped by instruments; indeed it may well be the instruments that produce the phenomena in the first place. And the instruments are nothing but theories materialized" (Bachelard 1984:13). In the 1930s, when Bachelard was writing, he was concerned that the scientific epistemologies of his age might tempt scientists to fall for the lure of the given, a lure that would be an impediment to the scientific imagination. But there is little danger of a failure of scientific imagination today. This imagination is shaped by commercial and media hype about breakthroughs, fictional narratives of potential futures and the dilemmas we might face within them, risk assessments of technologies that are inescapably futuristic, the work of the new specialists of "foresight," "horizon scanning," and "scenario planning" within and outside public and governmental agencies—images of the future that that seem obligatory if the hopes, research grants, investment strategies, and so forth that drive contemporary science are to be generated in the present. Practitioners of cultural studies have certainly had a role in the generation of these images that simultaneously appeal and alarm. But social theorists and bioethicists have also played their part in conjuring up a future landscape rich in moral dangers and slippery slopes that might hurtle us to a very different form of humanity.

These expectations, promises, hopes and fears about the powers of biomedicine are undoubtedly exaggerated. But they do point to something important. We can now at least contemplate, and sometimes achieve, the micromanipulation of many of the capacities of the body and the mind in the pursuit of secular desires and aspirations. Some, in the wake of Haraway's Cyborg Manifesto (Haraway 1991a), suggest that we are, in some way, becoming "posthuman"—in particular through our capacity to simulate and augment the capacities of the body with prosthetic devices (Hayles 1999).[2] But were we ever just "human"—were our capacities ever so natural? I doubt it: humans have never been "natural," and at least since the invention of language we have been augmenting our capacities through intellectual, material, and human technologies. Indeed the capacities that we take to be natural arise not from human nature but from distributed associations of humans, artifacts, and objects (I discuss this extensively in Rose 1996b, especially ch. 8). Further, as I have suggested in chapter 1, our current ways of being human do not make us less biological, on the contrary, it is as if, in the inescapable connections that have now been forged between human life and biotechnology, we have become more biological. This is why, rather than suggesting we have become posthuman, I ask to what extent we are inhabiting an "emergent form of life." The concept of a form of life has a long heritage, most notably, perhaps, in Ludwig Wittgenstein's *Philosophical Investigations* (Wittgenstein 1958). But my direct inspiration for using it in the current context comes from Stefan Beck and Michi Knecht, who took it as the title for a conference on these themes.[3] We can take the idea of a *form of life* to denote a certain way of thinking about and enacting one's existence according to certain rules and premises. But more than just a way of living, the idea of a form of life also refers to the *life form*, the entities that inhabit that way of living and their characteristics. Further, *emergence* suggests that the present, while not radically different from that which preceded it, may nonetheless be a moment within a process in which something novel is taking shape. Something novel that is taking shape as a consequence of the intersection of multiple contingent pathways, while not being a consequence of any individual development. Something novel that is arising from the intertwining of ways of thinking and acting in a range of practices—medical, legal, economic, political, ethical—while not being posited directly by any of them. Something whose characteristics may be identifiable, but whose productivity cannot be predicted. To what extent is ours an "emergent form of life"?

We might ask this question in relation to the familiar themes of globalization, the rise of information technologies, accelerating concerns about security, and much more. But here I focus on medicine. Previously, one might suggest, the role of medicine was not to transform human capacities but to restore a lost normativity. The body had its natural norms, illness was a loss of those norms, and medical interventions sought to restore those norms or to mimic them in some way. The hope was that, with a cured or at least treated body and mind, the individual might be able to live out his or her natural life in something like a normal manner. But these norms no longer seem so normalizing, these normativities appear in principle open to conscious manipulation, and new norms created by biomedical artifice are already a reality. New reproductive technologies overcome natural limitations of age, of infertility, of sexuality on procreation. Genetic testing *in utero* allows selection against certain hereditary illnesses or disabilities. Sperm selection can be used to increase the chances of having a child of the desired sex. Embryo selection using preimplantation genetic diagnosis enables parents and doctors to choose children who are tissue matches to siblings requiring bone marrow or other transplants. Hormone replacement therapy and treatment for age-related sexual dysfunction extend sexual and reproductive lives. Psychopharmaceuticals, while not as effective or predictable as many suggest, do allow some modification of mood, emotion, cognition, and volition. Some of the normativities once considered to be inscribed in the laws of organic life itself have already moved, at least in the wealthy West, into the field of choice, and are laden with all the demands that choice imposes. It has become possible for individuals to think of their embodied selves as open to modification in new ways and hence to acquire further obligations for the responsible self-management of their biological and somatic existence.

But we should pause before diagnosing an epochal shift. The sphere of medicine has long exceeded that of the identification of disease and the restoration of normality. Indeed Leon Kass and his colleagues list a series of aspects of human life that, over the last two centuries, have "become part of the doctor's business": "childbirth, infertility, sexual mores and practices, aspects of criminal behavior, alcoholism, abnormal behavior, anxiety, stress, dementia, old age, death, grief, and mourning" (President's Council on Bioethics [U.S.] and Kass 2003: 303). There is nothing new in principle about the use of technologies that seek to manipulate life rather than treat disease—we need only think of the cure-alls, potions, and strategies that have been adopted down the ages in many cultures to boost fertility, increase the likelihood of male children, preserve health and strength, and so forth, let alone those that alter mood. And yet, on the other hand, something does seem to be occurring. Indeed my use of the term *the politics of life* is intended to contrast our present to the configuration analyzed by Foucault in a paper written in the 1970s called "The politics of health in the eighteenth century" (Foucault 1999).[4] Foucault, in analyses that provided the empirical basis for his concepts of biopower and biopolitics, pointed to the ways in which, in Europe in the eighteenth century, many authorities were troubled by issues of illness as they arose at a range of sites in the social body, and developed strategies for intervening upon them, in the towns, hospitals, and in the family in the name of health. The problem space that demanded a politics of health began to consolidate in the nineteenth century, and the twentieth century saw the construction of a whole range of complex and hybrid social technologies for the management of illnesses, and the health of individuals and populations and of their sexual and reproductive activity. But, as I have argued in previous chapters, the field of our contemporary biopolitics is not defined by health and illness, or even by the parameters of sexuality and procreation. It is a space of problems concerning the optimization of life itself.

I want to explore this "optimization" in relation to two linked issues—"susceptibility" and "enhancement." In each, biomedical knowledge intervenes on human beings in the present with an eye to optimizing their future vitality. Susceptibility indexes the multitude of biomedical projects that try to identify and treat persons currently without symptoms, in the name of preventing diseases or pathologies that might manifest themselves in the future. Enhancement refers to the attempts to optimize or improve almost any capacity of the human body or soul—strength, endurance, longevity, attention, intelligence—to open it to artifice and include its management within the remit of biomedicine from bench to clinic and marketplace.

A New Epoch?

I have suggested that if our contemporary biopolitics has novelty, this is not the outcome of any single event. But I have also argued that one of the keys to contemporary biopolitics of life lies in the new molecular scale on which life is envisaged and acted upon.[5] It is linked to the possibilities that this molecular gaze opens up for the reverse engineering of life, its transformation into intelligible sequences of processes that can be modeled, reconstructed in vitro, tinkered with, and reoriented by molecular interventions to eliminate undesirable anomalies and enhance desirable outcomes. In principle, it seems, any element of a living organism—any element of life—can be isolated, its properties identified, mobilized, manipulated, and recombined with anything else. And I have argued that these developments are given particular salience in a regime of the self that accords bodily health great ethical value, and in which individual and collective forms of subjectification often focus around the somatic.[6]

Reflecting upon similar developments, Hans-Jörg Rheinberger has suggested that we have moved from an age concerned with *representing* organisms and their processes—an age concerned with discovery—into a technological age, one concerned with *intervention*, whose telos is that of rewriting and transforming life (Rheinberger 2000). I do not share Rheinberger's view that these changes have overcome some primordial or ontological distinction between nature and culture. But I do argue, as he does, that the central logic of our current engagement with the vital order is not to know but to transform it through technical intervention. Knowledge of life is placed, *ab initio*—or perhaps even earlier if that is possible—in the service of the work that life does upon itself in order to improve itself. Our very understanding of who we are, of the *life-forms* we are and the *forms of life* we inhabit, have folded *bios* back onto *zoë*. By this I mean that the question of the good life—*bios*—has become intrinsically a matter of the vital processes of our animal life—*zoë*. Since the form of *bios* is constitutively subject to contestation, life itself—not merely the health that made a good life possible, or the sickness whose experience might ennoble or instruct us in the ethics of living—is now at stake in our politics.

Claims of novelty can often be countered with evidence of historical precedents and continuities, and this is no exception. Indeed I have been stimulated to reflect on my suggestion that something new is taking shape by a remark about my work made by the excellent medical historian Charles Rosenberg.[7] While generally positive in his comments, he had a hesitation that, he said, arose from "the historian's predictable caution about the newness of the new." In a recent paper called "What is disease?" (in memory of another great medical historian, Owsei Temkin), Rosenberg himself suggested that contemporary disputes over the boundaries of disease have a long history (Rosenberg 2003). Conflicts among authorities: Who has the power to define, diagnose, and treat? What are the borderlines of illness: Are headache, insomnia, back pain, irritable bowel syndrome, and the like diseases, symptoms of diseases, or inescapable conditions of life itself? And who should decide—doctors, medical administrators, patients themselves? When are conditions deserving of medical treatment and when are they not? Where are the boundaries between conditions for which an individual is to be accorded responsibility, and those for which responsibility is to be located elsewhere—in the organs, in fate, in heredity? These boundary disputes are not novel, they have been endemic to medicine. And what Rosenberg calls "technocreep" also has a long history: diagnostic tools elicit signs that are taken as evidence of pathologies that were previously invisible—generating what Rosenberg terms "protodiseases." These questions have historically had an even greater salience for conditions whose manifestations are largely emotional or behavioral—depressions, compulsions, and so forth.

The historian's caution thus forces us to recognize that such blurred boundaries are not unique to our emergent forms of life. Nonetheless, I think, something is happening. Some examples may help us understand what.

Susceptibility

In his essay on the normal and the pathological, to which I have already referred, Canguilhem argues that the notion of the norm in medicine arises from the normativity of life itself: "It is life itself and not medical judgement which makes the biological normal a concept of value and not a concept of statistical reality" (Canguilhem 1978: 73). This does not only provide the basis for his distinction, which I have questioned, between vital and social norms. It also, for Canguilhem, places the doctor on the side of life, for life itself is normative, engaged in an "effort of defense and struggle against all that is of negative value" to re-establish and maintain its norms: Health is not normality but normativity—the capacity of the organism to adjust to changing conditions. Disease limits this normativity, even if pathology has its own norms. Medicine seeks to restore that normativity even if the norms achieved by therapeutics are not those that preceded them.

The notion of susceptibility in contemporary genomic medicine seems to escape this logic. It operates as a third term between the normal and the pathological (Novas and Rose 2000). Until its eruption into frank illness, the normativity of life seems to ignore such symptomless prediseases, yet genomic medicine reverses this disregard, and makes these hidden seeds of future troubles central to its diagnostic and therapeutic hopes. But in fact, the idea of hidden susceptibilities is not so novel. While the term "susceptibility" is used as a noun only in the early twentieth century to mean a person capable of contracting a disease or deficient in defenses against one, the OED traces a very similar term, "predisposition," to the early seventeenth century, defining it as a physical condition that makes a person susceptible to a disease. However another phrase of Canguilhem's is helpful here, one that he takes from René Leriche. "Health, says Leriche, is life lived in the silence of the organs." Conversely, "disease is what irritates men in the normal course of their lives and work, and above all, what makes them suffer" (Canguilhem 1978: 46).[8] From this perspective, disease is a state of suffering—a human matter. Thus for Canguilhem, "It is the individual who is the judge of this transformation [between the normal and the pathological] because it is he who suffers from it" (106). But what then of a predisease, whether it be an irregular heart rhythm or a substitution of one base for another in a coding sequence? Is one diseased if that defect lurks in a silence, if it exists outside the awareness of an individual? Perhaps these definitional or ontological questions are the wrong ones. It is better to ask, instead, How does it come about that susceptibility is something that calls out for medical attention? How does it come to be that, contrary to Canguilhem's humanism, the doctors and genetic counselors, and now the genomics researchers and biobankers, have acquired the right, against that of the individual who appears existentially well, to define and diagnose a state of potential disease, or protodisease, and in doing so, to render the person a "pre-patient."

Of course, the discourse of predispositions already addressed a liability lodged within the individual organism—the predisposed person. From the nineteenth century onward, predispositions were located in a lineage, and handed down in an inherited constitution from one generation to the next. One inherits a constitution, which if it is weakened or tainted in some way may predispose one to not merely any one of a number of illnesses such as tuberculosis, scrofula, syphilis but also to neuropathy, epilepsy, or madness, or to drunkenness, gambling, and any form of physical and moral viciousness. One's constitution might be affected in utero by the bad habits of one's mother, or by events occurring in pregnancy. It can be further weakened over one's lifetime by vicious habits, and passed down in an even more enfeebled form to one's offspring. And, when a weakened constitution is exposed to certain exciting causes, it may lead to frank pathology. Thus too much excitement may lead to an episode of mania, or poor habits such as masturbation might lead to weakness or frank physical illness.

The configuration of constitution, predisposition and, exciting cause leading to the expression of pathology was re-arranged, but not fundamentally altered, with the gradual distinction between inheritance and environment. Even into the twentieth century, the sense remained that the sickness lurking in an inherited, tainted, or defective constitution might be provoked into early or more severe activity by bad habits or vice. And, in the reverse direction, it was widely suggested that its expression might be inhibited or even prevented by living a pure and virtuous life. So the idea of inherited predispositions to certain illnesses, which rendered one susceptible to pathology in the presence of certain exciting causes, is an old one. And so is the idea that one should moderate one's habits, diet, morals, and forms of life in order to minimize the chance of provoking the onset of an inherited condition.

So what is it that leads me to include the idea of susceptibility as a key dimension of an emergent form of life? It does not, in itself, mark an epistemological transformation, it is true. But a mutation emerges nonetheless from the combination of a shift in scale, a shift in technical capacities, and a shift in the aspirations of expertise. The combination of the idea of susceptibility as genomic precision, the technology of genetic screening, and the promise of preventive medical intervention appears to offer a quantum leap in the capacity of biomedical expertise to bring a potential unwanted future into the present and make it calculable. In addition, the very fact of having done so seems to invite or even demand medical intervention on the susceptible individual in the present to direct his or her path to a different, and more desirable—less diseased—future.

As is well known, the first move here was epidemiological. It was marked by the use of risk scales based on calculations as to the probabilities of certain kinds of illness for those in certain groups established by family history or demographics—race, age, weight—in addition to lifestyle factors such as smoking. Such risk-based identification and treatment of individuals proliferates, now aided by increasingly sophisticated testing, screening, and imaging technologies.[9] And screening and risk profiling is linked to intervention—the prescribing of lipid-lowering drugs or antihypertensives to reduce the risk of heart attack and stroke, for example. As Rosenberg suggests, these symptomless conditions are "iatrogenerated—created . . . by the ideas, practices, and personnel of medicine" (2003: 503). They depend for their emergence on a whole series of developments, notably large-scale long-term research studies of morbidity and mortality. In the closing decades of the twentieth century, they were also fostered by the proselytizing activities of individual medics or subspecialties, sometimes coupled with disease awareness campaigns by patient's groups, to draw attention to a condition deserving of medical attention. These protodiseases have become the object of exceptionally fruitful exploitation by the pharmaceutical industry: for example two lipid lowering drugs—Lipitor (atorvastatin) and Zocor (simvastatin)—were among the top sellers in the United States in 2003, grossing $6.3 billion and $5.10 billion, respectively.[10] Note that high lipid levels, high blood pressure, and similar indications are not, themselves, diseases. What is treated by doctors and drugs here is not disease but the almost infinitely expandable and malleable empire of risk. While public heath medicine has long engaged in strategies of disease prevention and health promotion, these individualized and pharmaceuticalized practices or early and long-term intervention into corporeal processes in the name of the "treatment" of risk have come to be a central dimension of the politics of life in the twenty-first century.

But what susceptibility promises, in the age of genomics, is more than risk assessment and risk management—more, that is to say, than intervention based on a correlation between factors such as age, weight, or diet whose link to the disease process may be unknown or distant. Susceptibility, it is claimed, is something that can be defined at the level of the individual body itself—a variation within the sequence of DNA bases in an individual's genome that predispose that person to the development of a particular disease or disorder. A 2004 visit to the website of the Institute for Systems Biology—"revolutionizing science, enhancing life"—co-founded in 2000 by Leroy Hood, who was an early advocate for the Human Genome Project in which he played a key role, indicates something of the promissory culture that is taking shape.[11] "The mission of the Institute for Systems Biology is two-fold: revolutionizing biology through a systems approach; and predictive, preventive and personalized medicine," they tell us:

> The Human Genome Project has catalyzed two paradigm changes in contemporary biology and medicine—systems biology and predictive, preventive and personalized medicine. The Human Genome Project has provided access to the extensive human genome variability (polymorphisms) that distinguish each of us from one another (apart from identical twins). On average, 1 letter in 1000 differs between your DNA and mine. This means, on average, we differ from one another by approximately 6 million DNA variations. Most of these variations have no influence on our appearance or behavior. However, a few make some of us tall or short and others thin or fat. An additional few predispose to diseases such as cancer, cardiovascular disease, neurological disease, or metabolic diseases. Within 10–15 years, we will have identified hundreds, if not thousands of genes that predispose to disease. We will be able to analyze the relevant DNA sequences from these genes from a small amount of blood and use these to predict a probabilistic future health history for each individual. This is predictive medicine. Since it is an anathema in medicine to predict without being able to cure or prevent, we will use systems approaches over the next 15–25 years to place these defective genes in the context of their biological systems and learn how to circumvent their limitations. This is preventive medicine. The agents for preventive medicine will

include drugs, embryonic stem cell therapy, engineered proteins, genetically-engineered cells, and many others. Because each of us differs by 6 million DNA variations, we will each be predisposed to differing combinations of late onset diseases. Accordingly, physicians will have to treat each of us as individuals with our unique sets of predispositions. This is personalized medicine.

While any two randomly selected humans will share 99.9 percent of the three billion base pairs that make up their genome, the 0.1 percent difference between them amounts to millions of differences—usual estimates vary from two million to ten million but some put the number much higher—at the level of base pair differences (single nucleotide polymorphisms [SNPs]). It is upon these SNPs—or coinherited clusters of SNPs known as haplotypes or HAPs—that the search is currently concentrated for medically salient differences between individuals in their susceptibility to diseases, and in their amenability to certain kinds of treatments. The promise of such contemporary programs to identify variations at the genomic level is that they will enable clinicians to move beyond epidemiological characterizations of risk indicators, or trial and error use of different therapeutics, in order to identify the precise sequences of bases at particular loci responsible for increased susceptibility or variations in drug efficacy, thus enabling individual diagnosis and "tailored treatment." And here, unlike in earlier forms of risk assessment, the variation in the sequence in question is, at least in principle, integral to the mechanisms whereby that disease will develop.

The Institute for Systems Biology is only one of a range of different projects that are seeking to discover identify the SNPs or HAPs linked to disease susceptibility, with the aim of prediction and prevention. This is the promise that led to such investment, of hope, research, and capital, both public and private, in genome banking projects—from DeCode in Iceland and UmanGenomics in Sweden, to the UK Biobank and Celera Diagnostics. In each of these projects, tissues were collected from lineages of affected and unaffected individuals, DNA was sequenced, and an attempt was made to find correlations between particular SNP patterns and liabilities to develop diseases, especially common complex disorders. The belief, as I have said, is that once these correlations have been discovered, retrospectively as it were, the SNP patterns so correlated can be deployed prospectively in genetic tests to enable presymptomatic diagnoses and preventive intervention. There is already a growing industry of presymptomatic testing—for conditions from Alzheimer's to periodontal disease—using gene expression chips as high throughput platforms to identify genetic variation at the SNP level. The move to common complex disorders entails largely dispensing with the "gene for" model of an earlier thought style, in favor of multiple interactions between SNP variations at several different loci and their capacity to express themselves in certain environmental and biographical conditions. If successful, such endeavors will enable diagnostic tests to be used routinely, if not in the immediate context of the consultation in the doctor's surgery then certainly in reference laboratories, to enable presymptomatic diagnosis and preventive interventions on a scale previously unimaginable.

As in all such promissory endeavors, the "if" is big, especially in relation to the common complex disorders. Nonetheless a string of announcements emerge from such companies implying that a breakthrough is just around the corner. Take the press release issued on June 3, 2004, entitled "Celera Diagnostics Identifies Novel Genetic Markers Linked to Increased Risk of Heart Attack."[12] It reports:

> Scientists from Celera Diagnostics presented data linking genetic variations in two genes with increased risk for myocardial infarction (MI) or heart attack, at the International Vascular Biology Meeting in Toronto. Celera Diagnostics, a joint venture between the Applied Biosystems Group (NYSE: ABI) and Celera Genomics Group (NYSE: CRA) of Applera Corporation, conducted the study in collaboration with researchers from the Cleveland Clinic Foundation and the University of California, San Francisco. . . . The two genetic markers, or single nucleotide polymorphisms (SNPs), presented include a SNP in the aquaporin 10 gene (AQP10), which is associated with a two fold increased risk for MI. The second SNP is in KIAA1462, a gene of unknown function. Each of these SNPs confers a risk for MI comparable to conventional risk factors such as smoking, high blood pressure and elevated cholesterol levels. . . . These genetic markers were identified through an association study of more than 9,000 functional SNPs. Celera Diagnostics evaluated DNA samples from more than 1,400 individuals to compare patterns of genetic variation in people with a history of MI to those with no history of chronic heart disease. The results were replicated in a separate sample collection of over 1,000 individuals.

Kathy Ordoñez, President of Celera Diagnostics, is quoted in this press release as saying:

> Our association studies are broadening our understanding of the underlying biology of cardiovascular disease, and producing opportunities to create diagnostic and therapeutic value. To deliver on the promise of Targeted Medicine, we must translate these exciting discoveries into clinical practice. We are working with Quest Diagnostics (NYSE:

> DGX) and other collaborators to identify the most informative constellation of markers associated with MI. In conjunction with scientists at Celera Genomics, we are also considering the therapeutic potential of these and other markers.

But we might also note the closing paragraphs of the press release, worth quoting in full:

> Certain statements in this press release are forward-looking. These may be identified by the use of forward-looking words or phrases such as "believe," "plan," and "should," among others. These forward-looking statements are based on Applera Corporation's current expectations. The Private Securities Litigation Reform Act of 1995 provides a "safe harbor" for such forward-looking statements. In order to comply with the terms of the safe harbor, Applera notes that a variety of factors could cause actual results and experience to differ materially from the anticipated results or other expectations expressed in such forward-looking statements. These factors include but are not limited to (1) uncertainty as to whether Celera Diagnostics or Celera Genomics will be able to generate products, or therapeutic or diagnostic value from disease association studies conducted by Celera Diagnostics; and (2) other factors that might be described from time to time in Applera's filings with the Securities and Exchange Commission. All information in this press release is as of the date of the release, and Applera does not undertake any duty to update this information, including any forward-looking statements, unless required by law.

According to Martin D. Chitwood and Nichole T. Browning, the intent of Congress in passing the Private Securities Litigation Reform Act of 1995 was to implement procedural protections to discourage "meritless" securities class action litigation.[13] Chitwood and Harley tell us:

> Proponents of the PSLRA, consisting primarily of venture capitalists, corporate interests, and accounting firms, argued that these proposed changes to the federal securities law were necessary to eliminate "meritless" private lawsuits and to encourage directors and officers of public companies to publish statements about their companies' financial prospects without fear of shareholder lawsuits. Opponents of the PSLRA, however, including numerous consumer protection groups, argued that those who lobbied for the legislation were motivated solely by their desires to protect themselves from meritorious lawsuits.

President Clinton sought to veto the act, observing that it would "have the effect of closing the courthouse door on investors who have legitimate claims." Despite these concerns, Congress overrode President Clinton's veto and the act went into effect on December 22, 1995. The safe harbor that it produced allows corporations and individual defendants to avoid liability for "forward-looking" statements that prove false if "the statement is accompanied by meaningful cautionary statements identifying important factors that could cause actual results to differ materially from those in the forward-looking statement."[14] The promissory culture thus incorporates its uncertain nature and risky futurology into its own activities in such a way to secure itself against the risks that it inescapably generates and thrives upon.

Analogous promises of imminent personalization, though seldom taking advantage of a statutory safe harbor, are made in the field known as pharmacogenomics (see, for example, Roses 2002a, 2002b, 2004). Here, trials and studies are being carried out on many disease/drug combinations, collecting and sequencing DNA samples, and correlating drug response with genomic information. It has long been recognized that responses to drugs differ among individuals—the same drug can be efficacious in treating one individual for a condition, ineffective in treating another, and produce severe adverse effects in another, all apparently suffering from "the same" illness. This has usually been ascribed to variations in pharmacokinetics—that is to say in the enzyme systems responsible for the metabolism of the drug in question—however it may also relate to individual variations in the actual disease mechanism itself. And the aim is to identify the specific SNPs or HAPs correlated with efficacy, inefficacy, or adverse effects, and hence to develop genetic tests that will enable or perhaps even require clinicians to determine, having diagnosed a disease in a particular patient, which of the available drugs is most appropriate. Some predict the pharmaceutical industry will move away from "blockbusters" to "minibusters," others that conditions previously considered to be variants of a single disease will be "dissected" by the drugs themselves and reclassified as distinct conditions at the molecular level. The hope invested in such tailored medicine is that the efficacy of drugs can be increased by individualizing drug prescribing on the basis of a genetic test—to maximize "adherence" to drug regimes by ensuring that patients do not stop taking drugs because of unpleasant side effects, to minimize waste by ensuring patients are not prescribed drugs that are ineffective in light of their particular genomics, and to reduce the individual and public health costs of adverse reactions to drugs, sometimes estimated to be among the leading half dozen causes of hospital admission and mortality.

The science fiction writers and popular journalists have already imagined a "tomorrow's world" of genetic individualization, where fate can be read out of a gene chip at the moment of birth, in the womb, or in preimplantation genetic diagnosis.[15] Industry analysts predict a rapidly growing market for molecular diagnostics, ranging from the chips to the pharmaceuticals. The field of players in molecular diagnostics is large and growing: as early as 2002, a Frost and Sullivan report on U.S. Pharmacogenomic Markets identified sixty-seven key market players.[16] But we need to introduce some cautions. First, of course, most of these projects have yet to generate clinically significant findings. Despite the enthusiastic predictions of so many involved in pharmacogenomic research, very few diagnostic tests for drug response are currently on the market, and clinical demand for such tests is uncertain (Hedgecoe 2005). In any event, the language of personalization or tailoring is misleading, as the best such tests are likely to achieve is to allocate individuals to a group on the basis of the probability of response—thus informing the clinician that a person has say, a 90 percent chance of responding well to a drug, or less than 10 percent chance of responding at all. This may assist doctors in the initial choice of drug, but such probabilistic data is not likely to be compelling to patients refused drugs that may just offer them a chance of effective treatment, or to doctors whose prescribing habits are shaped by many factors other than such information (Nuffield Council on Bioethics 2003).

As far as disease susceptibility is concerned, we can already see some programs of national screening in place for single gene disorders. For example, in a little publicized development in England, the "NHS Sickle & Thalassaemia Screening Programme" was initiated in 2001 to implement "effective and appropriate screening programmes for women and children including a new national linked antenatal and neonatal screening programme for haemoglobinopathy and sickle cell disease by 2004"—this involves antenatal screening for thalassemia and sickle cell, aiming "to allow informed reproductive choice by identifying couples at risk of an affected infant at an early stage in pregnancy. Options include prenatal diagnosis and termination of affected pregnancy as well as continuation of affected pregnancies." There is also a program of screening of *all* newborns "as an integral part of the newborn bloodspot screening programme."[17] Sickle cell and thalassemia are, of course, simple genetic disorders showing Mendelian patterns of inheritance, and the results of genetic testing are relatively unambiguous.[18] Genetic screening, here, seems simply to be an extension of the kind of genetic counseling that has long been available for those prospective parents who believe themselves to have a family history of a severe inherited disorder. And yet it extends the requirements to calculate genetic risk, and the obligations of choice, potentially to everyone considering reproduction.

A number of other well-publicized genetic tests for susceptibility to relatively rare single gene disorders have now been brought to the market and the clinic.[19] But the first ten years of operation of the major commercial biobanking enterprises—for example DeCode in Iceland, UmanGenomics in Sweden, and indeed Celera Diagnostics in the United States, have largely failed in their attempts to find clinically useful markers for the genomic basis of susceptibilities to common complex diseases, disappointing both those who invested capital and those who invested hope. Despite the unbounded optimism of the researchers and the biotech companies, the genomics of common complex disorders remain elusive. Common complex diseases clearly involve the interaction of multiple susceptibility and protective loci, and attention is already shifting from genomic sequences to gene expression and from the DNA to the cell and to complex various epigenetic mechanisms regulating gene activity. Further, even where susceptibilities are identified for simple diseases with an identifiable genetic locus, there is no necessary symmetry between genomic diagnosis of susceptibility and molecular treatments. The gene for Huntington's disease was located on the short arm of Chromosome 4 in 1983, and sequenced some ten years later. But although individuals can take a genetic test that will reveal with very high probability whether they will, at some point, develop the disease, no treatment is yet available—hence in most jurisdictions less than half of the potential testees chose to know their HD status. And even where a positive result is obtained from a test for single gene conditions like HD, the reality is not a qualitative change in which a certain fate is written in, and readable from, the genes by scientific foretellers of the future. In fact, what is produced by such tests is not certainty, but probabilities and uncertainties. Even in the extreme case of a monogenetic and inevitably fatal disease like HD, a genetic diagnosis that reveals that a particular individual is carrying an expanded number of CAG repeats at the locus 4p 16.3 on the short arm of chromosome, does not predict when the disease will develop, how rapidly it will progress, and with what level of severity—indeed some with such a diagnosis may live to old age or die from unrelated conditions.[20] How much more so, then, for such current exemplars where the predictive power of the genetic test is much lower—for example the APOE e4 allele at 19q 13.2 for late onset Alzheimer's or variants of BRCA 1 at 17q 21 and BRCA 2 at 13q 12.3 linked to early onset breast cancer. In these and most other cases, the probabilities generated by the genomic diagnosis are much lower and only certain limited forms of the condition are involved at all.

Presymptomatic genomic diagnosis, that is to say, will mainly generate information that is not qualitatively different in predictive power from probabilities arising from family history. And yet something is happening, not so much in clinical practice, perhaps, but in the form of life that is emerging. For even if no revolutionary advances in treatment are produced, once diagnosed with susceptibilities the responsible asymptomatic individual is enrolled for a life sentence in the world of medicine—of tests, of drugs, of self-examination and self-definition as a "prepatient" suffering from a protosickness.[21] And, in the near future, perhaps, they will subject themselves to new forms of monitoring that engender a new ethical relation to the self.

Take, for example, the "Ubimon" project, which was part of "Ubiquitous Computing for Healthcare in the Community" based at Imperial College, London in 2004.[22] Ubimon—"Ubiquitous Monitoring Environment for Wearable and Implantable Sensors"—aimed to develop miniaturized sensors, perhaps using nanotechnology, that could be implanted into the body to detect early signs of disorder, say cardiac arrhythmia, or the fatty streak lesion that is an early indication of atherosclerosis; these would use wireless free communication techniques to send this data to the patient's PDA or to a central monitoring station. The hope was that this would, in the first instance, enable reliable baseline information to be gathered on individual function, and then, when such baselines were established, detect anomalies and either alert the patient to consult with their physician or, at a later stage, perhaps trigger the automatic release of medication from a depot implant in the patient's body. And, of course, who could possibly find such ubiquitous health monitoring anything but desirable, for its aim is simple, to avert disease and to prolong life.[23]

This is a world that is resisted at the risk of irresponsibility: as documented, for example, in Callon and Rabeharisoa's study of the difficult choice made by Gino (Callon and Rabeharisoa 2004). Gino, who suffers from Limb Girdle Muscular Dystrophy like many on the island of Reunion, refuses to subjectify himself in terms of his disease—while his brother Leon is a key member of the patient's association, lobbying for research funds and for political recognition, Gino prefers to drink wine and watch Marseilles in the local bar. His recalcitrance seems impossible to articulate; not only is he the target of numerous attempts by relatives, friends, and others to get him to act responsibly, but he refuses to justify himself when confronted by the sociologists. In this emerging form of life, then, the susceptible individual would be obliged to engage in responsible self-management, to debate and justify his or her choices with others, to enter into complex calculations of risks and benefits, to act now in the present in relation to probable futures brought into view. What moral judgments would be deployed for those who chose to live "in the silence of the organs"? Would the forms of life now lived by those with a family history of Huntington's, early Alzheimer's, or breast cancer now become those of each of us? Would each individual, in consultation with their doctor, have to become their own bioethicist, forming a relation with a new set of authorities? Would we each, in deliberating about our choices, have to become our own heath economists, calculating our own QALYs—those Quality Adjusted Life Years beloved by health economists and managers of health care organizations in determining rational treatment choice—assessing the impact of different presymptomatic treatments, and making our own evaluations of the costs and benefits of different forms of life?

So here, in the field of susceptibilities, one can observe new forms of subjectification taking shape, new self-technologies whose "ethical substance," to use Foucault's term is soma, and whose telos is the prolonging of healthy life. One can also see new biosocial communities assembling around these somatic identities—parents and families raising money, funding research, lobbying politicians, enacting their biological citizenship by demanding their rights for attention to their particular disorder—as in the Limb Girdle Muscular Dystrophy studied by Callon's group. The obligations of knowledge and choice are onerous enough. But subjectification can take another form. One can imagine susceptibility testing in a number of situations where it would be far from voluntary. Many fear susceptibility testing in insurance, or at least the demand, by insurance companies, that the results of any such tests be declared to them; right now, the fear seems greater than the reality. But one can also envisage testing of potential employees who might be exposed to chemical or environmental hazards, as in the earlier example of testing for sickle-cell traits in African American employees of some sections of the U.S. military. Some are proposing testing for susceptibilities in the serotonin system linked with inability to control conduct—say in disruptive school children or those convicted of aggressive or impulsive crimes. Here, in the neuroscientific and psychopharmacological field, responses purporting to be treatment are already at hand in the form of preventive administration of psychiatric drugs. In situations such as these, where other diagnostic and predictive tests are already being used, and preventive interventions are key elements in regulatory rationalities, the promise of molecular diagnosis is a potent one for researchers, private, and for the professionals responsible for predicting risks. That is to say, new potentials are emerging for "governing through susceptibilities," in the name of securing others from their consequences.

Enhancement

Let me turn to the second of my themes—enhancement—and return to the meditation on "biotechnology and the pursuit of happiness" proposed by Kass and his bioethical colleagues. One might expect them to base their attempt to establish the limits of the permissible on an idea of "human nature." This was the tack taken by Francis Fukuyama, in his *Posthuman Future*, who was certain that "our species-typical gamut of emotional responses . . . constitute a safe harbor that allows us to connect, potentially, with all other human beings" and that these were necessary to establish our "shared humanity" (Fukuyama 2002). But, while the account produced by Kass and company does embody such ideas, it rests on a rather different ideal, that of the human being as a creature "whose precise limitations are the source of its—of our—keenest attachments, whose weaknesses are the source of its—our—keenest attachments, and whose natural gifts may be, if we do not squander or destroy them, exactly what we need to flourish and perfect ourselves—*as human beings*" (President's Council on Bioethics [U.S.] and Kass 2003).

Whatever the reality of the developments that concern Kass and his colleagues, this *style of reflection*—on ourselves as human beings and the danger of our hubris—seems a significant characteristic of our emergent form of life: manifested in books, media debate, popular cinema, the deliberations of ethics committee, and much more. But before we diagnose this as an instance of the "reflexive individualism" characteristic of our own uncertain age, we should pause.

Philosophers, moralists, and novelists have long played out the disputes between those who feel humans are at their most noble when they recognize their natural limits, and embrace their inevitable finitude in the fallen world in which they live, and those who say that the most human form of life is to conquer, subdue, manipulate, and escape such natural limits. My favorite is "the great disputation on sickness and health" in *The Magic Mountain*, published in 1924 (Mann 1960).[24] For the humanist, and Freemason, Ludovico Settembrini, the triumph of modern medical science is the triumph of reason and humanitarianism, the triumph of the principles of health over sickness and virtue over vice, the triumph of a social morality of the normal man. For the Jesuit, and revolutionary, Leo Naptha, this ethic of the normal man and the triumph of reason is banal and vulgar. Human spirituality and human freedom were not bound to a veneration of the healthy body, but to an ethic that gives a positive value to bodily suffering. "There were those who wanted to make him 'healthy,' to make him 'go back to nature,' when the truth was, he had never been 'natural.' " For Naptha, all the propaganda carried on by the prophets of nature, the experiments in regeneration, the uncooked food, fresh-air cures, sunbaths, and so on stood, in fact, for the dehumanization, the animalization of man whereas "the more ailing he was, by so much more was he the more man" (1960: 466).

This debate on the value of natural limits to the human form of life thus seems to be an endemic feature of modern forms of life, dating back to at least European thought of the nineteenth century, perhaps even an aspect of the ethos of enlightenment itself. Of course any medical historian will tell us that the features of the "given humanness" enumerated by Kass and his colleagues have nothing given or natural about them: there is nothing natural about our present life span, our temporality of reproduction, our sense of ourselves as individual privileged actors realizing our secular life potential or any of the rest of it. But the historical relativists among us might as well save their breath, for such empirical and historical observations are radically insufficient to interrupt the debate. Perhaps it would be more productive to analyze these formulations as themselves elements in ethical technologies. How and in what ways has bioethics become so indispensable to our modes of governing life—to the linkages between ways of governing others and ways of governing ourselves?

Elsewhere I have suggested that we are living in an ethopolitical age (Rose 1996a), where issues as diverse as crime control and political apathy are problematized in terms of ethics. No longer posed in the languages of justice, welfare, or equity, ethopolitics here is about the value of different forms of life, styles of life, ways of living, and how these should be judged and governed. Nowhere is this ethicalization of politics more evident than in the value-driven debates over scientific developments, whether these concern global warming or reproductive technologies. It is in this context, I think, that we need to understand how bioethics has become a necessary supplement to the imperatives of political decision making concerning the life sciences, under conditions of moral uncertainty and lack of consensus, and where these are coupled with economic imperatives and aspirations, with clinical demands and ambitions, and with citizens' claims to treatment and rights to health (Gottweis 2002a, Rose 2002). Medical ethics was once inscribed in the persona of the medical agent—the doctor or nurse—that amalgam of wisdom, expertise, and judgment with all its archaic codes of conduct, ambiguous rights, and conflicting obligations. But today, the ethics of the medical personage seems insufficient. It is insufficient at the level of political technologies of regulation of legal and political subjects. And it also seems insufficient at the level of the pastoral technologies of the clinical encounter between doctor or genetic counselor and living human subjects. The encirclement of the medicine of the clinic—not merely by bioethics, but also by "evidence based medicine," by the demands of "patient choice," by the shadow of the law and the audits of the health management organizations—seems almost complete. We have yet to diagnose the costs and benefits of this reconfiguration.[25]

But let me turn directly to the aspect of optimization that concerns Kass and his colleagues. Their proposition is that we are on the cusp of a new age, in which we are no longer content with the restoration of sickened bodies and souls to their organic, vital norms. In this new age, it seems, we are able to reshape key aspects of the functioning of our bodies and souls more or less at will. For some bioethicists, and for social philosophers like Habermas, the genome has become the repository of our human nature, to be interfered with at our peril. But most disturbing for these American bioethicists is the manipulation of the soul itself: this is the field of concerns that has now benefited from being named—it is "neuroethics." Thus Kass and colleagues write:

> The awesome powers modern science has placed in our hands to control the external world increasingly enable us to control our inner experience . . . increasingly we can produce through drugs the subjective experience of contentment and well-being in the absence of the goods that normally engender them. In some cases . . . the new drugs can help return a person to the world and enable him to take responsibility for his life. But in many other cases, the growing power to manage our mental lives pharmacologically threatens our happiness by estranging us not only from the world but also from the sentiments, passions, and qualities of mind and character that enable us to live in it well . . . the creating of calmer moods and moments of heightened pleasure or self-satisfaction that bear no relation to our actual undertakings threatens to erode our sentiments, passions, and virtues. What is to be particularly feared about the increasingly common and casual use of mind-altering drugs, then, is not that they will induce us to dwell on happiness at the expense of other human goods, but that they will seduce us into resting content with a shallow and factitious happiness. (President's Council on Bioethics (U.S.) and Kass 2003: 266–7)

This suggestion, that happiness, whether shallow or not, can be obtained by the ingestion of a pill, without further work on the self, is misleading.[26] So, therefore, are the ethical dilemmas that it gives rise to. Consider, for example, the new generation of antidepressants that are often taken as the exemplars or forerunners for such "cosmetic psychopharmacology." A careful reading of Peter Kramer, who is credited with the invention of the term, and whose work seems to have had a major impact on Kass and company and on many other contributors to debates on neuroethics, shows that he does not actually espouse the view that mental processes can be reshaped at will by pills alone. In *Listening to Prozac*, his case histories present the unhappiness of his patients as the complex outcome of interaction between biographical experience, self-narratives and meaning systems, and the long-term shaping of neural pathways in the brain. Further, he acknowledges that the effects of the drug vary from person to person—some feeling "better than well" while others feel no change or become agitated and anxious (Kramer 1994).[27] As Elizabeth Wilson has argued, in her analysis of Kramer's work: "Nowhere in *Listening to Prozac* is neurology—with or without pharmaceutical assistance—deployed as a univocal determinant of human psychology" (Wilson 2004b: 27). Even in his most optimistic assessment of the effects of ingesting a psychopharmaceutical, the changes an individual undergoes are a consequence, not of taking the pill alone, but of therapeutic encounters with Kramer himself, of the narrative recasting a life history, of engagement in novel social situations, and much more.

The drugs for treating "attention deficit hyperactivity disorder" may seem an exception to this argument. The effect of methylphenidate (Ritalin) and dexamphetamine (Adderall) on the conduct of young children does appear to be rapid and self-evident. However it is important to note that the effects on attention, cognitive tasks, and working memory produced by psychostimulants are often exaggerated.[28] Children diagnosed with Attention Deficit Hyperactivity Disorder do benefit from methylphenidate when tested on narrowly conceived standardized tasks (Elliott et al. 1997). However, the benefits of methylphenidate are not pervasive; rather they appear to be mediated by baseline measures of cognitive performance and task novelty (Mehta et al. 2000). While there may be some improved functioning in limited cognitive ranges, such narrow and time-limited improvements in cognitive function and control hardly amount to the re-engineering of the human soul. Further, as Ilina Singh's detailed and careful empirical research in "Boys on Ritalin" has shown, parents do not feel that they are creating "calmer moods and moments of heightened pleasure or self-satisfaction that bear no relation to our actual undertakings," as Kass and company put it (Singh 2002, 2003, 2004). What is at stake, it appears, is something quite different—parents do not seek to create a false or modified self for their sons, but hope and feel that when their child takes the drug, his true self can appear.[29]

Popular hype aside, there seems no anthropological and sociological support for the suggestion that individuals use psychiatric drugs to enhance themselves. Even the quasibiographical or semifictional accounts of experiences with SSRI antidepressants do not indicate that they are either used for, or experienced as, transforming personalities or producing happiness.[30] At best, to adjust a familiar phrase, those who administer them, and those who take them, are trying to transform crippling misery into everyday unhappiness; some undesirable adverse effects, such as loss of libido, or suppression of other affects, may be the price that those who take these drugs are prepared to pay.[31] Indeed, like an earlier generation of "mothers little helpers," as evidence accumulates of anxiety, suicidal thoughts, and withdrawal difficulties associated with such pharmaceuticals, the suggestion that a little pill contains the alluring and dangerous possibility of pain-free happiness seems hard to take seriously. Even a cursory reading of the biomedical literature would be enough to show that SSRIs do not, in fact, allow manipulation of moods or personality at will—indeed they do so less, and less reliably, than such "dumb" drugs as alcohol or marijuana. And, of course, even drugs such as MDMA (Ecstasy), which is widely used by young people worldwide to change their perceptions and capacities in a limited forum and for a specific purpose, for example in a "rave party," do not contain these properties in the pharmaceutical product itself. Consider a simple example: alcohol. The same volume of alcohol will have quite different behavioral, emotional, and cognitive consequences depending on whether it is taken on a solitary and melancholic evening at home, at a celebratory fortieth birthday party, on the terrace of a football match, or in the controlled setting of a psychological experiment.[32] As sociologists have long demonstrated, these "effects" are never simply given in the drug: they are embedded in complex situations and the affects they generate require all manner of social and contextual supports.[33]

If one reflects on the telos of pharmacological ethics, by which I mean the ends, purposes, or objectives that are implicit in the work that one is enjoined to do upon oneself when taking these drugs, the suggestion that the drugs themselves inaugurate a new shallow regime of manipulation of the self begins to fade away. Those who market the new generation of "smart" pharmaceuticals for treating depression, anxiety, panic disorders, and the like are rather careful in the ethics they seek to embed in these drugs. As their advertisements show, they do not promise individuals the power to remake their self or soul at will. What is offered here, by drugs, indeed perhaps by almost all those biomedical technologies that concern Kass and company except those of competitive sport, is not actually the creation of some kind of super-being. On the contrary, these technologies operate within a regime of the self that is actually rather familiar. Individuals are prescribed them, and utilize them, in the hope of restoring themselves to a state they feel they have lost. "Hello me" is the badge worn by the smiling young woman who, in August 2004, greeted visitors to Glaxo Smith Kline's Paxil website, advertising controlled release Paxil (which treats social anxiety disorder, depression, and panic disorder).[34] The ethic of authenticity that runs through a previous generation of psychological and psychodynamic treatments is at work here as well (Rose 1989). In a regime of the self that stresses self-fulfillment and the need for each individual to become the actor at the center of his or her life, these drugs, which are often used to complement rather than to substitute for other therapies, do not promise a new self, but a return to the real self, or a realization of the true self. As Carl Elliot has pointed out, where "the significance of life has become deeply bound up with self-fulfillment" there is complex relationship between self-fulfillment and authenticity, such that "a person can seen an enhancement technology as a way to achieve a more authentic self, even as the technology dramatically alters his or her identity" (Elliott 2003: xx–xxi). The cultural hype about designer drugs, like that of designer babies, deserves analysis. But while cultural representations may be of "designer moods," what is sold to the patient is a dream of control. Take control of your moods, treat anxieties that are the symptoms of illness, feel like yourself again, get your life back: these are the hopes, and the narratives, that mobilize the relations between the drug companies, the prescribers, and the consumers of psychiatric drugs.

Indeed, despite the worries that this is an ethic that offers happiness without requiring an exertion of the will, getting your life back is seldom framed as a simple matter of taking a pill. A rapid trip around the websites for depression, for example, shows that the restoration of the self requires work. The subject must engage in self-analysis of the ebbs and flows of mood, thought, and emotion. The subject must learn new ways of self reflection, self-assessment, and insight. These forms of self-scrutiny are often materialized in the form of questionnaires to complete or diaries to maintain, meticulously charting the variations in feelings, moods, behavior, and thoughts in different situations.[35] Frequently the subject is recommended to engage in other forms of therapy, for example cognitive therapy, which requires that thought works on thought itself. Here, as elsewhere, we are obliged to work on ourselves to make ourselves what we really are. Paradoxically this is precisely the ethic of a real identity, of the work on ourselves that makes us truly human, that provides the touchstone for our bioethicists. The use of the phrase "cosmetic psychopharmacology" is thus misleading. While cosmetic surgery may sometimes—by no means always—promise a fabricated exterior, here, as in most contemporary practices for "governing the soul," none of the parties involved consider their aspirations to be the fabrication of an artificial self. No doubt both legal and illegal drugs have long offered to "take you out of yourself," but the new generation of psychiatric drugs no more offer designer moods than PGD offers "designer babies."

Is the issue different when it is not mood or emotion that is at stake, but cognition?[36] Recently concerns have shifted from the SSRIs to the issue of "cognitive enhancement." Debates focus on a number of examples that seem to indicate that we are already beginning to see the widespread use of pharmaceuticals to enhance mental functions. The main examples are the widespread use of Ritalin among students who have not been diagnosed with ADHD;[37] the use of Cephalon's Provigil (modafinil) developed for the treatment of sleep disorders in an attempt to increase alertness and mental energy;[38] and the prospect that drugs initially developed for the treatment of age-related memory loss, "Mild Cognitive Impairment, and early Alzheimer"s Disease, will be used "off label."[39] Harry Tracey, publisher of *NeuroInvestment*, is widely quoted as estimating in 2004 that at least 40 potential cognitive enhancers were currently in clinical development.[40]

Take, for example, Tim Tully's work with his company Helicon on the development of a memory-enhancing compound, targeted at a protein named CREB that appears to regulate neurotransmission linked to long-term memory (based on his work with fruit flies). This work received extensive publicity in April 2002, when it was dubbed Viagra for the Brain.[41] CREB is cAMP Response Element Binding Protein. This appears to be a cellular controller of gene function that regulates the signaling pathway that enables long-term memory formation both in flies and in all mammals, including humans. Subsequent functional genomic studies at Helicon identified numerous additional genes that control different and parallel memory pathways. Helicon is using this work as a basis to create drugs for use in memory disorders; they claim the genes they have identified are targets for the production of drugs that enhance and diminish memory.

Helicon's rival in this work is Memory Pharmaceuticals, a biopharmaceutical company founded in 1998 by Eric Kandel, University Professor of Neurobiology at Columbia University and Senior Investigator at the Howard Hughes Medical Institute, recipient of the 2000 Nobel Prize in Physiology or Medicine for his pioneering work in the field of learning and memory, and founding Director of the Center for Neurobiology and Behavior at the Columbia University College of Physicians and Surgeons. Memory Pharmaceuticals is focused on developing innovative drugs for the treatment of debilitating central nervous system (CNS) disorders such as Alzheimer's Disease, depression, schizophrenia, vascular dementia, Mild Cognitive Impairment, and memory impairments associated with aging. A 2002 article in *Forbes*[42] lays out the hype and hope:

> In 1998 Kandel's team injected aging mice with a failed antidepressant called Rolipram, which prevents the breakdown of cyclic-AMP by blocking an enzyme called phosphodiesterase-4. The hope was that the drug would boost old, tiring brain cells. Rolipram, developed in the late 1980s, never made it because it did not work well and caused nausea and vomiting. But old mice put on Rolipram began navigating mazes faster. . . . Kandel shared the amazing results with Walter Gilbert, a friend and Nobel laureate at Harvard who founded Biogen. Gilbert contacted the venture capitalist Jonathan Fleming of Oxford Bioscience Partners, who helped raise $38 million to form Memory

> Pharmaceuticals. Axel Unterbeck, then head of dementia research at Bayer, signed on as president. . . . "I was stunned. Never had I seen data like this," says Unterbeck, now Memory's chief science officer. . . . Now Kandel is devising a Rolipram-like drug that targets the brain's memory centers but avoids regions that control the vomiting reflex. It turns out that some of the 20 variants of phosphodiesterase-4 play different roles. Researchers at Memory Pharmaceuticals carefully mapped the regions in the brain where each variant is found. It is testing prototype drugs that block those present only in the hippocampus. In animal tests, the compounds duplicate Rolipram's success without the nasty side effects. The first human trials are about 18 months away, most likely first in Alzheimer's patients. Unterbeck says, "If it is safe, the market is incalculable."

Some ethicists follow Kass and Fukuyama in their concerns about the implications of the spread of such "doping practices" (Butcher 2003). The web pages of Life Enhancement Products Inc., a commercial company selling nutritional supplements, quote Howard Gardner, a professor of cognition and education at Harvard University, as a spokesperson for this opposition: "By the time that we are aware of it, it will be too late," Gardner said. "I think that this change is unlikely to be stoppable, but I believe that those of us who are opposed to cognitive enhancements of individuals within the normal distribution of the population should stand up and be counted. We might just make a difference." But for those at Life Enhancement Products Inc. the capacity to improve oneself is a matter of liberty: "Let those of us on the opposite end of the ethical spectrum stand up and be counted: Let people be free to choose to be smarter and to take whatever steps are necessary. Be proud that your interests have led you to this pro-cognitive conclusion long ago. Stand up and let others know. And don't fall prey to the "doping" argument. "Smarting" is the way to go."[43]

Cognitive enhancement is now one of the staple concerns of neuroethics (Caplan and Farah forthcoming, Farah 2002, Wolpe 2003). Paul Wolpe points out that humans have long used multiple ways to enhance their cognitive functioning: "We send our children to school, memorize poetry, develop training programs, medicate, enrich our word power, read novels . . . try to get a good night's sleep before exams, eat "brain food" such as fish . . . all actions that, to one degree or anther, are intended to create environments, inner states or improved functioning that will encourage or support a desired level of neurological performance" (Wolpe 2002: 391). For Wolpe, it is the direct engagement with the neurochemical, structural, or electrical components of the brain that sets contemporary enhancement techniques apart and raises new ethical challenges. But

Wolpe and colleagues are not convinced by the concerns expressed from those who worry about equity: "In comparison with other forms of enhancement that contribute to gaps in socioeconomic achievement, from good nutrition to high-quality schools, neurocognitive enhancement could prove easier to distribute equitably (Farah et al. 2004: 423). Indeed, they point out that, without waiting for the ethical issues to be resolved, the market for nutraceutical and other foods, additives, vitamins, and techniques to enhance cognitive performance, however disputable their claims of efficacy, is already huge.

As Wolpe points out, these developing neurotechnologies are significant, not so much because of what they will or will not be able to do, but because they force us to engage directly with the kinds of selves we think we are or could be (Wolpe 2002: 394). But, as I have already argued in relation to Canguilhem's distinction of social and vital norms, any distinction between treatment and enhancement grounded in a postulate of a natural vital order of the human body and soul, however desirable to the bioethicists, is bound to fail. The question of the proper limits of these technologies will not be resolved by an appeal to human nature, human dignity, or a rejection of artificiality. But this will not erase the significance of these styles of thinking, for they pose the question of who we are as human beings, what we should fear, and what we might hope for—what we might legitimately desire and what desires might legitimately be denied. Here, as elsewhere, the disputes about life itself, about the possibilities that are opened to us in our emerging form of life, and about the dilemmas generated by our increasing and inescapable responsibility for our own biology, will be worked out only in the messy interactions of science, technology, commerce, and consumption that are the territory of contemporary vital politics.

Mapping Emergent Forms of Life

Although I have focused here on only one dimension of our emerging form of life, that of optimization, I have tried to show that new configurations of knowledge, authority, technology, and subjectivity are emerging—emerging tentatively, it is true. Many of these hopes will be disappointed, these fears will prove ungrounded, and the hype will move on to some other domain—nanotechnology, perhaps. Freed from the weight of expectations, relieved of its promissory overload, susceptibility testing for a limited number of conditions and a limited number of drug responses will no doubt become routinized, integrated into the normal practice of medicine. Freed from the portentous bioethical discourse that surrounds it, the distinction between normalization and enhancement will come to seem spurious, replaced by the more modest and pragmatic questions of what interventions are available to whom, at what cost, with what efficacy, and with what safeguards.

Nonetheless, I think these disputes about susceptibility and enhancement have a wider significance. Medicine has long been central to our philosophical and ethical understanding of ourselves. Hence, as it mutates, those understandings mutate, generating a new ontology of ourselves. These new forms of life, these new ideas of what kinds of persons we are and could or should become, are emerging at the multiple intersections between the imperatives of the market and the drive for shareholder value, the new imaginations of the body and its processes that have been brought into existence, the drive of biomedical researchers for papers, prizes, and intellectual property, and the hopes of national governments for new economic opportunities. But in the United States, and to a lesser extent elsewhere in the anglophone world, in Europe and in Japan, they are energized by the illimitable demands of those who possess the financial and cultural wherewithal to their rights as consumers to medical resources in the name of the maximization of their somatic selves and those of their families. The multiple transactions between expertise and subjectivity, and the multiple injunctions and managed desires to reform and remake ourselves through calculated intervention in the name of our authenticity, self realization, and freedom, have been central to the "government of the self" in advanced liberal democracies (Rose 1989). What is new, perhaps, is the centrality accorded to the soma, to the flesh, the organs, the tissues, the cells, the gene sequences, and molecular corporeality to our individual and collective ways of understanding and managing ourselves as human beings.

Many years ago now, when feminists proclaimed "our bodies, our selves," they envisaged a very different form of politics, one in which the body was a natural object, which we must rescue from its alienated state, retrieve from the grasp of the experts, which we can and should each know and tend for ourselves (Boston Women's Health Book Collective et al. 1978). Now our bodies form the basis of many related but different projects, in ways that could not have been anticipated in the critiques of medical authority in the 1960s and 1970s.[44] The conscious and calculated management, maintenance, modification, and manipulation of our somatic existence throughout the course of our lives and through all vicissitudes has become the subject of so much noisy and incessant speech, the organizing feature of a complex of knowledge, power, and value, and the hesitant potential basis of a new ontology. In this sense, our bodies have become ourselves, become central to our expectations, hopes, our individual and collective identities, and our biological responsibilities in this emergent form of life.

Chapter 4
At Genetic Risk

Among the many consequences of recent advances in the life sciences and biomedicine are some mutations in "personhood."[1] These are not merely modifications of lay, professional, and scientific ideas about human identity and subjectivity, but shifts in the presuppositions about human beings that are embedded in and underpin many practices. In this chapter I focus on one of these: the human being who is "genetically at risk." This kind of person is born at the intersection of at least three trajectories. *First*, we see the growing belief that many undesirable conditions—physical illnesses or behavioral pathologies—have a genetic basis. This may be in the form of a "genetic mutation" for a particular pathology such as Phenylketonuria or Huntington's Disease, or it may be in a certain genetic makeup, which may involve small variations in many genes and their interactions with one another, which increase the likelihood that certain individuals will develop a particular condition such as breast cancer. *Second*, researchers claim that they have the capacity to characterize the genetic sequences or markers associated with the occurrence of many conditions at the molecular level and that this capacity will increase. This arises, in particular, from beliefs about the clinically relevant information that will be generated by large databases that integrate DNA analysis of blood or tissue samples with family histories and personal medical records.[2] *Third*, doctors claim that they are increasingly able to identify specific individuals with the genetic profile linked to the development of particular conditions prior to their onset through diagnostic tests. This identification may be precise, where genetic screening is able to identify the genetic variants or polymorphisms themselves. It may be probabilistic, where screening is based on the identification of genetic markers associated with increased likelihood of being affected, or where identification is through family histories or the identification of factors associated with the condition. These developments, in association with other characteristics of contemporary forms of personhood in advanced liberal democracies, have consequences for the ways in which individuals are governed, and the ways in which they govern themselves.

In such societies, the reorganization of many illnesses and pathologies along a genetic axis does not generate fatalism. On the contrary it creates an obligation to act in the present in relation to the potential futures that now come into view. The discourses and practices of genetics here link up with those of risk. While hereditary knowledges have long been associated with various forms of risk thinking, the availability of predictive genetic testing introduces a qualitative new dimension into genetic risk, creating new categories of individuals and according genetic risk a new calculability. As a result of these new knowledges, it is becoming possible for individuals to be identified as genetically at risk for a particular condition prior to any symptoms appearing. Many thus fear that such "genetically risky individuals" may then be treated, by themselves, and by others ranging from employers and insurance companies to future spouses and genetic counselors, as if their nature and destiny was indelibly "marked" by this genetic flaw, despite the fact that the "penetrance" of the genes may be unknown, that in most cases only a certain percentage of individuals in this class will suffer in this way, and that the timing of onset and severity of any disorder are unpredictable. Debate in many countries in Europe and North America has focused on whether such persons will suffer social stigma and exclusion from certain opportunities, services, or benefits. It is also possible that those diagnosed with predispositions to pathologies on the basis of a genetic test, despite the fact that they seem normal or healthy, may find themselves, voluntarily or involuntarily, placed under the aegis of the medical, psychiatric, or legal professions, and the subject of various forms of surveillance or treatment in the name of prevention.[3]

The new genetics also links up with contemporary practices of identity. In advanced liberal democracies—the geographical and political regions with which I will be concerned in this chapter—genetics takes its salience within a political and ethical field in which individuals are increasingly obligated to formulate life strategies, to seek to maximize their life chances, to take actions or refrain from actions in order to increase the quality of their lives, and to act prudently in relation to themselves and to others. As life has become a strategic enterprise, "the categories of health and illness have become vehicles for the self-production and exercise of subjectivities endowed with the faculties of choice and will" (Greco 1993). Over the past three decades many aspects of biomedical languages of description and judgment—high blood pressure, abnormal heart rhythm, raised blood cholesterol, and the like—moved from the esoteric discourse of science to the lay expertise of citizens. Today, new ideas about genetics are supplementing older notions of heredity and genes within these languages of self-description and self-judgment, inscribing genetic knowledge into the heart of corporeal existence (Kenen 1994). Like earlier vocabularies these genetic languages render visible aspects of human individuality to others and to oneself that go beyond "experience," not only making sense of it in new ways, but actually reorganizing it a new way and according to new values about who we are, what we must do, and what we can hope for. Like Ian Hacking's interactive kinds (Hacking 1986, 1995), this reshapes the self-descriptions and possible forms of action of the genetically risky individual him or herself. Thus genetic styles of thought not only give life strategies a genetic coloration but also create new ethical responsibilities. When an illness or a pathology is thought of as genetic, it is no longer an individual matter. It has become familial, a matter both of family histories and potential family futures. In this way genetic thought induces "genetic responsibility": it reshapes prudence and obligation, in relation to marriage, having children, pursuing a career, and organizing ones financial affairs.

The rise of the person genetically at risk is entangled with the spread of the molecular vision of life (Chadarevian and Kamminga 1998, Kay 1993). Of course, geneticists still gather information on family histories. But increasingly, for genetic researchers, this gross level of data is only a stepping-stone in the attempt to construct linkage maps that can then be the basis of DNA sequencing to identify the exact chromosomal location and sequence of the mutated gene in question and enable the development of highly specific diagnostic techniques.[4] Within this molecular thought style, illnesses are increasingly visualized in terms of sequences of DNA base pairs at particular locations on a specific chromosome. For example, one condition involving fronto-temporal Dementia and Parkinsonism is known as FTDP-17 because it is linked to a number of mutations in a specific region of Chromosome 17. Increased susceptibility to certain forms of breast cancer has been linked to the mutations known as BRCA1 and BRCA2 on Chromosome 13. Researchers have tried to link psychiatric disorders such as manic depression, and even variations in personality such as novelty seeking, with the synthesis or nonsynthesis of particular proteins or the characteristics of particular neuronal transmitters or neural receptor sites—Chromosome 11 once being a particular favorite.[5] As the body becomes the subject of a molecular gaze, susceptibility to disease has been molecularized, and genetic risk becomes a molecular matter.

In this chapter, I will first say a little more about personhood and its contemporary genetic mutation. Next, since the identification of persons in terms of their hereditary makeup and defects or propensities is not itself new, I provide a brief history of the ideas and practices of genetic risk. I then consider two sets of practices. Through an examination of recent controversies about genetic discrimination in education, employment, and insurance, I will argue that this marks the birth of a distinctive, although not completely novel, practice of subjectification that operates alongside, and intersects with, others that are operative in different situations. However I will try to show that this cannot be understood simply in terms of the birth or rebirth of genetic essentialism and, in particular, that it does not efface, but indeed links up with, prevalent regimes of subjectification that stress an ethics of enterprise, responsibility, and self-actualization. Second, drawing extensively on the work of Carlos Novas, who has studied the ways in which those at risk of Huntington's Disease debate the dilemmas that face them, I explore some of the ways that forms of subjectivity generated by genetic risk are bound up with new ethical problematizations and new ethical relations (Novas 2003).[6] Like Novas I argue that far from generating resignation to fate or passivity in the face of biological destiny or biomedical expertise, these new forms of subjectification are linked to the emergence of complex ethical technologies for the management of biological and social existence. And I follow Novas in suggesting that these are located within a temporal field of life strategies, in which individuals seek to plan their present in the light of their beliefs about the future that their genetic endowment might hold. These new modes of subjectivity produce the obligation to calculate choices in a complex interpersonal field, not only in terms of individuals' relations to themselves, but also in terms of their relations to others, including not only actual and potential kin, past and present, but also genetic professionals and biomedical researchers.

Somatic Individuality

A number of authors have suggested that we are witnessing a whole-scale geneticization of identity with the consequent reduction of the human subject to a mere expression of their genetic complement (Dreyfuss and Nelkin 1992, Lippman 1991, 1992). While these authors accept that genes play a role in all sorts of illnesses, in interaction with one another and with social, biographical, psychological, and environmental factors, they claim that "geneticization" is a determinism that asserts that genes "cause" disorders. They argue that these genetic narratives of health and disease orient the ways in which problems are defined, viewed, and managed within society. They suggest that this legitimates funding and support for the projects of the gene mappers, and hence defines more and more problems of health and disease as genetic disorders. Geneticization is seen as an individualizing tactic that redirects scarce resources away from social solutions to social problems, and represents a threat to doctrines such as equal opportunity, as well as to ideas of free will, intentionality, and responsibility. "The individual affixed with a genetic label can be isolated from the context in which s / he became sick. . . . The individual, not society, is seen to require change; social problems improperly become individual pathologies" (Lippman 1992: 1472–73). Hence the application of genetic knowledge in the diagnosis, assessment, and treatment is associated—wittingly or unwittingly—with reactionary and stigmatizing political strategies.

These arguments makes some significant points. But taken as a whole, I find them misleading.[7]

The geneticization argument implies that to ascribe genetic identity to individuals or groups is to objectify them, hence denying something essential to human subjectivity. But to make human individuality the object of positive knowledge is not "subjection" in the sense of domination and the suppression of freedom—it is the *creation* of subjects that is at stake here. Today, as at the birth of clinical medicine, the sick person bears their illness within their corporeality and vitality—it is the body itself that has become ill. But this somaticization of illness did not, in fact, mandate the perpetual passivity of the patient. In fact, clinical medicine, increasingly over the last half of the twentieth century, constituted the patient as an "active" subject—one who must play a part in the game of cure (Armstrong 1984, Arney and Bergen 1984). While not denying that illness is inscribed in the body, contemporary medical practice has required the patient to offer up his or her voice in the diagnostic process in order to permit the disease itself to be identified, has asked that they commit themselves to the practice of the cure as part of a therapeutic alliance, and has advised them to conduct themselves prudently prior to illness, in the light of information about risks to health. The same is true of the fabrication of the persons genetically at risk in contemporary medical genetics. The patient is to become skilled, prudent, and active, an ally of the doctor, a protoprofessional, and to take a share of the responsibility for getting themselves better. No wonder, then, that as ideas of genetic risk spread, individuals themselves request genetic testing, and commercial organizations have been set up that offer testing for predispositions, online, direct to the consumer, for a small fee. Thus, for example, by early 2005, DNA Direct, based in San Francisco, was offering patients genetic testing for predispositions to Alpha-1 antitrypsin disorder, breast cancer (using the tests developed and patented by Myriad Genetics), ovarian cancer, cystic fibrosis, haemochromotosis, infertility, and recurrent miscarriage and thrombophilia.[8] "Our tests and services can empower you, your family and your healthcare team," says their website, also quoting a satisfied customer: "I liked being able to test myself at home. Finding out helps explain the past. . . . It also will help keep me healthy in the future."

Patients at genetic risk and their families are thus not passive elements in the practice of cure. The studies carried out by Paul Rabinow (Rabinow 1999), by Vololona Rabeharisoa and Michel Callon (Callon and Rabeharisoa 2004, 1998a, 1998b), and by Deborah Heath, Rayna Rapp, and Karen Sue Taussig (Heath et al. 2004) have all shown that such persons—the ill patients themselves, those "asymptomatically ill," and their families—are increasingly demanding control over the practices linked to their own health, seeking multiple forms of expert and nonexpert advice in devising their life strategies, and expecting medics to act as the servants and not the masters of this process.[9] These persons defined by genetic disease have an investment in scientists fulfilling their promises and discovering the basis of, and the cure or treatment for, genetic conditions. Medicine, including medical genetics, notwithstanding its resolutely somatic understanding of the mechanisms of disease, has been one of the key sites for the fabrication of the contemporary self—free yet responsible, enterprising yet prudent, conducting life in a calculative manner by making choices with an eye to the future and with the aspiration of maintaining and increasing his or her own well-being and that of their family.

Critics also tend to suggest that the new medical genetics leads to a focus on the individual as an isolate. I disagree. Within such practices, individuals are subjectified through their location within networks of relatedness and obligation. Consider, for example, the practice of genetic counseling, which I discuss in more detail later. In a study of genetic counseling consultations, Armstrong and his colleagues show how the genetic identity of the counseled individual is established by locating him or her within a system of relations—mapping a set of remembered relations of lineage onto a remembered web of illnesses—at the same time as those social and familiar relations are reworked in genetic terms (Armstrong et al. 1998, Konrad 2005). The illness or condition becomes a "family" matter. The patient's problem might be traced to a family member in a previous generation; the diagnosis in one person has all kinds of implications not only for that individual but also for their relatives. New questions are raised about the connection, or lack of it, between family members' ideas of kinship and the "natural" relations traced out by DNA itself—notably for adopted children, those conceived with donated sperm and, of course, those who are not the biological children of the person considered to be their father. Genetic identity is thus revealed and established within a web of genetic connectedness overlaid upon a web of family bonds and family memories, with the burden of mutual obligations and caring commitments, and with all the ethical dilemmas they entail. In becoming part of such a genetic network, the subject genetically at risk may re-think his or her relation to current family—lovers, potential and actual spouses, children, grandchildren, and so forth—in terms of these issues of risk and inheritance. They may reshape their form of life—lifestyle, diet, leisure activities, alcohol, smoking—in these terms, thus also reshaping relations with those with whom they interact. They are also brought into relation with novel associations—not those of "society," but of "community"—organizations and groupings of those similarly at risk; patients at particular hospitals or clinics; participants in trials of new therapies; subjects of documentaries and dramas on radio, television, and the movies.

The mutations in personhood associated with the new life sciences and biomedical technologies of life are multiple and they are not exhausted by genetics. For example, new reproductive technologies have split apart categories that were previously coterminous—birth mother, psychological mother, familial father, sperm donor, egg donor, and so forth—thus transforming the relations of kinship that used to play such a fundamental role in the rhetorics and practices of identity formation (Franklin 1997, Strathern 1992, 1999). Developments in psychopharmacology have transformed the ways in which individuals are understood, as the very features that seemed to constitute their individuality—such as personality or mood—now appear to be amenable to transformation by the use of specially engineered drugs such as Prozac (Elliott 2003, 1999, Fraser 2001, Kramer 1994, Rose 2003, Slater 1999). New visions of personhood are coming to the fore associated with the growing interest and sophistication in brain imaging techniques, which localize the features of the personality, affects, cognition, and the like in particular regions of the brain (Beaulieu 2000, Dumit 2003). Practices of subjectification that operate in genetic terms—in terms of genetic forms of reasoning, explanation, prediction, and treatment of human individuals, families, or groups—find their place within this wider array of ways of thinking about, and acting upon human individuality in "bodily" terms—the general "somaticization" of individuality in an array of practices and styles of thought, from techniques of bodily modification to the rise of corporealism in social and feminist theory and philosophy.

In any event, the geneticization of identity has to be located in a more complex field of identity practices. Advanced liberal democracies are traversed by multiple practices of identification and identity claims—in terms of nationality, culture, sexuality, religion, dietary choice, lifestyle preference, and much more. Only some of these ascriptions of, and claims about, identity are biological or biomedical. Indeed biomedical identity practices and identity claims, including those that operate in terms of genetics, find their place among a bewildering array of other identity claims and identificatory practices. Subjects or others sometimes do rewrite aspects of their identity in biological terms, but they sometimes contest vehemently such a rewriting. Identities are plural and multiple: one is identified as a gay man within some practices, as a Muslim within others, as a carrier for sickle-cell disease within others. Even when regulatory practices utilize biological conceptions of personhood, genetic identity is rarely hegemonic. In insurance, as we shall see, genetic information is considered alongside other nongenetic aspects of personhood—medical history, habits such as smoking, risks associated with lifestyle choices, and so forth. In the courtroom, a range of biological evidence, including that from brain scans,is now entering in the determination of aspects of personhood such as capacity to stand trial or responsibility—but courts have proved remarkably resistant to arguments that responsibility or intentionality at law should be reconceptualized in terms of evidence from genetics (Rose 2000b).[10] Ideas about biological, biomedical, and genetic identity will certainly infuse, interact, combine, and contest with other identity claims; I doubt that they will supplant them.

A Brief History of Genetic Risk[11]

While hereditary thought has a long history, the twentieth century saw the emergence of a range of practices that defined individuals in relation to their genetic constitution. Here, drawing on the work of Carlos Novas (Novas 2003), I focus on one of these practices: genetic counseling, in which individuals, under the guidance of authorities, are encouraged to reflect upon their inherited constitution, with the explicit aim of affecting their conduct. Novas has described the historical variability of the ways in which the subject of the genetic consultation has been understood, and the related changes in the types of reflection and forms of conduct that counselors have tried to engender. This enables him to identify the distinctiveness of the person genetically at risk as this conception, and the practices to which it is linked, has developed from the early 1970s to the present.

Genetic counseling links up the objectifying knowledge of genetics, which operates at the level of the soma, and the subjectifying knowledge of the human sciences, which works upon the conduct of human conduct. Novas has proposed the term "technologies of genetic selfhood" to characterize the ways in which the practices of genetic counseling incite an individual, couple, or family to reflect upon their genetic constitution with the aim of affecting their conduct in light of this knowledge

(Novas 2003). He suggests that we might think of these as heterogeneous assemblages that involve a combination of forms of knowing, expertise, and diagnostic techniques. Knowledge forms an integral component of such technologies and practices of self-government and this knowledge itself is heterogeneous, ranging from Mendelian genetics to Rogerian psychotherapy, and is usually further tailored to the perceived needs of the consultand. Novas shows that the provision of genetic advice is a highly technical process that makes use of family pedigrees, clinical observation, risk and probability analysis, serological analysis, tremometers, electroencephalographs, prenatal diagnosis, and predictive genetic testing (either through linkage or mutation) in order to help visualize, diagnose, or communicate genetic status. Many different experts have performed genetic counseling—physicians, pediatricians, geneticists, neurologists, psychiatrists, or psychotherapists—and there have been many contests over the appropriate form of professional expertise and the training required to perform it. And technologies of genetic selfhood vary greatly in the norms of subjectivity they embrace.

Genetic counseling needs to be located within more general biopolitical rationalities. Following Novas, I think it is possible to make a broad heuristic division into three periods that differ in the objectives, problematizations, normative orientation and practices of genetic advice giving.[12] The first "eugenic" period runs through the 1930s and 1940s. The eugenic forms of thought that spread across Europe, North America, and elsewhere in the first half of the twentieth century gave rise to attempts to reshape "voluntary" individual reproductive decisions in the light of eugenic considerations. In the 1930s genetic counseling took its place alongside eugenic strategies of public education and the use of film and propaganda. Thus the genetic counseling of the early 1930s to the late 1940s largely operated within a concern to shape individual reproductive decisions, and to limit the reproduction of those with hereditary diseases or defects, in order to improve the quality of human stock in the population as a whole. Genetic advice in the eugenic age required an assessment of good and bad genetic qualities, and an evaluation of the seriousness of the heredity defect. Couples who were intelligent and of good physique were encouraged to bear greater numbers of children, while those that were not were encouraged to restrict their childbearing or family size. While some subjects were thought to possess the moral capacity to control their reproduction, most of those who carried hereditary pathologies were construed as passive individuals who often unwittingly bore more children than they should, either because they failed to acknowledge the genetic risks they faced or because their judgment was impaired by the effects of the pathology itself.

From the 1950s to the early 1970s, Novas suggests, genetic counselors seek to dissociate the field from the negative eugenics associated with the Nazi regime. They argue that the prevention of genetic disease in democratic societies can take place only through voluntary measures (Dice 1952). Its aim should be the optimization of population health, mostly through the prevention of birth defects (Fine 1993). Those who sought to develop this "nondirective" approach grounded in preventive medicine believed that parents wanted healthy children. Thus they believed that those who were at risk of having defective children, once given accurate knowledge by their genetic counselor, would make "responsible" use of that knowledge and make their own prudent decisions concerning reproduction—they would chose not to have children, to limit their family size, or to adopt (Dice 1952, Herndon 1955).

Novas argues that during this period genetic counseling was redefined as a form of guidance to help relieve anxieties, fears, and inner tensions of being provided with genetic risk information (Falek and Glanville 1962; Kallmann 1956, 1961; Roberts 1961). Further, genetic counseling itself had to use psychology in order to optimize the assimilation of genetic risk information, as individuals confronted by genetic risks might resort to immature patterns of behavior, neutralizing the genetic realities through pathological defense mechanisms such as repression, displacement, rationalization, and projection. Genetic counseling was thus reconceived as a kind of short-term psychotherapy that could instill a sense of responsibility within consultands who would come to recognize the value of a well-planned family (Kallmann 1962: 253).

During the 1970s, Novas suggests, a new form of "psychosocial counseling" became dominant, and this has remained the model up to the present. In this model, the identification of genetic risks has become bound up with a concern to maximize life chances and improve quality of life; in order to do this, genetic counseling was no longer to be exclusively concerned with the prevention of genetic disease, but must be involved in the communication of genetic risk (Kenen 1984). This was the time when there were major developments in genetic knowledge, and a range of new techniques such as prenatal testing and preimplantation diagnosis, together with the possibility of the abortion of fetuses thought to carry, or potentially to carry, genetic pathologies. Those undergoing genetic consultation were confronted with the range of new choices that developments in biomedicine had placed before them; they had to make complex decisions concerning their own life and the lives of their actual or potential offspring. And increasingly they were addressed as autonomous individuals making informed and responsible choices for their own futures.

As presymptomatic and predisposition genetic testing became more widely available, Novas points to an increasing concern to identify those who were also at risk for adverse psychological outcomes, such as depression or suicide as a result of the genetic testing process. Psychometric testing was one resource that was drawn upon, and those undergoing genetic testing were also evaluated in terms of their coping resources, sources of support either from spouses, friends, family, or participation in support groups (Bloch et al. 1993: 370, Decruyenaere et al. 1996, 1999). Most recently, psychosocial genetic counseling has come to focus on the modification of lifestyle (Biesecker and Marteau 1999) and, in particular, on promoting the autonomy and self-directedness of the client (Elwyn et al. 2000). Through techniques such as shared decision making, the client is him- or herself a party to the relevant information, and takes a portion of the responsibility for any decision, including, for example a decision to disclose genetic information to kin who may also be genetically at risk. And such disclosure is itself seen as vital, in the light of the need to "give them the opportunity to plan their lives, to make informed decisions about reproduction, and to seek surveillance for early signs of a complication for which medical intervention can be effective . . . [a] genetic counseling session must not be construed as the passive transference of information from expert to layman, but as a dynamic, or ongoing, process in which the counselor and counselee both have a role to play in determining what occurs and what is understood" (Hallowell and Richards 1997). They must be active planners and decision makers, and reconfigure their identity in terms of a genetic past, a genetic present, and a genetic future (Ogden 1995). Genetic counseling thus links with other injunctions to women in relation to reproduction that urge them to take informed choices over their own reproduction in the light of expert knowledges and techniques. On the one hand, this opens the possibility for new strategies of control, in which psychology plays a key role in attempts to modify the behavior of those deemed genetically at risk. But in the context in which each individual life has become conceived as a kind of project to be planned, this reconfiguration of identity reshapes the ethical field within which genetically risky individuals must govern themselves and their lives.

Genetic Discrimination

On February 8, 2000, at a meeting of the American Association for the Advancement of Science, the then President of the United States Bill Clinton signed an executive order that prohibited every federal department and agency from using genetic information in any action involving hiring or promotion (White House 2000). On the same day, Clinton endorsed the Genetic Nondiscrimination in Health Insurance and Employment Act of 1999, which extends such protections to the private sector, and to individuals purchasing health insurance. This was in the context of a 1996 study that had shown that 25 percent of the respondents or affected family members believed they were refused life insurance, 22 percent believed they were refused health insurance, and 13 percent said that they or a family member had been denied a job, or was fired from a job, because of a genetic condition in the family; many of those questioned had refused genetic tests because of fear of genetic discrimination, or had not revealed genetic information to insurers or to employers (Lapham et al. 1996). Other studies had found a widespread fear that employers would, in the future, demand that potential employees take genetic tests, and that many would be deterred from taking such tests in the present if employers could access the results. Clinton reasserted the view he had expressed since 1997: the effort to find genetic cures for diseases should not undermine protections for patients.

The potential of genetic knowledge to be used in discriminatory ways has been a concern for many years. In 1992 Paul Billings and his colleagues defined genetic discrimination as "discrimination against an individual or against members of that individual's family solely because of real or perceived differences from the "normal" genotype" (Billings et al. 1992), and found some evidence of such discrimination in the health and life insurance industries and elsewhere. They warned of the possible growth of stigmatization and denial of services and entitlements to individuals who have a genetic diagnosis but are asymptomatic, and pointed to the dangers of creating "a new social underclass based on genetic discrimination ('the asymptomatically ill')" (476). Since that time further studies have argued that there is evidence of genetic discrimination—that is to say, discrimination against currently healthy individuals on the basis of genotype alone, sometimes based on misunderstandings, such as when an individual is a carrier for a mutation for a particular disorder but will not develop the disease. Some have suggested that the rise of genetic testing blurs the boundaries that were previously used to define "pre-existing conditions": an individual carrying a detectable mutation or mutations that predispose them to develop a particular disorder in particular circumstances might be considered to have a "pre-existing condition"—even if they are unaware of it. Studies have also suggested that the fear of genetic discrimination is widespread among those who have family histories of particular disorders, and that this is leading to reluctance to take genetic tests, concern about disclosure of information on such tests by medical practitioners, and in some cases to the falsification or nondiscourse of relevant aspects of medical histories to insurers, employers, or others.[13] And many have argued that existing antidiscrimination provisions are insufficient to deal with these issues.

In the 1990s, many states in the United States enacted laws that prohibit the use of genetic information by insurance companies in the pricing, issuing, or structuring of health insurance (Hall 2000). A study published in 2000 concludes that such laws do not seem to have had much impact, for there were in fact almost no well-documented cases of health insurers asking for or using presymptomatic genetic test results either before or after the passage of such laws, or whether such laws exist in the state in question. Perhaps, as Hall says, industry norms are more significant than laws in organizing practices of this type. And, one might add, perhaps the fear of such discrimination is more real a force than actual discrimination; thus despite an absence of evidence, the U.S. House of Representatives Education and Workforce Subcommittee on Employer-Employee Relations examined the issue again in July 2004. "Lawrence Lorber . . . representing the U.S. Chamber of Commerce, told the subcommittee that legislation is not needed. 'Employer collection and misuse of genetic information remains largely confined to the pages of science fiction.' . . . Even in cases in which discrimination does take place, he said, existing laws like the Americans with Disabilities Act and the Health Insurance Portability and Accountability Act 'are more than ample to deal with the misuse of genetic information.' " However the Chair, Senator Judd Gregg, who helped steer a genetic discrimination bill through the Senate by a vote of 95 to 0 in October 2003 urged the House to act. He noted in a written statement that his committee "heard from a broad range of civil rights, employment and insurance experts, the overwhelming majority of whom agree that there are gaps in current law with respect to genetic information."[14]

These debates have continued in many countries. France, Norway, Australia, Denmark, the Netherlands, and Austria have all passed laws that either severely limit or outright forbid the use of information derived from genetic tests for anything other than medical or scientific purposes, and such legislation was also being considered in Germany at the end of 2004.[15] In 2004, the U.S. Coalition for Genetic Fairness, an alliance of groups supporting legislation banning genetic discrimination, part- sponsored by Affymetrix (whose chips, of course, are part of the technology that makes such testing possible) published a brochure detailing cases where they allege such discrimination occured—for example when children were denied health insurance because they carried "the gene for" alpha-1-antitrypsin deficiency, despite advice from medical professionals that they would never develop the disease themselves.[16] In the light of these concerns the U.S. Senate, in February 2005, passed by 98 votes to 0 yet another piece of legislation, the Genetic Information Nondiscrimination Act of 2005, which would amend a series of existing health and employment acts, outlaw such discrimination whether by employers, health plans, or labor organizations, and extend medical privacy and confidentiality rules to cover genetic information.[17] Whether or not there is, or will be genetic discrimination, the injection of debates over genetics into practices of educational assessment, employee recruitment, and actuarial calculation is significant in its own right. Such debates are themselves acting as vectors for the spread of genetic conceptions of personhood, for the generation of "genetic responsibility," and for the partial reshaping of ethical dilemmas in molecular terms.

Education is another key surface of emergence for these new ways of thinking. Medical diagnoses of educational difficulties and explanations of failure in biological terms have a long history. Schooling exposes children to the gaze and evaluation of professionals: the regime of the school and the norms of conduct and performance it establishes have long served to individuate children not able or willing to conform to those norms, and opened them up to treatment "in their best interests." Notable here have been the diagnoses of minimal brain dysfunction, hyperactivity, attention deficit disorder, and attention deficit hyperactivity disorder: these have been the sites of medical and psychiatric expansion for some decades, especially in the United States, with the widespread use of psychostimulant drugs, methylphenidate (Ritalin) and dexamphetamine (Adderall), as a response. More recently there is evidence of the widespread diagnosis of depression, posttraumatic shock disorder, and conduct disorder in children, coupled with prescribing of antidepressant medication, notably serotonin selective reuptake inhibitors such as Prozac, Zoloft, and Paxil to children.[18] The molecular vision of the new genetics adds the possibility of predictive tests to such practices, and hence of genetic screening and presymptomatic intervention.

Of course, strategies of screening and preventive intervention have long been advocated in relation to juvenile delinquency and criminality. Direct genetic screening for pathologies of conduct together with preventive intervention was first advocated in the late 1960s, when the claim was made, later disproved, that the XYY condition—the possession of an extra Y chromosome—was linked to immaturity, inadequate control of aggressive instincts, and hence to an increased probability of violent crime.[19] In the recent examples, screening of schoolchildren would not merely be for genes linked to future psychiatric disorders, but for genes that are claimed to link directly to problems arising within the schoolroom. As early as 1987, reports in popular science journals claimed to locate a particular gene on Chromosome 15 possessed by members of a family in which there was a history of reading disability, holding out the hope of predictive tests and presymptomatic treatment to avoid the problems of dyslexia (Vellutino 1987, cited in Hubbard and Ward 1999). In 1995, articles appeared in *Developmental Brain Dysfunction* reporting pilot projects to screen school children for major chromosomal abnormalities (duplicate X or Y chromosomes, fragile X syndrome), with the aim of providing genetic counseling and other forms of medical intervention (Callahan et al. 1995, Staley et al. 1995). Recent research claims that childhood hyperactivity is highly heritable, and that specific genes encoding aspects of the dopamine transport and reception system (DAT1 and DRD-4) may be implicated, though admitting that the findings need further replication and validation (Plomin and Mcguffin 2003, Thapar et al. 1999).

Dorothy Nelkin and Laurence Tancredi suggest that, over time, such claims will lead to the use of genetic and biologically based tests as part of the standard testing regime in schools, and that educational authorities and parents will come to see them as objective assessments that have predictive value (Nelkin and Tancredi 1989). This illustrates the two faces of genetic individualization. On the one hand, those in favor of such tests would argue that they were necessary to provide such children with special attention and tailored regimes of learning. On the other, they may also lead to discrimination in the acceptance of children by certain schools, they are likely to be transmitted along with other information to post-school institutions such as colleges or potential employers, and may thus generate a long-lasting spoliation of identity at the molecular level, and a life sentence to existence under the gaze of the therapeutic professions. Whatever their direct consequences, however, the unique visibility that the practices of schooling confer upon children, which is both universalizing and individualizing, will play a key role in the dissemination of these new molecular visions of conduct and its determinants.

A further surface of emergence for these new forms of thinking is the workplace. This is a particular issue in the United States, where most health insurance is obtained either through private schemes or workplace-based group schemes (for a good early review, see Gostin 1991; for a comparative study focusing on Canada, see Lemmens and Poupak 1998). Individual insurance is individually underwritten on the basis of risk factors—such as age, medical history, occupation, health related habits such as smoking, and the like—hence, the relevance of individual information on genetic risk. Group insurance is underwritten on the basis of the risk characteristics of the group, such as type of industry, age and gender characteristics, experience of prior claims, and the like. As employers invest more in the recruitment, training, and retention of workers, and are faced with growing insurance premiums and other costs associated with accidents, injury, and ill health among their workforce, they are seeking increasingly to adopt hiring practices that screen out those employees most at risk of future ill health and disability. The debate focuses on whether it is legitimate for genetic screening to join older techniques—such as the requirement for information on personal medical history, the use of psychological tests and examinations, and the taking of family histories—in the attempt to identify individuals who may be "accident prone" or are particularly liable to develop physical illness or psychological conditions rendering them unable to work effectively, or who may be particularly vulnerable to certain features of specific working conditions.

In Britain the situation is different, because heath insurance is obtained through a national scheme, the National Health Service, funded through universal compulsory taxation set at levels unconnected with medical history or risk category. Unlike the private and group systems in the United States, a universal system must cope inescapably with the aggregate levels of genetically related diseases in the population—levels that are unlikely to be directly affected by genetic testing (Low et al. 1998, O'Neill 1998). This distinction is, however, only partial. This is not only because increasing numbers of people in the United Kingdom are opting for private health insurance, or being given entry into such schemes through their employment, and hence having to undergo medical screening prior to the issuing of the employment contract. Nor is it because some other, perhaps more optional, insurance products are also individually underwritten, such as travel insurance, and require disclosure of relevant medical information. It is also because life insurance, in the United Kingdom as well as the United States, is individually underwritten, and life insurance in the United Kingdom is virtually obligatory for those wishing to obtain a mortgage or loan for house purchase.

Where insurance is individually underwritten, insurers are concerned about accurate and full disclosure of information first to correctly allocate the applicant to a risk pool, but second because of the fear of antiselection or adverse selection: where an applicant deliberately withholds information that indicates they are at increased risk—for example, cigarette smoking—in order to take advantage of premiums set for those with a lower level of risk. Thus, for example, they worry about the possibility that an individual who has genetic knowledge unknown to his or her insurers will conceal the test result and buy life insurance at the standard rate, knowing that their life expectancy is short, hence greatly enhancing their inheritance in the short term and driving up premiums for other policy-holders in the long term—or driving others to companies that could offer lower rates because they had more effective methods of screening out those at high risk. Insurers are concerned that the impact of such shifts might be exacerbated by those who have tested negative for various genetic conditions deciding that they do not need insurance coverage; these were the people whose premiums would have provided the resources—the unwitting subsidization—for the demands made by those who did need to claim.

Historically insurers and applicants have largely operated on the basis of mutual genetic ignorance. However, insurers in the United States by and large argue that there is no significant distinction between genetic information and other health-related information concerning risks, and that the distinction between genetic information and other information on medical risk is unsustainable (Pokorski 1997). Pokorski, an executive of a leading U.S. reinsurance company, points to a host of evidence in support of his view. Textbooks routinely suggest that almost all medical conditions have a genetic basis; the National Center for Human Genome Research has observed that "For policy purposes, it will become increasingly difficult to distinguish genetic from non-genetic diseases, and genetic information from non-genetic information" (National Center for Human Genome Research 1993), and an editorial in *The Lancet* argued that "it will soon be impossible to talk of medical and genetic tests as separate creatures" (Lancet [Anon] 1996). Indeed, as technology companies develop inexpensive biochips to screen for hundreds of genetic defects, drop-in shops and centers are envisaged where individuals can obtain information on genetic testing and common genetic disorders and even offer "walk-in testing," mail-order kits, and home test kits, it seems likely that genetic information may come to be as widely known to individuals as information on other insurantially relevant risk factors such as raised blood pressure, high cholesterol levels, abnormal heart rate, and high body-mass indices.

In the light of these developments, it is easy to understand the view of The Association of British Insurers, in the code of practice it issued in December 1997—that it is not only conceptually unwarranted, but also practically impossible to distinguish genetic information from other medical information of relevance to insurers (Association of British Insurers 1997). Yet on the same day, the Human Genetic Advisory Commission set up by the British Government published a report recommending precisely the opposite: a moratorium for at least two years on any requirement for an applicant for life insurance having to disclose the results of any genetic test to a prospective insurer, until sound actuarial evidence supporting the use of specific tests in relation to specific insurance products has been scientifically validated and is publicly available (Human Genetics Advisory Committee 1997). Critics of the insurers' demands for disclosure of purportedly predictive genetic information argued that those making use of such information are often ignorant of the exact implications; that genetic predictions have rarely been validated and therefore do not allow accurate assignment of an individual to a risk pool; that there is a high level of uncertainty as to the age of onset and severity of most genetically related conditions; and that there is lack of knowledge as to the interaction between genes and environment, and between genes and lifestyle (e.g., O'Neill 1998). Newspapers warned of the dangers of the creation of a "genetic underclass" (Daily Telegraph 2000). In March 2005, the U.K. Government and the Association of British Insurers agreed a Concordat, including an extension of the voluntary moratorium on the use of predictive genetic test results by insurers, which had been due to expire in 2006, by five years to 2011.[20] The Concordat also enshrined agreements with the companies that no one will be required to disclose the result of a genetic test taken as part of a research study, that insurers will not put pressure on those seeking insurance to take a predictive genetic test, and that they will not ask individuals to reveal another person's test results. This Concordat was designed, as the government press release makes clear, not because such discrimination is actually occurring, but in order to reassure people about taking genetic tests.[21]

In the United States, some authors have argued that health care providers are under an obligation *not* to disclose genetic information to insurance companies or employers who might use it to discriminate against them; genetic counselors are being advised to counsel their clients on how to minimize potential insurance problems; some are urging that genetic testing is carried out on an anonymous basis wherever possible to limit the potential for identification of the testees; others are advising individuals to buy all the life and health insurance they require before they undergo genetic testing; individuals are being advised to apply for a large number of small life insurance policies that are less carefully scrutinized rather than applying for a single substantial policy; and physicians are being urged to keep two sets of records, one complete set to be used for health care and one, without adverse test results, which can be accessed by insurers (all examples cited from Pokorski 1997).

No doubt this dispute will be resolved differently in different jurisdictions, at least in the short term. Genetic information has significance, in part, because of a more general shift that is occurring in insurance toward risk segmentation (see Ericson et al. 2000 for an excellent discussion). While insurance can act so as to socialize risk, the current tendency has been to utilize knowledge about populations and information about individuals to "unpool" risks and to allocate individuals to tightly defined risk categories. Such tightly defined risk pools will then determine not only the cost of insurance and its benefits and exclusions for those included, but may also lead to the exclusion of those whose risks make them unfit for inclusion into a commercially viable risk pool. This strategy is justified in terms of twin obligations: not to burden prudent consumers with the cost of risks incurred by the imprudent, and to safeguard and maximize profitability for shareholders by incorporating all known relevant information into risk calculations. But the result is not only to encourage moral hazard (lying about one's history in order to obtain insurance at preferable rates, or even at all), but also to drive those whose risks are deemed too high (those with poor driving records or in areas with high rates of burglary) to specific niche market insurance companies, usually not only charging high rates but also offering limited cover. Within this logic, it is clear that genetic risk is a key factor for such segmentation and unpooling practices. These practices can undoubtedly be seen as congruent with the emphasis on personal responsibility and prudence in advanced liberal practices of government. But where high risk is a matter of inheritance, it seems that the old adage still applies: if you want to be successful, choose your parents carefully. In a regime of fragmented communities of risk, private insurance no longer redistributes accident to mitigate the arbitrariness of fate and the hazards that, but for the grace of God, may face us all. Where back-dated genetic prudence is an impossibility, the individual and his or her family has few options when required to take personal responsibility for the governance of their genetic risks. For the wealthy or articulate in this situation, this may enhance the development of risk communities that take their own measures to ensure their security. However, especially in the United States, where upwards of forty million people have no health insurance whatsoever, those who are excluded from this option are likely to find themselves abandoned to whatever residual provision the "facilitating" state may offer.[22]

The actual penetration of genetic reason into insurantial practices will clearly depend on these contestations. The truth of certain genetic claims will be disputed. Reductive genetic arguments are likely to be rejected by all parties. Legal measures may restrict the occasions where an individual may face compulsory genetic testing at the demand of some authority. Factors such as cost may limit the general use of predictive screening. Irrespective of these actual outcomes, however, disputes over insurance are acting as key vectors for the spread and proliferation of genetic reasoning and its penetration into conceptions of personhood. In these struggles over the way in which the body should enter insurantial practices of calculation, individual fate, and hence individuality itself, has acquired a genetic dimension. And perhaps the key sites of innovation here are to be found neither in the calculative practices of the insurers, nor in the vociferous defenders of privacy and rights, but in the forms of life of the actors themselves, the subjects both of genetic risk and of insurantial practice. Somatic individuals do not relate to themselves as simply the expression of an underlying genetic identity. Even when genetically at risk, such individuals consider themselves to be creatures of rights, legal subjects whose somatic personhood grants them entitlements as well as obligations. As we shall see presently, among the rights that may be claimed is the right to know one's genetic status, one's level and pattern of genetic risk. And, further, the somatic individual, incorporating his or her genetic status, is also a subject of self-actualization, responsibility, choice, and prudence—ethics that can be only operative in the light of a knowledge of one's bodily truth. Individuals themselves are faced with questions as to whether to take genetic tests in order to predict their own future and act prudently within it, in relation, say, to their obligation to their family, the need to make provisions by way of insurance in the event of their death or incapacity, and their wish to conduct their affairs in the world in the light of a knowledge of their genetic status. Genetic identity, that is to say, induces "genetic responsibility."

Genetic Responsibility

The advances in the life sciences associated with molecular genetics and the mapping of the human genome create new possibilities for thinking about and acting on the conduct of human beings as somatic individuals. The merger of the language of genetics and risk provides a rich vocabulary by which to render intelligible our identities, our conceptions of health, and our relations with others. This opens up distinctive fields for ethical self-problematization.[23] The language of genetic risk increasingly provides a grid of perception that informs decisions on how to conduct one's life, have children, get married, or pursue a career. With the emergence of the genetically at risk person, genes themselves have been constituted as an "ethical substance" (Foucault 1985: 26)—that one works in relation to the self (genetic identity, reproduction, health) and in relation to others (siblings, kin, marriage, children). And this ethical work framed in terms of genetics intersects with, and becomes allied with, a more general style of work on the self in contemporary advanced liberal democracies that construes life as a project, framed in terms of the values of autonomy, self-actualization, prudence, responsibility, and choice.

Carlos Novas has illuminated some aspects of these new forms of personhood through his study of webforums and chat rooms related to Huntington's Disease (Novas 2003).[24] These are sites on the Internet where subjects at risk and others can discuss their own ways of understanding and responding to issues related to risky genes and genetic disease. Of course, these are not "representative"—only a tiny minority of those genetically at risk are involved in such activities, probably disproportionately drawn from those who are young, relatively wealthy, and better educated. Yet they clearly exemplify the formation of a new ethics of biomedical subjectivity that is emerging out of the plethora of new movements organized around the new sciences of medicine and life. This is a form of life in which actual and potential patients—in other words, all of us—have come to be "passionately curious about their health, happiness and freedom" (Rabinow 1994: 63). Novas argues that, like earlier practices of confession and diary writing, the practices of posting, reading, and replying to messages in these webforums and chat rooms can be seen as techniques of the self, entailing the disclosure of one's experiences and thoughts according to particular rules, norms, values, and forms of authority. Through these practices of disclosure, individuals develop a language to narrate and reflect on their genetic identity, seek advice on how to conduct their lives appropriately, and assume responsibilities for the management of genetic illness. In HD, key issues concern the decision as to whether to have children, the decision to get married, and disclosure to other family members that they face the prospect of developing a debilitating neurological disorder. These informal practices of mutual disclosure around such issues among those who identify themselves with a virtual community are significant because they constitute a novel form of authority—an authority based not on training, status, or possession of esoteric skills, but on experience. And, like those older forms of authority, experiential authority, the experiential authority of others, can be "folded" into the self (Barry et al. 1996b, Diprose 1998, Rose 1996). In the process, relations with older forms of authority, such as medical and genetic expertise, mutate. These small, yet important mutations are starting to shape the ways in which novel life strategies are formulated and developed. Governing oneself in the light of one's risky genes is intimately tied to identity projects, the crafting of healthy bodies, and the management of our relations with others, in relation to a wide range of authorities.

Novas identifies four key dimensions of these technologies of the self. The first is the location of a "*molecular-genetic identity*" through mapping a lineage in genetic terms, a form of medico-genetic biographical narration that can encompass several generations such as grandparents, and include kin such as aunts, uncles, and cousins and which draws on a knowledge of genetics and its mode of transmission for a particular genetic illness. This constructs individuals, by themselves and by others, as being at genetic risk for HD and enmeshes family members, whether they carry the mutation or not. Much of the debate focuses around the decision to undergo a predictive genetic test, which is discussed as a potentially life-altering decision that one makes for oneself, in relation to one's genetic legacy, and for significant others. These tests—like all predictive genetic tests—introduce new ways of understanding, describing, and experiencing oneself as a person at risk. Key aspects of self-identity come to be defined in terms of the sequence of bases at a particular location on the short arm of Chromosome 4. The number of CAG repeats at a particular locus on this chromosome is significant as they not only indicate the presence of HD, but may be indicative of age of onset and the severity of symptoms experienced. And once tested, this genetic knowledge becomes important in shaping ethical decisions concerning reproduction, and introducing new norms of reproductive health shaped in terms of a concern for others. The very availability of predictive testing creates new categories of individuals: those at risk but not tested; those HD positive; those HD negative; and, significantly, those who are HD intermediate. Testing positive or negative for HD thus can form a new genetic identity, no longer "at risk" the tested individual is either "not at risk" or "a person who will develop HD." But the predictive test can also be indeterminate when the number of CAG repeats falls in a borderline between the normal and the pathological. In these instances, the identity of the molecular genetic self is put into question. Whatever the risk status, however, it is clear that it has become possible, perhaps necessary for those at risk of HD to think of themselves as molecular genetic beings.

The second dimension identified by Novas involves the formation of a *domain of ethical problematization.* He found that one of the most frequently posed ethical questions in the HD webforum concerned the decision to have children in light of knowledge of being genetically at risk or presymptomatic. This is one example of the way in which quite complex molecular genetic knowledge has begun to permeate the field of reproductive decision making as it concerns paternal transmission of HD. Epidemiological research has shown that fathers at risk for HD have the potential to transmit more severe and earlier onset forms of HD to their offspring. This clearly serves to complicate the decision to have children for such men, and also for their actual or potential partners, who must take into account the small risk of transmitting earlier and more severe forms of HD. Once the field of reproductive decisions becomes structured by knowledge of molecular risk, each individual becomes obliged to bring the genetic future—the quality of life that potential offspring will have in terms of genetic illness—into the present field of ethical concerns. But, in addition, a new problem space of communication takes shape. When should the at risk person tell siblings or children who are also at risk that they have decided to undergo predictive genetic testing? When should they tell children or other family members that they face the prospect of inheriting a severe neurological disorder? In our age of authenticity, the norm of truthful speech increasingly infuses familial relations. How, then, should we shape our communicative conduct with regard to potentially life-altering information? In this context, where autonomy and choice are paramount, and where genetic information is thought of as containing the potential to transform one's life, the disclosure of genetic risk information gets framed in terms of the language of rights: the right to know of one's kin and children, so that they may have the right to choose versus the right not to know, the right not to be known, the fear of the consequences that that knowledge may bring for one's conduct of one's own life and for one's treatment by others—friends, employers, teachers or insurers.

Novas identifies a third dimension—a new *relation to expertise*. At least for some of those who participate in these Web debates, the combination of genetic and enterprising forms of selfhood creates new relations with expertise, reconfiguring power relations in significant ways. The obligations of autonomy and responsibility here mean that individuals are not content with being the passive subjects of medicogenetic knowledge and treatment. The responsible and autonomous genetic subject becomes a layexpert in the management of HD, gaining as much knowledge as possible about the disease and applying it to him or herself, with the aim of optimizing health and improving quality of life. These lay experts use the webforum not only to educate one another in advanced genetic knowledge, but also in the mundane business of care, such as the side effects of medications, the use of feeding tubes to prevent the deterioration of a person whose HD produces difficulty swallowing, and the alleviation of muscle pains caused by the choreic movements. Hence the lay experts are also "experts by experience" as they generate and authorize their own knowledge, and the Web communities become mediators, organizers, compilers, and editors of knowledge, not only about the disorder, but also about the forms of life required to live with it. Relations with professional expertise are often conducted "at a distance." Partly, this distance is established by the use of the Internet as a source of information. But, more significantly, it is because professional experts are no longer regarded as the sole authority of truth. The genetic counseling session becomes merely one source of information for the individual to make choices about his or her life, to be connected with that distributed across multiple sites, requiring active discovery and assimilation. Somatic individuals engage with this knowledge as consumers, aware of the range of knowledge products on the market, and demand that their choice is constantly expanded. This places new obligations on geneticists and clinical researchers to produce new forms of knowledge that can be useful to those suffering from this disease. Somatic individuals who are at risk thus become active in shaping the enterprise of science, engaging their hope in political lobbying for support of biomedical research, donating a portion of their income toward finding a cure, fund-raising to support research, donating tissues and blood for genetic analysis, and being willing to take part in experimental clinical trials for potential therapies to cure HD (Heath et al. 2004, Rabeharisoa and Callon 1998a, Rabinow 1999).

A fourth dimension identified by Novas concerns the formation of what he terms "*life strategies*"—ways of thinking about and acting upon our lives in the present in relation to some future goals. Life strategies are formulated in a complex ethical field, as the question of how one should conduct one's life arises at the intersection of many, often conflicting, ethical problems. Only a finite set of forms of life are at our disposal: the practices and techniques that we have to shape the self and mould our lives are contoured by dominant cultural practices and are historically specific. In the webforums that Novas studied, life seemed to be understood along a temporal axis, in terms of stages when one would like to accomplish certain objectives, or in terms of tasks, aims, or objectives to be accomplished in the time remaining on this earth. These are understood in languages incorporating elements assembled from genetic counselors, psychologists, support groups, Internet discussion groups, advice columns, television programs, and conversations with friends and family members. The life strategies adopted for the government of genetic risks—the take up of predictive genetic testing, genetic counseling, the disclosure of the self in the webforum and to one's family and kin, family planning methods combine with other obligations—to family, to work, to religion—that are also organized across a life course and are shaped by the desire and the obligation to govern life in the present in the name of the future and in the light of genetic risk.

Genetic Personhood—A New Ontology?

I have argued that the critics of biological determinism, genetic reductionism, geneticism, and the like have considerably oversimplified the shifts in forms of personhood associated with the rise of ideas and practices of genetic risk. There is little evidence that modern genetic biomedicine reduces the genetically risky person to a passive body-machine that is merely to be the object of a dominating medical expertise. Even if biomedicine did frame genetic subjectivity in these terms, the genetic subjects that inhabit contemporary advanced liberal democracies are very different. Genetic forms of thought have become intertwined with the obligation to live one's life as a project, generating a range of ethical conundrums about the ways one might conduct one's life, formulate objectives, and plan for the future in relation to genetic risk. The genetic axis of personhood is assembled together with all those other axes that construct the subject as autonomous, prudent, responsible, and self-actualizing. At least for some, the power of biomedical expertise is itself reconfigured in this assembly, and placed in the position of a resource to be drawn on in life planning, rather than as a master discourse in arbitrating forms of life or decisions as to procreation. It is located among multiple other forms of expertise, notably that developed in the virtual communities of at risk persons themselves, and their families and allies. The virtual community of the HD webforum analyzed by Novas is one example of the growing array of virtual communities of somatic individuals organizing key axes of their forms of life around their sickly, risky, improvable, or manipulable corporeality.

More fundamentally, criticisms posed in terms of biological and genetic determinism fail to recognize a significant change that is occurring in conceptions of life itself.[25] The explanatory form of the genetics they criticize is that of a depth ontology. For these critics, biologists are thought to construe the genetic code as a deep inner truth, the cause of sickness or health, merely expressed in the surface of corporeality, conduct, character, etcetera. Explanatory structures that operate in terms of depths and surfaces clearly characterize much modern thought—political economy with its arguments about the hidden hand of the market or the causal powers of the mode of extraction of surplus value; the depth ontologies of the human subject associated with psychoanalysis and all the dynamic psychologies. I would not wish to deny that such explanatory forms are also prevalent in biological thought, especially in its semipopular forms (such as Richard Dawkins or the writings of the sociobiologists). The spontaneous philosophy of the biologist is undoubtedly "modern" in this sense—reflecting on their practice and representing it to others, they tend to ascribe a deep ontological reality to their concepts and portray them as the hidden truths that produce and determine a realm of observable effects. But, as a whole tradition of philosophers of science from Bachelard onwards has taught us, one should not mistake the spontaneous philosophy of the scientist for the operative epistemology or ontology of scientific activity. In this sense, despite popular and pseudophilosophical accounts, I suggest that contemporary genetics is beginning to operate in a "flattened" world, a world of surfaces rather than depths. In the developing explanatory schemas of postgenomics, the genetic code is no longer thought of as a deep structure that causes or determines, but rather as only one set of relays in complex, ramifying, and nonhierarchical networks, filiations, and connections (Deleuze 1988b).

I would not claim that the metaphysics of the gene has been abandoned. But I would argue that such an ontology is not unchallenged, and that these challenges and alternatives will accumulate over the next decade. Perhaps, that is to say, we need to analyze the way that genetics and genetic risk might figure in the forms of personhood associated with a post-ontological conception of life, a vitality not of depths and determinations but of surfaces and associations.

Chapter 5

Biological Citizens

A new kind of citizenship is taking shape in the age of biomedicine, biotechnology, and genomics.[1] This is a shift in what I term "biological citizenship."[2] Since Marshall's classic essay (Marshall 1950) it is conventional to think of a kind of evolution of citizenship since the eighteenth century in Europe, North America, and Australia: the civil rights granted in the eighteenth century necessitated the extension of political citizenship in the nineteenth century and of social citizenship in the twentieth century.[3] This perspective is useful, to the extent that it breaks with political-philosophical considerations of citizenship and locates citizenship within the political history of "citizenship projects." By citizenship projects, I mean the ways that authorities thought about (some) individuals as potential citizens, and the ways they tried to act upon them in that context. For example: defining those who were entitled to participate in the political affairs of a city or region; imposing a single legal system across a national territory; obliging citizens to speak a single national language; establishing a national system of universal compulsory education; designing and planning buildings and public spaces in the hope that they would encourage certain ways of thinking, feeling, and acting; developing social insurance systems to bind national subjects together in the sharing of risks. Such projects for creating citizens were central both to the idea of the national state, and to the practical techniques of the formation of such states. Citizenship was fundamentally national.

Many events and forces are placing such a national form of citizenship in question. As anguished debates about "multiculturalism" indicate, the nation can no longer be assumed to be a cultural or religious unity, nor can citizenship be so easily tied to hopes for a single national identity. As debates about "globalization" indicate, the idea of a single territorially bounded national economy has become problematic. As debates about economic and political migration indicate, the capacity of states to delimit citizens in terms of place of birth or lineage or race is placed in question. Discussions of these challenges rarely touch on issues of biology, bioscience, or biomedicine. But developments in these areas also challenge existing conceptions of national citizenship, and intersect with all these other developments in significant ways. Indeed I would make a wider claim: biological presuppositions, explicitly or implicitly, have underlain many citizenship projects, shaped conceptions of what it means to be a citizen, and underpinned distinctions between actual, potential, troublesome, and impossible citizens.

Of course, there have been many discussions of the importance of biological beliefs for the politics and history of the nineteenth and twentieth centuries. But the biologization of politics has rarely been explored from the perspective of citizenship. Yet histories of the idea of race, degeneracy, and eugenics, of ideas and policies around women, motherhood, and the family, and of demography and the census show how many citizenship projects were framed in biological terms—in terms of race, blood lines, stock, intelligence, and so forth. I use the term "biological citizenship" descriptively, to encompass all those citizenship projects that have linked their conceptions of citizens to beliefs about the biological existence of human beings, as individuals, as men and women, as families and lineages, as communities, as populations and races, and as species. And like other dimensions of citizenship, biological citizenship is undergoing transformation and is reterritorializing itself along national, local, and transnational dimensions.

Inevitably, in discussing these issues, the specter of racialized national politics, eugenics, and racial hygiene is summoned from its sleep. Such biological understanding of human beings were clearly linked to notions of citizenship and to projects of citizen building at the level of the individual and of the nation-state. Nonetheless, contemporary biological citizenship, in the advanced liberal democracies of "the West" that are my principal focus here, does not take this racialized and nationalized form.[4] The forms of biological citizenship that I discuss here are differentially territorialized. As analyses of bioprospecting and biopiracy show, not all have equal citizenship in this new biological age. Nonetheless, the links between biology and human worth and human defects today differ significantly from those of the eugenic age.[5] Different ideas about the role of biology in human worth are entailed in practices of selective abortion, preimplantation genetic diagnosis, and embryo selection. Different ideas about the biological responsibilities of the citizen are embodied in contemporary norms of health and practices of health education. Different citizenship practices can be seen in the increasing importance of corporeality to practices of identity, and in new technologies that intervene on the body at levels ranging from the superficial (cosmetic surgery) to the molecular (gene therapy). A different sense of the importance of the "bare life" of human beings as the basis of citizenship claims and protections is bound up in contemporary transnational practices of human rights. And while it is true that many states are, once more, regarding the specific hereditary stock of their population as a resource to be managed, these endeavors are not driven by a search for racial purity. Instead, they are grounded in the hope that certain specific characteristics of the genes of groups of their citizens may potentially provide a valuable resource for the generation of intellectual property rights, for biotechnological innovation, and for the creation of biovalue.[6]

However an analysis of biological citizenship cannot merely focus on strategies for "making up citizens" imposed from above. The languages and aspirations of citizenship have shaped the ways in which individuals understand themselves and relate to themselves and to others. Projects of biological citizenship in the nineteenth and twentieth century produced citizens who understood their nationality, allegiances, and distinctions, at least in part, in biological terms. They linked themselves to their fellow citizens and distinguished themselves from others, noncitizens, partly in biological terms. These biological senses of identification and affiliation made certain kinds of ethical demands possible: demands on oneself; on one's kin, community, society; on those who exercised authority.

It is this sense of biological citizenship that is most clearly developed by Adriana Petryna in her study of post-Chernobyl Ukraine (Petryna 2002). The government of the newly independent Ukraine based its claims to a right to govern on the democratically expressed will of its citizens. And those citizens who had, or who claim to have, been exposed to the radiation effects of the nuclear explosion at the Chernobyl reactor, believed that they had rights to health services and social support that they could claim from that government in the name of their damaged biological bodies. In this context, she argues "the very idea of citizenship is now charged with the superadded burden of survival . . . a large and largely impoverished segment of the population has learned to negotiate the terms of its economic and social inclusion using the very constituent matter of life" (5). Biological citizenship can thus embody a demand for particular protections, for the enactment or cessation of particular policies or actions, or, as in this case, access to special resources—here "to a form of social welfare based on medical, scientific, and legal criteria that both acknowledge biological injury and compensate for it" (4). Life acquires a new potential value, to be negotiated in a whole range of practices of regulation and compensation. This is not a unique situation. We can see something similar in campaigns for redress for the victims of Bhopal (Kumar 2004), and in numerous American examples of fights for compensation for biomedical damage, portrayed in semifictionalized accounts in films such as *Erin Brockovich* and *A Civil Action*. Of course, there are very different political, legal, and ethical framings in these different locales. But in each case claims on political authorities and corporate entities are being made by those who have suffered biological damage, in terms of their "vital" rights as citizens.

Biological citizenship is both individualizing and collectivizing. It is individualized, to the extent that individuals shape their relations with themselves in terms of a knowledge of their somatic individuality. Biological images, explanations, values, and judgments thus get entangled with other languages of self-description and other criteria of self-judgment, within a more general contemporary "regime of the self" as a prudent yet enterprising individual, actively shaping his or her life course through acts of choice (Novas and Rose 2000). The responsibility for the self now implicates both "corporeal" and "genetic" responsibility: one has long been responsible for the health and illness of the body, but now "somatic individuals" must also know and manage the implications of one's own genome. This responsibility for the self to manage its present in the light of a knowledge of its own future can be termed "genetic prudence": a prudential norm that introduces new distinctions between good and bad subjects of ethical choice and biological susceptibility (on prudence, see O'Malley 1996).

Biological citizenship also has a collectivizing moment. Paul Rabinow proposed the concept of "biosociality" to characterize these forms of collectivization organized around the commonality of a shared somatic or genetic status, and has drawn attention to the new ethical technologies that are being assembled around the proliferating categories of corporeal vulnerability, somatic suffering, and genetic risk and susceptibility (Rabinow 1996a). Biosocial groupings—collectivities formed around a biological conception of a shared identity—have a long history, and medical activism by those who refuse the status of mere "patients" long predates recent developments in biomedicine and genomics. Many of these earlier activist groupings were fiercely opposed to the powers and claims of medical expertise. Some remain implacably antimedical; others operate in a manner that, while not explicitly opposed to established medical knowledge, prefers to remain complementary to it. Nonetheless collectivities organized around specific biomedical classifications are increasingly significant. The forms of citizenship entailed here often involve quite specialized scientific and medical knowledge of one's condition: one might term this "informational biocitizenship." They involve the usual forms of activism such as campaigning for better treatment, ending stigma, gaining access to services, and the like: one might term this "rights biocitizenship." But they also involve new ways of making citizenship by incorporation into communities linked electronically by email lists and websites: one might term this "digital biocitizenship."

Both in its individualizing and its collectivizing moments, contemporary biological citizenship operates within the field of hope. Hope plays a fundamental yet ambiguous role in contemporary somatic ethics.[7] Sarah Franklin introduced the idea of "hope technologies," in the context of her study of assisted reproduction: within such technologies, professional aspirations, commercial ambitions, and personal desires are intertwined and reshaped around a biosocial telos (Franklin 1997). The maintenance of hope has become a crucial element, not merely in the reproductive technologies, but more generally within the therapeutic armory of nursing care for patients with cancer and other life-threatening illnesses; it is also, some suggest, threatened by the contemporary obligations not to deceive patients, but to give them full information about the nature of their condition and the statistics of recovery and mortality (Hickey 1986, Hinds 1984, Hinds and Martin 1988, Mikluscak-Copper and Miller 1991, Peräkyla 1991, Ruddick 1999). It was this dilemma that Mary-Jo Delvecchio Good, Byron Good, and their colleagues pointed to in 1990, discussing not only the ways in which, in American treatment practices for cancer, physicians sought to instill and maintain the hope of their patients in the possibility of cure or remission as a therapeutic tool, but also the way in which physicians maintained their own hopefulness through commitment to the progressive efficacy of biomedical therapeutics (Good et al. 1990). Although they focused on the patient-physician relationship, and the norms of disclosure of information to patients, they suggested that this was part of a "political economy of hope," because hope linked these elements of therapeutic practice to the funding of "research and treatment institutions, . . . the patterns of availability and promotion of particular anti-cancer therapies, [and the] search for treatments and cure by patients and their families" (60).[8] Nik Brown explored similar dynamics in his study of hopes around xenotransplantation, usefully pointing to the semantic complexity of the contemporary term: In a world imbued with a drive to master the future and still clinging to an ambivalent belief in progress, hope draws our gaze to a horizon upon which things are imagined that we expect with desire, or desire with expectation (Brown 1998). Hope, as Brown points out, is not merely a set of beliefs, but is infused with affects that structure the actions of many of those involved in working in situations where illness, and the fear of illness, can generate despair, desperation, and terror in the face of the future. And deeply emotional representations of the fears and hopes of sufferers, and their expectations that new medical technologies will deliver them from their suffering, structure many popular representations of patients and their illnesses, and are often deployed by medical charities, support groups and others in seeking to raise funds to keep that hope alive (Brown 1998, especially chapter 4).

Carlos Novas has expanded and deepened the sense in which we might understand the political economy of hope that has taken shape around contemporary biomedicine (Novas 2001). He shows how this economy has been constituted by interlinking hopes of many different types and of diverse actors: the hope of patients and their families for effective treatment; the use of hope as a therapeutic instrument by medical practitioners; the hope of those managing health services that might minimize or mitigate the impact of common disorders such as stroke, heart disease, or cancer; the hope of those with a family history of genetic disease for children who are not suffering from debilitating conditions; the hope of us all for an old age not marred by Parkinson's or Alzheimer's disease; the hope of the pharmaceutical industry and biotech companies for treatments that will generate increased profits and shareholder value; the hope of scientists and researchers for career advancement and fame. Thus many new experts and forms of professional expertise have been involved in the generation, modulation, instrumentalization, and management of these hopes, and of the fears, anxieties, and disappointments that provide them with such a powerful dynamic.

Biological citizenship is a more general version of what Deborah Heath, Rayna Rapp, and Karen-Sue Taussig have termed "genetic citizenship": a way of understanding human differences, especially those related to health, in terms of genetic influences. They argue that developments in genetics are not just engendering new ways in which individuals and authorities are seeking to manage procreation according to genetic criteria,[9] but also giving rise to "new forms of democratic participation, blurring the boundaries between state and society, and between public and private interests" (Heath et al. 2004: 152). This is manifested in a range of struggles over individual identities, forms of collectivization, demands for recognition, access to knowledge, and claims to expertise. It is creating new spaces of public dispute about the minutiae of bodily experiences and their ethical implications. It is generating new objects of contestation, not least those concerning the respective powers and responsibilities of public bodies, private corporations, health providers and insurers, and individuals themselves. It is creating novel forums for political debate, new questions for democracy, and new styles of activism. Heath, Rapp, and Taussig argue that "it is "genetic citizenship" that connects discussions of rights, recognitions, and responsibilities to intimate, fundamental concerns about heritable identities, differential embodiment, and an *ethics of care*" (57, emphasis in original). And, in a context in which all of us, not merely those already enmeshed with disease, are potentially subjects of genetic screening for susceptibilities, they suggest that these movements and contestations may indicate something of shape of a future genetic citizenship for us all (166).

The organization of citizenship in genetic terms is undoubtedly significant, and it is already possible to identify programs to extend genetic education beyond those immediately involved, to educate parents, schoolchildren, and citizens in genetic literacy (see, for example, Jennings 2003).[10] But I think that genetics contribute only one dimension to contemporary biological citizenship, one axis of the ways in which the biological makeup of each and all can become an issue for political contestation, and for recognition and exclusion, and for demands for rights and the imposition of obligations. Biological citizenship has taken different forms in different national contexts, and in relation to different types of disease, disfigurement or disability. Its form is shaped by many factors, notably their biopolitical histories and modes of government, their traditions of activism, and their presuppositions about persons and their rights and obligations. In the remainder of this chapter, I will explore these issues in relation to some empirical examples. My aim is descriptive and diagnostic—to begin to map the new territory of biological citizenship and to develop some conceptual tools for its analysis.

Making Up the Nation

Paul Gilroy has suggested that gene-oriented constructions of "race" are very different from "the older versions of race-thinking that were produced in the eighteenth and nineteenth century" (Gilroy 2000: 15). As the relations between human beings and nature are transformed by genomics, the meaning of racial difference is changed; he believes that this provides the possibility of challenging the tainted logic of raciology. His assessment may be optimistic, but it points to the way in which certain presuppositions about biology bound together thinking about nation, people, race, population, and territory from the eighteenth century onwards. To think of individual and collective subjects of European nations was to think in terms of blood, stock, physiognomy, and inherent moral capacities. Those over whom Europeans would exercise colonial dominion were also thought of in these terms. In short, citizenship was grounded on what, from the early nineteenth century onwards, would be termed "biology." Distinctions within nations as to those more or less worthy of, or capable of, citizenship, and distinctions between peoples as to their respective capacities to rule and be ruled, were built on an explicit or implicit biological taxonomy inscribed in the soma of both individual and collective and passed down through a lineage.

This is not the place to review the various ways in which people, race, nation, history, and spirit were linked in the blood, divided and placed into hierarchies and patterns of descent. These can be traced from the philosophers of eighteenth-century liberalism, such as Locke and Mill, through eighteenth- and nineteenth-century raciology, into the political debates about racial deterioration and degeneracy in the second half of the nineteenth century, and concerns about the consequences of the size and fitness of the population for the fate of nation-states in imperial rivalry. Ideas of character and constitution, of blood, race, and nation, remained inextricably intertwined in the eugenic arguments of the first half of the twentieth century, which shaped the political imagination of the nation-states of North America, the Nordic countries, Australasia, South America, and elsewhere. Such ideas were translated into many different strategies to preserve the biological makeup of the populations of states. Some focused on outside threats, such as those posed by immigration from lower races. Others focused on threats from within, such as the dangers posed by the breeding of defective, insane, sick, or criminal individuals and their kin. Conceptions of the biological basis of national identity and national unity underlay many legal definitions of nationhood and citizenship in terms of descent. In Germany, the citizenship law of 1913, which was framed in these terms and defined citizenship in terms of the line of descent, survived the Nazi experience and remained in force until 1999.[11] In the 1920s, Chinese citizenship was built on the presupposition of a single lineage of blood of "the yellow race" (Dikötter 1998).[12] In the same period in Mexico, some attempted to argue that it was the fusion of blood that gave the Mexican race its defining characteristics (Stepan 1991). In each version, the nation was not only a political entity, it was a biological one. It could be strengthened only by attention to the individual and collective biological bodies of those who constituted it.

Within these twentieth-century projects of biological citizenship, there were clear differences between those who felt that their objectives could only be reached by strategies involving compulsion and those who opposed compulsion in the name of liberty. But this distinction did not map onto a simple division between strategies of reproductive control and strategies of health education and public health. Emphasis on the need to educate individuals so that they take personal responsibility for the genetic implications of their reproductive decisions is not new. For example, the genetic education of the citizen was a constant theme in the eugenic period, and early eugenicists developed all sorts of events to encourage individuals and families to reflect on themselves, their marriage partners, and their past and future lineage in eugenic terms, with a view to enhancing healthy procreation. Through education, the genetic citizen was to be enabled to take responsibility for his or her own heredity. I shall return to this question presently.

What, then, of the present? It would be too simple to believe that such concerns with the biological and / or genetic makeup of the population and the individual citizen have ceased to be matters of national political concern. The very existence of state-supported public health measures indicates that the vital biological existence of the citizen remains an issue within the political rationalities of the present. The very existence of certain practices that have now become routine in medical care—ultrasound, amniocentesis, chorionic villus sampling, and more—shows that judgments of value concerning certain features of the bodies and capacities of citizens have become inescapable, even if it is the individual citizen and her family who must carry the responsibility for the choice now rendered calculable for them. And successive state-funded health promotion programs show how the biological education of the citizen remains a national priority, although it is now supplemented by a host of other forces seeking to shape the reflexive gaze though which the citizen views his or her past, present, and future biological corporeality.

From another perspective, national genetic peculiarities have become a key resource for biomedicine and for commercial exploitation. This has involved the search for lineages with a high incidence of particular diseases and the belief that the study of such pedigrees would provide the key to unraveling the genetics of disease. Finland can serve as an initial example.[13] It has long been recognized by geneticists that sectors of the Finnish population are attractive for gene hunting because of a combination of low geographical mobility, relatively high rates of "inbreeding," good genealogical and health records, and high rates of prevalence of certain diseases.[14] For example, many claims about the discovery of genes linked with schizophrenia, manic depression, alcoholism, and other disorders were based on genetic research in Finland. In the age of genomics such conditions, once seen as burdens on the national population and its health service, have become potentially valuable resources: hence, they are included in the Finnish proclamation of biotechnology as a national imperative. As we discuss in detail later, the national population has become a resource not only for understanding particular pathologies, but also for profitable biomedical exploitation.

CHAPTER FIVE

Making Biological Citizens: From Public Value to Biovalue

Over the past decade, in many countries, there have been attempts to educate citizens so that they will be better able to engage in informed debate over the complex ethical and democratic dilemmas brought about by scientific and technological progress. The "public understanding of science" is seen one way of regaining the confidence and trust of lay members of the public in the regulatory mechanisms that govern science. It is also seen as a way of redressing a kind of "democratic deficit" that is said to exist when citizens do not actively participate in shaping scientific and technological futures. Such arguments concerning the need to enhance the scientific understanding of citizens have a long history. As far as biology and biomedicine are concerned, I have already commented on the attempts by eugenicists and similarly minded educators in the 1920s and 1930s to inculcate a particular version of scientific literacy—in this case the capacity to reflect in a eugenically informed manner on reproductive and marital choices. This is only one way in which the capacities of the individual for citizenship have been linked to his or her understanding of "advances in science."[15]

Attempts educate the public about science and technology are part of strategies for "making up" the biological citizen.[16] "Making up citizens" has involved the reshaping of the way in which persons are understood by authorities—be they political authorities, medical personnel, legal and penal professionals, potential employers, or insurance companies—in terms of categories such as the chronically sick, the disabled, the blind, the deaf, the child abuser, the psychopath. These categories organize the diagnostic, forensic, and interpretive gaze of different groups of professionals and experts. Classification of this sort is both dividing and unifying. It delimits the boundaries of those who get treated in a certain way—in punishment, therapy, employment, security, benefit, or reward. And it unifies those within the category, overriding their specific differences. New biological and biomedical languages are beginning to make up citizens in new ways in the deliberations, calculations, and strategies of experts and authorities: for example the emergence of categories such as the child with attention deficit hyperactivity disorder, the woman with premenstrual dysphoric disorder, or the person who is presymptomatically ill because of genetic susceptibilities.

Making up biological citizens also involves the creation of persons with a certain kind of relation to themselves. Such citizens use biologically colored languages to describe aspects of themselves or their identities, and to articulate their feelings of unhappiness, ailments, or predicaments. For example, they describe themselves as having high levels of blood cholesterol, as vulnerable to stress, as being immuno-compromised, or as having a hereditary predisposition to breast cancer or schizophrenia. They use these phrases, and the types of calculation to which they are attached, to make judgments as to how they could or should act, the kinds of things they fear, and the kind of lives for which they can hope. In part, of course, the languages that shape citizens' self-understandings and self-techniques are disseminated through authoritative channels: health education, medical advice, books written by doctors about particular conditions, documentaries on television that chart individuals coping with particular conditions. Thus, for example, Bruce Jennings of the Hastings Center, in a 2003 concept paper pointed to the work of groups from Genetic Alliance through the March of Dimes to the Oregon Health forum that are working to "bolster the social capital and genetic citizenship of ordinary citizens and particularly of cultural and ethnic minorities"—the kind of work he considers vital to ensure that genetic literacy and genetic citizenship are part of a movement of civic renewal and democratic empowerment (Jennings 2003: 4–5). Indeed, whatever may be said about their general level of scientific literacy, in biomedical arenas individuals are actively engaging with biological explanations and forming novel relations with scientific or medical authorities in the process of caring for, and about, health. But the contemporary biological citizen sits at the intersection between these more or less authoritative endeavors and a variety of other flows of information and forms of intervention. Or perhaps, "sits" is the wrong term, for even while sitting, an active scientific citizenship is increasingly enacted, in which individuals themselves are taking a dynamic role in enhancing their own scientific, especially biomedical, literacy. The active search for scientific knowledge is particularly marked in that of health and illness, of medicine, genetics, and pharmacology—in what Rabinow has termed "the third culture" (Rabinow 1994)—where what is at stake is each individual's own life, or that of those for whom they care. In engaging with such issues, the language with which citizens are coming to understand and describe themselves is increasingly biological.

For those suffering directly or indirectly from illness or disability, reading and immersing oneself in the scientific literature of the illness that oneself or a loved one suffers from can be a key technique. This knowledge can be used to gain a better understanding of the disease process, to provide better levels of care to those suffering from an illness, and to discuss and negotiate with the doctor a range of therapeutic possibilities. Over the last decade the Internet has come to provide a powerful new way in which those who have access to it, and who are curious about their health or illness, can engage in this process of biomedical self-shaping. But a key feature of the Internet is that it does not only give access to material disseminated by professionals, it also links an individual to self-narratives written by other patients or carers. These accounts usually offer a different narrative of life with an illness, setting out practical ways of managing a body that is ill, the effect and harms of particular therapeutic regimes, ways of negotiating access to the health care system, and so forth. That is to say, these narratives provide techniques for the leading of a life in the face of illness. They have a further distinctive feature, which relates to truth itself. Strategies for making up biological citizens "from above" tend to represent the science itself as unproblematic: they problematize the ways in which citizens misunderstand it. But these vectors "from below" pluralize biological and biomedical truth, introduce doubt and controversy, and relocate science in the fields of experience, politics, and capitalism.[17]

In response to the perceived power of such problematizations from below, those whose investment in biomedicine is measured in terms of capital returns and shareholder value—the biotech, biomedicine, and pharmaceutical companies—now actively engage themselves with the self-education of active biological citizens. They set up and sponsor many of the consumer support groups that have sprung up around disorders from attention deficit hyperactivity disorder (ADHD) to epidermolysis bullosa (EB). In doing so, they seek to represent their activities and their products as beneficial, to counter the claims of the critics, and to educate actual or potential consumers of their products. In the United States pharmaceutical companies are permitted to engage in "direct to consumer advertising" and television advertisements for the benefits of different brands of drugs are widespread: notably drugs for treating experiences of mental malaise, now coded as depression, anxiety, and panic disorders. But, across all jurisdictions, such companies are now using the Internet for this purpose. It is thus worth considering one example from this domain in some detail.

Eli Lilly's Prozac website, in 2001, was emblematic of techniques to promote a particular version of scientific or biological literacy.[18] The home page of this site was titled "Your Guide to Evaluating and Recovering from Depression." Prozac.com thus represented itself as a resource center where individuals could learn more about depression, its treatments, and ways to securing a recovery. It claimed—characteristic of all such direct to consumer practices—that the information and knowledge provided on this website were not intended to supplant the authority of the health professional, but rather to encourage the person suffering from depression to form an "active" alliance with the medic in the realization of a program of care. But, of course, this activity was to take a specific, brand related, form: a form supported through the provision of information on how Prozac can aid in recovery from depression.

In part, this is a matter of forming the problem in a particular manner. The Prozac.com website used a biological explanation of depression, couched in terms of the action of neurotransmitters. Text and animated images were used to provide a way for individuals to understand their depression at a molecular level, in terms of chemical imbalances and the action of neurotransmitters, and to imagine the ways in which Prozac could directly target and correct these molecular imbalances. It was, it seems, important for the depressed individual to learn about the action of Prozac at the neurochemical level. This was not because taking the drug was all that was required. On the contrary, it was because the individual should know "what to expect while you work toward your recovery."[19] The process of recovering from depression did not simply require compliance with a drug regime: "You can and should be an active participant in your recovery from depression."[20] This process of recovery enlisted a whole range of techniques of the self: practicing self-discovery, liking yourself, being kind to yourself, reducing stress, engaging in physical exercise, eating well, writing lists and keeping diaries, building self-esteem, joining a support group, and reading the Prozac.com newsletter. This website thus was clearly a node in a political economy of hope: it sutured together hopeful personal beliefs that one can recover from depression if one knows how to manage it, with the commercial hopes embodied in the marketing of Prozac itself.

The role of biomedical authority here is not to encourage the passive and compliant patienthood of a previous form of medical citizenship. Citizenship is to be active. The actual or potential patient must try to understand his or her depression, to work with doctors to obtain the best program of medical care, to engage in self-techniques to speed the process of recovery—and, of course, to ask his or her doctor to prescribe Prozac by name. Indeed, as the daily form of Prozac was out of patent, the website sought to maintain market share. On every single web page, a banner advertised a free trial of Prozac® Weekly™—which was in patent—and told patients that they could ask their doctor about this new formulation. Another page suggested that there may be differences between brand name Prozac and its generic equivalent, fluoxetine hydrochloride, explaining to potential customers that there is no such thing as "generic Prozac"—for example the generic version comes in different packaging—and that if they feel uncomfortable about changing to a generic, they should ask their doctor to prescribe brand name Prozac.[21] What kind of scientific literacy is being promoted here? What kinds of active biological citizens are being shaped, and to what ends? This is the citizenship of brand culture, where trust in brands appears capable of supplanting trust in neutral scientific expertise. The weaving together of Eli Lilly's commitment to education and brand marketing suggested the title of this section of the chapter —from public value to biovalue—for this is just one example of the way in which biovalue is entwined with, and sometimes supplants, public value in the biological education of citizen-consumers.

Biosociality: Active Biological Citizens

Perhaps the account so far has given the impression that biological citizens are individualized, required to understand their nature and cope with their fate alone or with their own family, accompanied only by the ministrations and advice of experts, the solitary reading of informative material, or seated alone at their computer searching the Web. Undoubtedly such isolation is the condition of many. But it is not the destiny of the biological citizen to be an isolated atom, at least in circumstances where the forms of life, ethical assumptions, types of politics, and communication technologies make new forms of collectivism possible. Early exemplars of these new forms of biological and biomedical activism were the campaigning groups that arose around AIDS, especially in the English-speaking world. AIDS activists organized themselves into groups, and constituted those who were actual or potential sufferers from the condition as "communities"—communities *for which* they would speak, and *to which* they were responsible. These groups had a number of functions: to spread information about the condition; to campaign for rights and combat stigma; to support those affected by the illness; to develop a set of techniques for the everyday management of the condition; to seek alternative forms of treatment; and to demand their own say in the development and deployment of medical expertise.

The case of HIV and AIDS activism is exemplary for another reason: while initially relations between the activists and the conventional biomedical community were antagonistic, gradually an alliance developed. The HIV / AIDS community, and the identifications it fostered, came to provide key elements for the government of the virus. That is to say, it was through their identification as members of this community, that those in "high risk groups" were recruited to their responsibilities as biological citizens; health educators came to realize that it was only by means of the pathways provided by AIDS activists that they would be able to gain the allegiance of the active gay men who were their primary target. In allying itself with the health establishment in promoting the message of safe(r) sex, AIDS activists, in return, would have their say in the organization and deployment of social resources, and indeed gain the resources necessary for their activities. This was not a matter of cooption, although some saw it as such, but of alliances and translations. And "governing through community" produced its own problems. Most notably, that of shaping the conduct of a younger generation of gay men who did not identify themselves in the same terms as the previous generation, and that of governing the conduct of "men who had sex with men" but who did not identify themselves as part of any gay community.

Since the 1980s, biosocial communities following a roughly similar form have proliferated, and, since the advent of the World Wide Web, they have found the Internet a congenial host territory. Take, for example, the issue of manic depression. Until quite recently, in the United Kingdom at least, in addition to physicians and medics, those with such a diagnosis or their families (if they were not among the very few actively allied to the antipsychiatry movement) could access only one other organized source of information and support: the National Association for Mental Health (MIND). Things began to change in the 1980s. In 1983, the Manic Depression Fellowship (MDF) was founded, which described itself as a "user led" organization whose aims are to "enable people affected by manic depression (bi-polar) to take control of their lives" through the services this organization offers.[22] These services include: MDF self-help groups, information and publications, employment advice, the MDF Self Management Training Programme, a 24-hour Legal Advice Line for employment, legal, benefits, and debt issues, and a travel insurance scheme. MDF also seeks to combat the stigma and prejudice experienced by those affected by manic depression, raise awareness of the disease, and develop partnerships with other organizations concerned with mental health.[23]

Throughout the 1980s, the MDF was joined by a host of other user- and survivor-led organizations, some local and some national. True that those based in the United Kingdom are somewhat few and far between, but outside the United Kingdom such biosocial communities are proliferating. For example, Pendulum Resources is a website that presents itself as a "Bipolar Disorders Portal," a gateway to comprehensive quasi-medical and other information. It urges people with bipolar disorder to participate in the NIMH-funded Bipolar Genome Study at the Washington University School of Medicine and in other similar projects in the hope that "this kind of study will enable medical researchers to find safer, more effective treatments for Mental Illness and brain disorders."[24] Pendulum also provides links to at least twenty-four home pages of people diagnosed with, or living with, bipolar disorder who describe, in very different ways, their modes of living with the condition. These include, for example, "A Better Place to Be" which contains, among other things, a diary of the website author's "personal struggle with bipolar disorder," a journal and a link that enables readers to ask questions.[25]

These new forms of citizenship are not always premised on genetics. Many of these biosocial communities do indeed refer to genetics, but its significance varies. While in single gene or single substitution disorders such as Huntington's, PXE or Canavan's disease, genetics clearly plays an organizing role, in the biosociality forming around other conditions, genetics is not dominant. In the case of "A Better Place To Be," under the page entitled "Sources of My Depression," the author writes under the heading of "serious reasons": "a genetic heritage that comes from being half Finnish," and "other genes in my DNA that tend toward improper chemical balance," but also cites her "need for more vocational satisfaction and personal fulfillment," her "lack of recovery from a dysfunctional childhood," and what she terms "whiney reasons" such as "nobody loves me," "everyone hates me," "tendency at times to identify any negative feeling as depression," and "lack of disposable income to purchase all the fun and necessary things I must have!" Indeed, in relation to psychiatry, the pertinence of genetics, and of biomedicine, is the subject of intense biopolitical dispute.[26] But nonetheless, these websites do demonstrate some significant features of contemporary biosociality.

Rayna Rapp, writing about women and men facing complex reproductive decisions brought about by the technology of amniocentesis, designates them as "moral pioneers" (Rapp 1999). Her argument—which would include AIDS activists (Epstein 1996, 1997; Martin 1994)—captures something crucial. These women and men were pioneers because, in their relation with their bodies, with their choices, with experts, with others in analogous situations, and with their destiny, they had to shape new ways of understanding, judging, and acting on themselves, and also to engage in a kind of reimagining of those to whom they owe responsibilities—their progeny, kin, medical helpers, cocitizens, community, society. Perhaps the language of pioneering implies too much heroic struggle, when many of those engaged in such issues do so through small-scale achievements in the everyday realities of their lives. Nonetheless, the new biosocial communities forming on the Web and outside it can be seen as moral pioneers— or perhaps "ethical pioneers"—of a new kind of active biomedical citizenship. They are pioneering of a new informed ethics of the self—a set of techniques for managing everyday life in relation to a condition, and in relation to expert knowledge. While some might deride these techniques of the biomedical self as a kind of narcissistic self-absorption, in fact they show an admirable ethical seriousness. Like those techniques Foucault found among the Greeks (Foucault 1978, 1985, 1986), they identify an aspect of the person to be worked on, they problematize it in certain ways, they elaborate a set of techniques for managing it, and they set out certain objectives or forms of life to be aimed for.

Of course, in a certain political, cultural, and moral milieu, this idea of activism in relation to one's biomedical condition can become a norm. Activism and responsibility have now become not only desirable but virtually obligatory—part of the obligation of the active biological citizen, to live his or her life though acts of calculation and choice. Such a citizen is obliged to inform him or herself not only about current illness, but also about susceptibilities and predispositions. Once so informed such an active biological citizen is obliged to take appropriate steps, such as adjusting diet, lifestyle, and habits in the name of the minimization of illness and the maximization of health. And he or she is obliged to conduct life responsibly in relation to others, to modulate decisions about jobs, marriage, and reproduction in the light of a knowledge of his or her present and future biomedical makeup. The enactment of such responsible behaviors has become routine and expected, built in to public health measures, producing new types of problematic persons—those who refuse to identify themselves with this responsible community of biological citizens (Callon and Rabeharisoa 1999, 2004).

These obligations, and the forms of biosociality with which they are linked, are specific to certain times and spaces. Despite the much-vaunted global span of the Internet, Manuel Castells has documented the national and regional variation in Internet access, which is dependent on the availability of telephone lines and other basic communication technologies, as well as the penetration of the computer hardware and software necessary to access it (Castells 2000). While young travelers of the world may be able to log on to their Internet connection from almost anywhere, the same is hardly true of those who are the prime potential subjects of biosociality. The kinds of biosociality found in the United States, Europe, and Australia are not merely products of the availability of certain technological means of communication, but linked to particular conceptions of citizenship and personhood. In particular, they connect up, in various ways, with the history of previous forms of political activism, with the varieties of identity politics, and with the existence of a vociferous politics of rights and recompense. But such forms of biosociality described here have no visible presence in many geographical regions. AIDS biosociality in sub-Saharan Africa is very different from that of Paris, San Francisco, or London.[27] Biological citizenship in Ukraine is not a matter of contesting the power of medical expertise, nor of sculpting an autonomous life in which collectively shaped self-understandings are a pathway to self-fulfillment: it takes the form of demanding redress from the state for certain ills, in the form of benefits, and activism is oriented toward demanding medical recognition for a condition and obtaining expert judgment as a credential to obtain state benefit (Petryna 2002).

CHAPTER FIVE

Organizing Hope

Citizenship has long associations with forms of local political activism: involvement with the local work of political parties, working in charitable organizations, and for causes such as reducing inner city poverty or improving literacy, as well as small-scale activities such as charity bake sales, car washes, or raffles in order to support the local church, school, or community center. These aspects of citizenship are constantly reshaped in relation to new causes and are often inventive in their styles of organizing and activism. As already noted, since the 1980s there has been an upsurge in citizen activism and political inventiveness around issues of health and illness. But while patients' organizations and support groups have been around for many years, today there is one notable innovation: the formation of direct alliances with scientists. Patients' organizations are increasingly not content with merely raising funds for biomedical research but seek an active role in shaping the direction of science in the hope that they can speed the process by which cures or treatments are developed. Thus, in the context of the hopes attached to recent discoveries in the fields of genetics and the neurosciences, patient groups have engaged in new forms of political activism, seeking to act directly on the truths and technologies generated by biomedical research. Contemporary biological citizenship thus both depends on and hopes that the science of the present will bring about cures or treatments in the near future (Novas 2001, 2003). Hope, here, is not mere wishing and anticipating—it postulates a certain achievable and desirable future, which requires action in the present for its realization.

Novas exemplifies his argument with reference to Huntington's Disease and in particular to the role of an on-line email discussion list called Hunt-Dis.[28] He points to the extraordinary personal advocacy work of some people involved on Hunt-Dis such as Carmen Leal, whose ex-husband Dave suffers from HD, and who is still actively involved in his care. She engages in a whole variety of activities for HD, editing a collection of stories and poems about people's experiences with the disease (Leal-Pock 1998), using her speaking and singing skills to provide inspiration to others,[29] and maintaining a website alongside others called the Huntington's Disease Advocacy Center.[30] Her hope, as Novas points out, is for "this to be the last generation to have to worry about Huntington's Disease. Thanks to researchers, there is now tremendous hope that this will definitely be the last generation."[31] And she urges those with the disease to contribute actively to this work: they should educate themselves and read about the disease; talk to others about the disease; save money and donate it to research; and participate in fund-raising activities, not only to contribute to the cure, but also to help to raise awareness of the disease. Biological citizenship here is financial, ethical, public, and active: it tries to shape a new public arena in which those who have a stake in the suffering wrought by a disease can participate. In such a public arena, the hopes and responsibilities of citizens are closely tied to their biology.

Hope is also manifested in the Huntington's Disease Advocacy Center website, again studied by Novas. It too encourages visitors to educate themselves, and provides links to scientific articles on HD, articles that will also generate hope as they demonstrate the advances made in understanding HD, the dedication of the scientists and researchers to the search for a cure, the evidence linking HD to other neurodegenerative diseases also under active investigation, the development of mouse models to enable further research, and the funding provided by the Huntington's Disease Society of America, all provide a rationale to look toward the future with expectation.[32]

Biological citizenship requires those with investments in their biology to *become* political. The HD website also provides advice on how to achieve this: how to lobby elected officials, how to prepare for a meeting with a political representative, how to build coalitions, a list of who to contact, and samples of correspondence written to political officials. Campaigns by such organizations against President George W. Bush's ban on stem cell research show that politics, as it is enacted by biological citizens in a political economy of hope, involves profoundly normative judgments about values and ethics concerning the uses and ends of life itself. We can see such political activism around technologies of hope in many other sites. For example, in Switzerland in November 2004, under the Swiss system of direct democracy, more than two-thirds of voters in a referendum approved a law allowing scientists to experiment on human stem cells taken from embryos: the "yes" campaign that prevailed campaigned under the slogan "Hope." Launching the government's campaign, Interior Minister Pascal Couchepin had argued that stem cell research offered "hope for incurable illnesses such as Parkinson's, paraplegia, diabetes and heart disease," and the result of the vote was welcomed by scientists who said the result was "a vote of confidence in medical research and a positive sign for Switzerland as a centre for science and research," and by the pharmaceutical industry, which said it "gave patients hope for new cures."[33]

Producing Biovalue: Materializing Ethics, Health, and Wealth

As politics begins to take on more "vital" qualities, and as biological life itself accrues even more personal and economic significance, the vitality of each and all of us becomes a potential source of value. The biological life of individual and collective subjects has long had a value that is as much economic as political—or rather, that is both economic and political. From the nineteenth century onwards, the preservation of this vital value and its enhancement became a matter of state: political authorities took on the obligation and responsibility for preserving, safeguarding, and enhancing the biological capital of their population. Along this dimension we can place a whole sequence of developments from clean water and sewage, registration of births and deaths, child welfare and maternity services, medical inspection of school children, and indeed the development of state-organized national health services. Of course, private enterprises played a key role in producing the food, services, and pharmaceuticals that would simultaneously generate private profit and public good. A market economy of health came into being. Over the twentieth century, this market was increasingly shaped by the activities of the "social" state—regulating purity and hygiene of foodstuffs, and the production and marketing of pharmaceuticals. But the regulated political economy of health—consisting of relations between the state apparatus, scientific and medical knowledge, the activities of commercial enterprises and the health-related consumption of individuals—is being reshaped, as the potentialities embodied in life itself become sources of value.

Bodily processes have long been productive of economic value, from the fees charged by physicians and hospitals to their patients, the market in medical technologies and, most notably, the commercialization of pharmaceuticals. However, contemporary biomedicine, by rendering the body visible, intelligible, calculable, and manipulable at the molecular level, has generated novel relations between life and commerce, and enabled older health technologies such as those of social citizenship, to link in new ways to circuits of capital.[34] Two examples of this logic can be found in Sweden and Iceland.

In 1999, an article in *Science* opened with the following lines: "Sweden and some other Nordic countries are sitting on a genomic gold mine. Their long-standing public health care systems have been quietly stockpiling unique collections of human tissue, some going back for decades. . . . The samples were originally stored for possible therapeutic or diagnostic uses for the patients themselves, but researchers now realize that they could contain valuable information about inherited traits that may make people susceptible to a variety of diseases" (Nilsson and Rose 1999: 894). In many Nordic countries, census data, patient records, and tissues samples maintained in the process of providing health care in the past—through a heritage that runs from the pastoral government of the church, through that of the strong state to that of the social state—have been combined with large-scale genomic analysis in order to transform their citizenry into a resource for the production of wealth and health. Perhaps the best-known example comes from Iceland, where deCODE Genetics was given an exclusive license to create and operate such a database by an Act of the Icelandic parliament in 1998.[35] They declare in their mission statement that they are "Making the Map of Life . . . a Blueprint for Health."[36] Similarly, the Swedish firm UmanGenomics, studied by Klaus Høyer, describes the "unique resources" that are available to it, including a "unique collection of blood samples and data in the Medical Biobank of Umeå," derived from records of health examinations of the local population amalgamated in a 1985 epidemiological study of the population combined with samples from state-supported medical examination and blood donation (Høyer 2002, 2003).[37] Despite the origins of these samples in public health, "UmanGenomics has the exclusive rights to commercialize information derived from these samples."[38] In the 1930s, Nazi propaganda focused on the cost of genetic illness to the German Reich. But today, the genomic disease heritage of a population, far from being conceived of as a drain on national resources, is seen as a potential resource for the generation of wealth and health, and one that provides great economic opportunities for the novel alliances of state and commerce taking shape within contemporary circuits of biovalue.

This transformation, in which high levels of certain genetic illness become potential sources of biovalue, need not be driven from above, from the state and private enterprise. It can also come from below, from patients' organizations themselves. Novas, along with a number of other researchers, have explored this in the case of a patients' organization called PXE International, founded by Patrick Terry and Sharon Terry in 1995, after their two children, Elizabeth and Ian, were diagnosed with pseudoxanthoma elasticum (PXE).[39] The Terrys played an important role in forming networks of support among affected families, getting researchers interested in studying the disease, organizing conferences for scientists and patients, and lobbying the U.S. government for more funding to be directed toward the study of PXE, but also of skin diseases more generally. PXE International also established a blood and tissue registry in order to create a central repository, and to avoid the need for patients suffering from the disease to donate multiple samples.[40] By maintaining this registry PXE International not only wished to exert an influence on how this material was used, but also was able to demand a share of the intellectual property rights that arises from it.

The productivity of this blood and tissue registry for the generation of biovalue was demonstrated in 2000, when the gene for PXE was discovered by researchers at the University of Hawaii and announced in two consecutive articles in *Nature Genetics*, each with Sharon Terry as a coauthor (Bergen et al. 2000, Le Saux et al. 2000).[41] This discovery not only generated new insights into the pathology of the disease, but also had significant potential for the exploitation of intellectual property rights. The technology transfer unit at University of Hawaii was initially reluctant to yield patent rights to PXE International, but as they had previously negotiated the terms and conditions of access to the registry, which included Sharon Terry being named as a coinventor, they were able to work out a process of sharing royalties and deciding on licensing deals.

The "PXE model" for lay advocacy and active lay engagement with research is one that has been advocated by other groups with similar concerns around rare genetic conditions.[42] The Terrys argued that PXE International's stake in the patent for this gene sequence was not driven by a logic of commercialization, but rather to serve the values and interests of persons suffering from genetic diseases. Thus Patrick Terry defended the potential of patient-controlled patents and asserted that, "We're not interested in lining our pockets. We just want a cure" (Coghlan 2001). Nonetheless, the Terrys hoped that knowledge of this single and quite rare condition could have wider implications for the health of those who suffer from apparently unrelated disorders. For example, it appeared that studies of this particular genetic pathway on Chromosome 16 might shed light on hypertension and cardiovascular research, since the mineralization of the midsize arteries in PXE mimics the general aging of the arteries (Fleischer 2001). PXE might also provide clues to macular degeneration; this affects the eyesight of many individuals suffering from this disorder, but another sixty million Americans are thought to be at risk for this condition due to ordinary aging.[43] Clearly, PXE International has the chance of making significant financial gains if a broader use for the sequence is found. However, Sharon Terry says that PXE International will resist the temptation of profiteering: she claims in an article in *The American Lawyer*, "It's been suggested that we could make a killing because who cares if we're making the cost of cardiovascular treatment huge. We always say, we don't just represent people with PXE, we represent anybody, who has anything."[44]

The new circuits of biovalue, and the new markets they inhabit, not only create new possibilities for the generation of wealth, but also embody and generate new ethical values.[45] As life itself is penetrated by market relations and becomes productive of wealth, the morality governing some forms of economic exchange is being reconfigured. In an economy where the vitality of biological processes can be bought and sold, ethics becomes both a marketable commodity and a service industry in its own right. UmanGenomics in Sweden, for example, trades on the fact that all the blood samples in its collection have been collected with full informed consent (Høyer 2002, 2003). It proclaims that "Correct ethical handling of human tissue and medical data is essential," and highlights how it has been "internationally recognized for its ethical stance and procedures."[46] Ethics, in this instance, is not only a means of access to a valued resource, but is also a marketable asset that a company can trade on in its commercial relationships with other enterprises and in its relation with public bodies and medical practitioners (Rosell 1991). Some additional aspects of the Terrys' work can illustrate this further. At the same time as the discovery of the PXE gene, in August 2002 Patrick Terry joined with Randy Scott—who earlier had founded the biotechnology company Incyte—in establishing a biotech company called Genomic Health (in Redwood, California), initially funded with seventy million dollars of venture capital, and later attracting major investment from merchant banks and pharmaceutical corporations, for the development and marketing of genetic tests and targeted therapies that would individualize treatment to patient's genomics.[47] This does not undermine the ethical claims of the Terrys, but illustrates the new relations between ethics and bioscience in the production of biovalue. The recruitment of Patrick Terry to Genomic Health, and his title of "Director of Consumer Advocacy," was intended as an outward and visible sign of the ethical nature of the company and its commitment to patients, which went alongside its ambition to bring affordable genomics to the consumer. Thus Randy Scott, writing as Chief Executive Officer in 2001, insisted that the future of the genomics industry rests on the "education, trust and support of the consumer" and that the uptake of these new products will only take place on "a foundation built on bioethics"—a foundation "critical to engaging consumers . . . whether for research, for medical treatment, or for business" (Scott 2001). Ethics and biovalue are inextricably intertwined, contributing to the commercial value of products through demonstrating a commitment to the ethical values of the biocitizen-consumer of health and his or her requirements for trust.

The values embedded in new genomic artifacts are polyvalent: indeed they bear upon the meaning of life itself. Once more, Randy Scott illuminates this point. It is critical, he suggests "for industry to begin to create an open public dialogue with all stakeholders in order to facilitate understanding and to build trust." And while this dialogue will be both difficult and complex, "our quest to cure disease and prolong life will ultimately lead to much deeper questions—the very definition of what it means to be human" (Scott 2001). That is to say, judgments as to the nature and value of different forms of life—different ways of being human—are actually inscribed in the outputs of the life sciences and biomedical research today, not just in their conditions and consequences. The process of generating biovalue embodies and transforms conceptions of human life: to that extent biomedical artifacts themselves have ethics.[48]

CHAPTER FIVE

Biological Citizenship in Advanced Liberal Democracies

While citizenship has long had a biological dimension, new kinds of biological citizens—with new subjectivities, new politics, and new ethics—are forming around contemporary developments in biomedicine. As aspects of life once placed on the side of fate become subjects of deliberation and decision, a new space of hope and fear is being established around genetic and somatic individuality. In the advanced liberal nations of the West—Europe, Australia, and the United States—this is not taking the form of fatalism and passivity, and nor are we seeing a revival of genetic or biological determinism. While in the residual social states in the post-Soviet era, biological citizenship may focus on the demand for financial support from state authorities, in the West novel practices of biological choice are taking place within a different "regime of the self." Here, each individual is engaged as a prudent yet enterprising individual, actively shaping his or her life course through acts of choice, activities that extend to the search for health in the face of the fear of illness, and the management of the risks—now the genetic susceptibilities—of disease. Some readers may remember a United Kingdom BBC television documentary series some years ago entitled *Your Life in Their Hands* about the heroic work of doctors and medical staff working to save their patients. Perhaps, then it is emblematic of this new ethics that a BBC radio series in 2004, which charted the activities of those who found themselves diagnosed with particular disorders—for example women with a family history of breast cancer finding themselves having to choose whether to take the test for BRCA 1 or BRCA 2, or how to live with a positive test result—was entitled *Our Lives in Our Hands*.

This shift from implacable abnormalities to manageable susceptibilities is entirely consistent with the wider reshaping in contemporary practices for the government of persons. Today, we are required to be flexible, to be in continuous training, life-long learning, to undergo perpetual assessment, continual incitement to buy, constantly to improve oneself, to monitor our health, to manage our risk. And such obligations extend to our genetic susceptibilities: hence the active responsible biological citizen must engage in a constant work of self-evaluation and the modulation of conduct, diet, lifestyle, and drug regime, in response to the changing requirements of the susceptible body. In tracing out, experimenting with, and contesting the new relations between truth, power, and commerce that traverse our living, suffering, mortal bodies, and challenging their vital limits, such active biological citizens are redefining what it means to be human today.

Chapter 6

Race in the Age of Genomic Medicine

"Genetics for the Human Race" was the title of a special supplement of *Nature Genetics* published in November 2004.[1] This was based on a meeting held at Howard University on May 15, 2003, entitled "Human Genome Variation and 'Race'—the State of the Science." A month earlier, on April 14, 2003, announcing the release of the "essentially completed" human genome sequence, Francis Collins, Director of the National Human Genome Research Institute had said, "All of this study of genotypes will have profound consequences on our understanding of race and ethnicity," and told scientists they were obligated to translate what they learn about race, ethnicity, and genetics into "information that will be usefully included as part of the often-contentious dialogue about race in our society."[2] At the heart of these concerns lay a crucial question: Would the genomics of the twenty-first century resurrect, or finally lay to rest, the scientific racism that has played such a formative and bloody role in the history of our present?

The contentious debate about race, genomics, and health to which Collins referred has taken a new turn since the launch of the Human Genome Project at the start of the 1990s. It has been most fractious in the United States, given the salience of arguments about racism there, but genetic and biomedical researchers from many different countries have also been caught up in these controversies. The debate has focused on whether, in a genomic age, racial or ethnic categories have any biological meaning or are "merely cultural." The view of most "Western" sociologists and anthropologists, since at least the mid-twentieth century, has indeed been that race and ethnicity are social and cultural phenomena. However salient notions of ethnicity and race they may be to individuals and groups in a social context, the argument went, these are not biologically meaningful and cannot map in any significant way onto differences at the genomic level. More radical critics suggest that consideration of race and ethnicity in genomic medicine mark a potential shift toward a racialized medical practice, presage the reawakening of a dangerous racial science, and represent a further turn in "genetic reductionism" and the "geneticization" of identity. These debates have taken on a specific significance in the racially fragmented polity of the United States, with its fundamentally inequitable health care system that disadvantages black and Hispanic Americans disproportionately. Thus, in 1998, the American Anthropological Association issued a statement on "Race" that concluded:

> The "racial" worldview was invented to assign some groups to perpetual low status, while others were permitted access to privilege, power, and wealth. The tragedy in the United States has been that the policies and practices stemming from this worldview succeeded all too well in constructing unequal populations among Europeans, Native Americans, and peoples of African descent. Given what we know about the capacity of normal humans to achieve and function within any culture, we conclude that present-day inequalities between so-called "racial" groups are not consequences of their biological inheritance but products of historical and contemporary social, economic, educational, and political circumstances.[3]

For many who thought this way, the attention in the 1990s to the genomics of population differences in susceptibilities to illness, and the investment of funds and research efforts in research on racial or ethnic differences in disease susceptibility and treatability was, at the least, a distraction from the obvious causes of inequity in health care. More worryingly, it might legitimate those inequities by locating them, not in social conditions but in the genome itself.[4] At the worst, some thought that it was a reactivation of the same kinds of racism that underpinned Samuel Cartwright's infamous 1851 invention of the diagnostic category of Drapetomania to characterize a disease peculiar to Negroes, manifested in a behavior evident in blacks but absent in whites: the tendency to run away from slave plantations.[5]

Yet many population geneticists were convinced that the mapping of the human genome has confirmed that the world population can be divided into five major groups—Africans, Caucasians, Pacific Islanders, East Asians, and Native Americans—defined by their "out of Africa" date (e.g., Rosenberg et al. 2002). Consequently, some argued that there were significant differences between these population groups at the genomic level, that medically significant genomic variations relating both to disease susceptibility and to treatability of complex diseases (not merely single gene disorders) occurred in different frequencies in different population groups (e.g., Bowcock et al. 1991, Bowcock et al. 1994, Calafell et al. 1998).[6] Different methods were used to assess these variations across populations and account for their emergence and spread through mutation, selection, drift, nonrandom mating, and admixture (Cavalli-Sforza and Feldman 2003, Cavalli-Sforza et al. 1994). Understandably, those developing such arguments in the United States tried to formulate methods for the genomic analysis of populations thought to have high levels of admixture though migration, immigration, and intermarriage. Thus Shriver and his colleagues applied admixture analysis to African Americans, Mexican Americans, Cubans, and Puerto Ricans in the United States, Africa Caribbeans, and others, and argued that the potential of such work for the identification of "genes causing complex diseases" was "immense": they held out particular hopes in relation to obesity, type 2 diabetes, prostate cancer, dementia, and hypertension (Hoggart et al. 2004, Holder and Shriver 2003: 60, Shriver et al. 2003). The supporters of genomic research on the relation between race and disease, including a number of leading African American organizations and research groups, argued that race-blind genomic medicine not only ignored the genomic evidence of variation in disease and death that mapped onto common understandings of race but were also unable to address national and international racial and ethnic divisions in morbidity and mortality. They contended that the move to racial classification was necessary to target the particular diseases suffered by disadvantaged racial groups, and to redress their underrepresentation in medical research, drug development, and access to effective health care. They hoped that, used at the aggregate level at which most public health programs operate, genetically based medicine could play a key role in reducing race-based health inequalities.

The need for genomic research on the relations of ethnicity and disease, and genomic explanations of differences in morbidity and mortality between groups defined in terms of ethnicity was put most strongly, perhaps most controversially, by Neill Risch and his colleagues. In a number of papers, they argued that self-identified ethnicity was the best guide to divisions between major population groups along medically significant genetic lines (Burchard et al. 2003, Mountain and Risch 2004, Risch et al. 2002, Risch 2000). Despite migration, they claimed, mating patterns had preserved many of these distinctions, although they accepted that the situation was more complicated for recently separated or admixed groups (notably Hispanics). Further, they considered that self-identified race, often criticized as based on partial and inadequate folk knowledge of kinship, was actually a very good guide to genetic makeup, that individuals were well aware of the ethnicity of their various forebears, and were therefore able to characterize themselves in genetically meaningful ways. Self-defined ethnicity was the best form of categorization, they argued, precisely because it was this subjective and social identification, rather than the genetic marker alone, that correlated with other factors such as diet, housing, employment, and income. Hence they concluded that racial self-categorization was vital for genomic based medical research.

Many European and American medical geneticists have been critical of these views. Some have sought a middle position, claiming it is possible to identify genomic differences that have some correlation with population groups and ethnicity, but that any characterization of such differences should not use terms like "race" or "ethnicity," but focus on the genetic markers themselves (Wilson et al. 2001). However most publicly espoused the view that, whether self-defined or socially defined, categories of race or ethnicity were a poor proxy for genetic variation and should not be used as such, as they risked reactivating old and stigmatizing ideas of biological differences between racial groups and feeding racism (see also, for example, Cooper et al. 2003). Editorials in leading U.S. biomedical journals such the *New England Journal of Medicine* asserted that while there are genetic distinctions between population groups, within-group differences exceeded between-group differences, and most genetic differences were "only skin deep"—relating to the small number of genes governing skin pigment, physiognomy, etcetera (Schwartz 2001b).[7] And many geneticists argued that disease prevention and health improvement strategies based on categories of race or ethnicity would be misleading and ineffective (Braun 2002).

These developments took place in the context of well established and relatively uncontentious arguments that some single gene disorders do occur in different frequencies in different population groups delineated in terms of race or ethnicity, and that these have clinically significant implications—well-known examples include cystic fibrosis (see, for example, the estimates for Caucasians, Ashkenazi Jews, Hispanics, African Americans, and Asians in American College of Obstetricians and Gynecologists 2001) and the haemoglobinopathies such as sickle-cell anemia and thalassemia.[8] Further, in many countries the suggestion that there are important relations at the genomic level between race, ethnicity, disease susceptibility, and drug response is not controversial. For example, in China, Japan, Vietnam, India, and many other countries outside "the West" biomedical researchers routinely consider that the genomic specificity of their population is relevant to the patterns of disease susceptibility that they encounter.[9] Similarly, those involved in the regulation of pharmaceuticals in these regions believe that there are genetically based differences between populations in the activity of many enzymes involved in drug metabolism, and that these must be taken into account in the process leading to the licensing and use of particular drugs in their countries.

In any event, basic genomic laboratory research on disease uses such classifications, and research papers routinely report associations between race or ethnicity, disease susceptibility, and genetic variation (for a few examples, see Aviles-Santa et al. 2004, Criqui et al. 2005, Farrer et al. 1997a, 1997b, Helgadottir et al. 2005, Itakura 2005, Koivukoski et al. 2004, Marsit et al. 2005, Moore et al. 2004, Mori et al. 2005). Such classification is also routine in clinical trials, especially those related to pharmacogenetic drug development—that is to say, research that tries to identify the genomic basis of variations in drug responses between individuals and groups in order to differentiate those for whom a drug will be efficacious from those for whom it will be ineffective or cause adverse effects (for a review examining the evidence in relation to minority Americans, see Burroughs et al. 2002). Most national and international genomic resources, such as biobanks and electronic databases of sequence information, also categorize DNA samples using classification schemes in terms of race, ethnicity, or population group. Such dividing practices—African American, Japanese, Caucasian etcetera—whether construed in terms of race, population, descent, or ethnicity, are integral to the style of thought and modes of analysis of contemporary biomedical genomics.

Nonetheless, at the start of the twenty-first century, many biomedical researchers in the United States remained unsettled by arguments suggesting genomic explanations for differences in disease profiles or drug responses between ethnic groups. In 2003 Rick Kittles, from the National Human Genome Center of Howard University, writing with Kenneth Weiss, an anthropologist, after reviewing the dismal history of race thinking and racialized eugenics nonetheless concluded that even when other variables were controlled, contemporary research had shown that there were differences in disease prevalence associated with race (Kittles and Weiss 2003). They argued that "labels chosen to correspond to historical geographical ancestry also statistically capture some genetic information, and that is the sense in which race can refer to something with biological meaning" (40); the "concept [of race] can be used constructively in genetic research but not all appropriate uses of race are genetic"; "population classification is an indirect but potentially useful ascertainment strategy for identifying genes or variants of particular interest, or for designing more effective risk-screening programmes" (34, 49). But they immediately qualified this limited acceptance of the language of race in genomic medicine: "Because race categories are heavily affected by sociocultural factors, even where genes are also involved, environmental changes may have a bigger impact on health. If improved health is our objective, we should put our effort proportionately to where the problems really lie" (58). Yet Howard University, as we shall see later, has played a key role in compiling genomic data on African Americans for biomedical and other research.

In this chapter, I will consider this ambivalence in more detail. Unlike most sociologists and anthropologists who have commented on these matters, I will conclude that the resurgence of interest in race and ethnicity in contemporary genomic medicine is not best understood within the trajectory of "racial science." Nor is it simply the most recent incarnation of the biogenetic legitimation of social inequality and discrimination. Indeed, even in the United States throughout the twentieth century, the linkages of race, biology, and medicine have taken very different shapes at different times and in relation to different conditions. As Melbourne Tapper argues, in his enlightening account of the intertwined genealogies of sickle-cell anemia and racial politics in the United States, "sickling," like any other disease, exists only as it is shaped within certain formations of knowledge, belief, explanation, and intervention: as these formations have been reconfigured throughout the twentieth century, so have the meanings and implications that have been accorded to the relation between sickling and race. In the 1920s to the 1940s, for example, sickling was used "to call into question the racial identities of whites afflicted by the phenomenon"; in the 1940s and 1950s colonial doctors and anthropologists used data on the geographical distribution of sickling to argue that "tribes were biologically defined entities and that East Africans, certain Southern Indian populations, and Sicilians and Greeks shared a common racial heritage"; in the 1970s, the link between race and sickling was an explicit element within the U.S. government's antidiscrimination policies, and in the demands of African American activist organizations, both arguing that medical resources should be directed toward the African American population affected by sickle-cell anemia in the name of social justice and civil rights (Tapper 1999: 3–4). Indeed Tapper traces the constitutive role of sickling in the United States of the 1970s in shaping the idea that African Americans were, or could become, responsible communities taking care of the health of their own members. As with sickling, so more generally in the relations between race and genetics: these links have no given or intrinsic politics; they take very different forms as they are entwined with distinct styles of thought about health, illness, and the body at different times and places. And, as with sickling, the relation between these terms today, in the United States at least, is intrinsically linked to the delineation and administration of biosocial communities, formed around beliefs in a shared disease heritage, demanding resources for the biomedical research that might reveal the genomic bases of these diseases, and mobilized by the hope of a cure. Perhaps, then, we might understand the contemporary allure of race in biomedicine in terms of the hopes, demands, and expectations of such communities of identity, as both subjects and targets of a new configuration of power around illness and its treatment.

Further, the debate needs to be located, not within the biology of the past but that of the present—a molecular genomic biology that is probabilistic not deterministic, open not closed, not identifying an essential racial truth that determines individuals to different fates but opening up the possibilities of intervention and transformation. In the racial science that took shape throughout the eighteenth and nineteenth centuries, whatever differences there were in its different versions, race was understood at a molar level. It was thought of as an inherited constitution that shaped the whole character and all capacities of each racialized individual. This constitution could be observed in visible characteristics, not merely skin color but physiognomy, body morphology, and the like. And as racial science encountered genetics, it was this molar notion of racial specificity and racial difference that was encoded in genes, seen as equally molar, determining these individual physical, mental, and moral characteristics and severely limiting the scope for modification by environment or experience. But the molecular gaze of contemporary genomics transforms this perception.

As we shall see in the debates discussed in this chapter, race now signifies an unstable space of ambivalence between the molecular level of the genome and the cell, and the molar level of classifications in terms of population group, country of origin, cultural diversity, and self-perception. It is in this new space of ambivalence that a new genomic and molecular biopolitics of race, health, and life is taking shape.[10] These ways of thinking and intervening certainly have their own ethical and sociopolitical difficulties, as for instance in the pressures placed on parents from particular countries of origin, ethnicities, religious, or racial groups to use genetic tests for SNPs or markers that are associated with an increased probability of a child developing an illness associated with that group. But they are not undertaken in the name of constituting and legitimating a hierarchy of differences, or with the hope of the improvement of the quality of the population. What, then, are the links of race, biopower, and the biopolitics of health and medicine in our postgenomic age?

Race and Biopower

Classifications of humans in terms of race, with differential values attached, have a long history in European thought. This is not the place to rehearse that history in any detail or to undertake the difficult task of transforming linear histories into a genealogy of race thinking in our present. Most historians agree that racial classifications arose in seventeenth-century Europe from travelers' encounters with people of strange lands and that, by the eighteenth century, an initial idealization of the primitive had given way to the identification of the primitive with blackness, and the location of such peoples at a lower stage of development in the "great chain of being." By the mid-eighteenth century, the division of humankind into four, five, or more varieties of races and the differential valuation of each in terms of habits, intellect, and beauty, were routine among anthropologists, philosophers, and in political thought, although the precise classification systems to be used, and the forms of differential valuation, were subject to much debate. In state building projects from the eighteenth century, and in the so-called "war of nations," conceptions of race formed a prism not just for the imagination of "the nation," but also for the political management of national health and vitality, and of international competitiveness. Darwin himself, in *The Descent of Man*, published in 1871, wrote that

> [m]an has been studied more carefully than any other animal, and yet there is the greatest possible diversity among capable judges whether he should be classed as a single species or race, or as two (Virey), as three (Jacquinot), as four (Kant), five (Blumenbach), six (Buffon), seven (Hunter), eight (Agassiz), eleven (Pickering), fifteen (Bory St. Vincent), sixteen (Desmoulins), twenty-two (Morton), sixty (Crawfurd), or as sixty- three, according to Burke. This diversity of judgment does not prove that the races ought not to be ranked as species, but it shews that they graduate into each other, and that it is hardly possible to discover clear distinctive characters between them. (Darwin 1871)

Notwithstanding Darwin's caution, in nineteenth century Europe there was a concerted biologization of race, linked to pre- and post-Darwinist evolutionary thinking, deployed within states and in their rationalities of colonial government, and underpinning the late-nineteenth-century obsession with degeneracy and race suicide (e.g., Mosse 1978).

It is, therefore, entirely understandable that many have been tempted to place contemporary concerns with race and biology within this history, and to argue that racialization was central to modern state formation and remains so within contemporary technologies of government (Goldberg 2001). Michel Foucault certainly thought that the link between race and biology was central to the genealogy of contemporary biopower. In his lectures of 1976, thinking largely, it seems, of the European history of racial science and its culmination in eugenics, he argued that "the emergence of biopower . . . inscribes [racism] in the mechanisms of the State . . . as the basic mechanism of power, as it is exercised in modern States

. . . racism justifies the death-function in the economy of biopower by appealing to the principle that the death of others makes one biologically stronger insofar as one is a member of a race or a population" (Foucault 2002: 254, 258). As we have seen, the idea of populations as races that could be undermined from outside by infiltration from lower races, and degraded from inside by excessive breeding of degenerates, underpinned the strategies of eugenics. This biologized racism linked the characteristics of individuals to those of the population as a whole, seeking to constrain or restrict the rights, reproduction, numbers, or powers of groups defined by their racial origins. Such ideas of race, linking individual constitution to population quality, also underpinned early twentieth-century attempts by many nations to control the racial character of their population through restrictions on immigration, informed by those such as Francis Amasa Walker, director of the 1870 and 1880 U.S. Census, who, in 1899, warned that Americans were in danger of being overrun by hordes of "degraded" immigrants from southern and Eastern Europe: "beaten men from beaten races" (I discuss Walker in Rose 1999: 222–224; see also Walker and Dewy 1899).

Undoubtedly, colonial projects also inspired racial classification and differentiation. Ian Hacking has pointed to the ways in which categories become institutionalized, especially by censuses and similar official practices for identifying and enumerating subjects, and has reminded us that the first European censuses were carried out on the colonies, from the Spanish census of Peru in 1548 to the British census of India of 1871–72 (Hacking 1990: 17, 2005: 112). The latter enumerated the population by many different overlapping classificatory systems, notably religion, nationality, language, and caste. In this enumeration, casual references to race abound; thus it was estimated that the population of British India contained:

> 140½ millions of Hindoos (including Sikhs), or 73½ per cent., 40 ¾ millions of Mahomedans, or 21½ per cent., and 9 ¼ millions of others, or barely 5 per cent., including under this title Buddhists and Jains, Christians, Jews, Parsees, Brahmoes, and Hill men of whose religion no census was taken or no accurate description can be given. . . . Although nearly the whole of the inhabitants of British India can be classed under one or other of the two prevailing religions, it will be found that, when arranged according to nationality or language, they present a very much greater variety. The population of the single province of Bengal contains many races and tribes. Bengal proper, and some of the adjacent districts, are inhabited by the Bengali, living amid a network of rivers and morasses, nourished on a watery rice diet, looking weak and puny, but able to bear much exposure, timid and

> slothful, but sharp-witted, industrious, and fond of sedentary employment; the Bengali-speaking people number some 37 millions. Allied to these, both in language and descent, even more timid, conservative, bigoted, and priest-ridden, are the Ooryas, or people of Orissa, numbering four millions. The Assamese, of whom there are less than two millions, speak a language very similar to Bengali, but have a large mixture of Indo-Chinese blood; they are proud and indolent, and addicted to the use of opium. The Hindustanis of Behar are hardier and more manly, have a less enervating climate, and use a more substantial diet; their language is Hindee, and, they number (in Bengal) some 20 millions. Besides these, there are the Sonthals, Koles, Gonds, and other aboriginal tribes in Chota Nagpoor, the wild mountain races in Julpigoree, the inhabitants of the Garo, Cossya, Jyntea, and Naga Hills, and those in Tipperah and the Chittagong Hill tracts. (Waterfield 1875: 16)[11]

Historians of colonialism, notably of prisons, madhouses, and the government of illness, have shown how, throughout the nineteenth century, such "dividing practices" underpinned the different rationales and techniques of government thought appropriate for those allocated to different categories, in part because of their perceived capacity to carry the burdens of self-control, foresight, and responsibility necessary for civility (Arnold 2002, 2005, Vaughan 1998). These classification schemes also shaped the practices of self-identification among European colonists themselves, from the meaning of "whiteness" to the obligations of colonial governors and military personnel (Osborne 1997, Stoler 1995). Such practices persisted into the twentieth century, not only in the classification mania that characterized the race science of eugenics and national socialism but elsewhere.

In their discussion of race classification in apartheid South Africa from 1950, Bowker and Star point out that "race classification and reclassification provided the bureaucratic underpinnings for a vicious racism" but also that this classification system became "enfolded into a working infrastructure" that led to a "naturalization" of political categories such that they could be enforced by bureaucrats in their routine everyday practices and decisions (Bowker and Star 1999: 195–196) As they point out, this racial classification "sought to divide people into four basic groups—Europeans, Asiatics, persons of mixed race or coloreds, and "natives" or "pure-blooded individuals of the Bantu race," the latter being subdivided further into eight main groups (197). Classifications, as originally derived from categories checked on the 1951 census returns, were encoded in laws and inscribed in the passbooks that nonwhite South Africans were required to carry; they regulated all aspects of life, from mobility to sexuality. But even so, despite the brutal consequences of classification, its basis was always unstable. They quote from a legal study of race classification published in 1969: "the absence of uniformity of definition flows primarily from the absence of any uniform or scientific basis of race classification. Any attempt at race classification and therefore of race definition can at best only be an approximation, for no scientific system of race classification has as yet been devised by man. In the final analysis, the legislature is attempting to define the indefinable" (Suzman 1960, quoted in Bowker and Star 1999: 202–203). Racial classification, it seems, was essential for the practices of government, yet always elusive for those who would seek a basis for that government in the objectivity and legitimacy generated by scientific knowledge.

We scarcely need reminding that racial classification was central to the forms of segregation that persisted in the United States into the 1960s, and to the contemporary politics of race in that country. Bowker and Star refer briefly to the controversies over the U.S. Census in the 1990s, but race has been central to the enumeration of the U.S. population since its nineteenth-century beginnings. Thus the 1840 census, which notoriously counted slaves as three-fifths of a person, appeared to show that the rate of insanity among free blacks was eleven times higher than that of slaves, and was used to argue that blacks were constitutionally unfit for freedom. The 1850 census classified respondents by "color," and "civil condition"—that is, whether free or slave—and also included a category for Mulatto.[12] The 1890 census, counting Chinese and Japanese and American Indians, subdividing Mulattos into Quadroons and Octoroons, and identifying the country of origin of immigrants, was a direct response to growing concerns about the changing racial makeup of the American population. As I have already noted, the results were used by those such as Francis Amasa Walker to argue for restrictions on immigration to preserve the existing "racial balance" of the nation.[13]

After a long history of post-war debate over the correct forms of racial classification—which among other things moved the onus of categorization from observation of physical traits and skin color by the enumerator, to self-identification on unspecified grounds—in 1977 the U.S. Office of Management and Budget (OMB) established an official classification standard for the measurement of race in the American population that was first implemented in the 1980 Census: (a) American Indians, (b) Asians and Pacific Islanders, (c) Non-Hispanic Blacks, (d) Non-Hispanic Whites, and (e) Hispanics. The OMB directive, mandating all agencies, contractors, and grantees to use these categories when collecting data on race, explained that "the classifications should not be interpreted as being scientific or anthropological in nature" (OMB 1977 quoted in Snipp 2003: 573). These categories were immediately politicized: throughout the

1980s and 1990s groups contested some categories and sought the inclusion of others to demonstrate inequities, to argue for resources, or in order to give official recognition to their own identity (Snipp 2003). As Kenneth Prewitt put it, in the politics of numbers at the end of the twentieth century, "being "measured" is to be politically noticed, and to be noticed is to have a claim on the nation's resources . . . political visibility follow[s] on the heels of statistical visibility" (Prewitt 1987).

No wonder, then, that so many, especially in America, would echo Donna Haraway's assessment that race "is a fracturing trauma in the body politic of the [American] nation—and in the mortal bodies of its people. Race kills, liberally and unequally; and race privileges, unspeakably and abundantly. . . . Race, at once an uncanny irreality and an inescapable presence, frightens me, and I am not alone" (Haraway 1995: 321). For those who read contemporary biopolitics through the lens of Giorgio Agamben or Antonio Negri, it is the negative, exclusionary, murderous dimension of race that lies at the heart of references to race in genomics (Agamben 1998, Hardt and Negri 2000, 2004). For those who read biopower through the familiar tropes of sociocritique, the denial of social determination, the erroneous individualization of the causes and solutions to problems of ill health, underpins all resorts to biological explanations; attempts to differentiate by race within genomics perpetuate a worldview that serves to bolster, mask, and legitimate structural inequality and racism. Yet more recent critical analysts, even in the United States, have recognized that, as Troy Duster puts it, "purging science of race—when race and ethnic classifications are embedded in the routine collection and classification (from oncology to epidemiology, from hematology to social anthropology, from genetics to sociology)—is not practicable, possible or even desirable" (Duster 2003: 258). Race as an organizing force in social relations, in stratifying practices, in self-identification, cannot be wished away, especially in a racially fractured polity such as the United States, and Duster argues that in some cases at least "we must conduct systematic investigations . . . into the role of the interaction of race (or ethnicity or religion)—however flawed as a biologically discrete or coherent taxonomic system—with feedback loops into the biological functioning of the human body, and then again in relation to medical practice. . . .Understanding what remains problematic about the concept of race—the complex interrelationships between sociopolitical processes and scientific knowledge claims or biomedical practices—is also fundamental to understanding clearly the natural-cultural relations that shape our lives" (272–73).[14]

Races may be constructed, shaped by arbitrary classifications, politically delineated and redelineated, historically and situationally variable, and much more. So, in less highly cathected ways, are many of the other categories we live by—and they are no less real for that—such, inescapably, is the nature of the classification schema that shape our forms of life, our ways of governing others, and our ways of relating to and governing ourselves.[15] To illuminate these issues, then, we need to turn here, not so much to history as to genealogy. We should not place the contemporary ensemble of race, genomics, and medicine within a linear history, trying to find its essence in some origin and singular line of development. Instead, we need to locate the current debates over race and genomics firmly within the transformed biopolitics of the twenty-first century. This is a biopolitics organized around the principle of fostering individual life, not of eliminating those that threaten the quality of populations; it is a biopolitics that does not operate under the sign of the sovereign state; it is a biopolitics that does not seek to legitimate inequality but to intervene upon its consequences. Crucially, it is a biopolitics in which references to the biological do not signal fatalism but are part of the economy of hope that characterizes contemporary biomedicine. That is to say, perhaps the links between race, medicine, and genomics are not exceptions, but integral parts of today's politics of life itself.

The Genomics of Difference

After World War II, claims of scientific basis for racial differentiation were officially discredited: by 1963, for example, the United Nations Declaration on the Elimination of All Forms of Racial Discrimination took as one of its premises "that any doctrine of racial differentiation . . . is scientifically false, morally condemnable, socially unjust and dangerous . . ."[16] Of course, racist practices hardly subsided, and race remained extremely salient as a socioeconomic category, a mark of discrimination, and a mode of identification. Nowhere was this more so than in the identity politics of the United States. But when, over the 1960s and 1970s, African Americans sought to trace their "roots" to Africa, they seldom related this genealogy to a biological substrate.

In part due to the persistent interventions of radical critics, the link between biological understandings of distinctions among population groups and their sociopolitical implications seemed broken. Many biologists and biomedical researchers still believed they encountered genetic differences, not least when they considered variations in the prevalence of particular diseases in different regions or the efficacy of medicines in different national populations, but these explanations seldom had a high public profile. More publicity was given to those few individuals and groups who still claimed that there was a politically significant correlation between human qualities and racially differentiated biology—but Jensen,

Shockley, and others were largely marginalized and vilified, and often discredited (for some examples of the disputes, see Jensen and Miele 2002, see Kamin 1974, Lewontin et al. 1984, Shockley and Pearson 1992).

At one point it appeared that genomics itself would mark the end of biological racism. Not only did humans share over 98 percent of their genetic complement with chimpanzees, but within group variations in DNA sequences were greater than those between groups: the idea that population groups identified by race were genetically homogeneous and genetically distinct from one another seemed untenable (Marks 2002).[17] However a new molecular conception of population difference rapidly emerged out of genomic thinking. From the start of the Human Genome Project, population geneticists argued for an attention to diversity. The initial proposer of the Human Genome Diversity Project, L. Luca Cavalli-Sforza argued that it would "explore the full range of genome diversity within the human family" and it was hoped that the project would "help combat the widespread popular fear and ignorance of human genetics and . . . make a significant contribution to the elimination of racism."[18] While there were many critics of this project, by 1991 it was "adopted" by HUGO, the Human Genome Organization, which provided US$1.2 million to set up workshops to develop the technical and organizational aspects of the project, to consider the social and ethical implications, and to conduct a pilot study (M'charek 2005, Reardon 2001, 2005).

As the sequence of bases in the human genome began to be mapped within the Human Genome Project itself, variations at the molecular level emerged that gave a new significance to the issue of genomic difference. Genome mapping led to the conclusion that while the three billion base pairs that make up the DNA sequence of any two randomly selected individuals will be 99.9 percent identical, there are many variations at the Single Nucleotide Polymorphism level where, say, a T is substituted for a C. On average, it was claimed, one letter in 1,000 differed between two individuals—which made a total of many million variations between them (estimates of the number ranged from 6 million to 15 million). It began to be argued that these SNP differences between individuals were very significant, notably in relation to susceptibility to particular diseases and treatability by specific drugs. Most significantly, it appeared that these differences occur with different frequencies in different populations, and might hold the key to the long recognized differences in disease susceptibility.

Further, it appeared that sets of nearby SNPs on the same chromosome were inherited in blocks called haplotypes, and that these could be identified by tag SNPs. "HapMapping," as this method was termed, seemed to promise a more economical way of identifying SNPs relevant to disease. In 1999 the United Kingdom-based charity, the Wellcome Trust, announced a consortium with ten pharmaceutical companies to find and map 300,000 common DNA sequence variations, and NIH, Wellcome, together with labs in Japan and China, collaborated in an international HapMap Project, collecting DNA from blood samples in Nigeria, Japan, China, and the United States (from U.S. residents with ancestry from northern and western Europe). From the start, the Project was equipped with a bioethical arm—the International HapMap ELSI Group.[19] This ELSI group was involved in the decisions as to how to sample, which groups to include, and how to name them. The conclusion was that the samples would be anonymous, but would be identified by the population from which they were collected. The Project described its approach as follows:

> The International HapMap Project will be carried out by a consortium of public and private institutions in five countries—Canada, China, Japan, the USA and the UK. . . .The map will be based on DNA samples obtained from hundreds of people in geographically distinct populations: Nigerian Yorubas, Han Chinese, Japanese, and US residents of European origin. These populations have been selected for their diverse population histories, which may result in differences in haplotype structure and frequencies, and are not meant to be representative of different ethnic or racial groups.[20]

It thus emphasized population groups rather than race or ethnicity, and justified the choice of comparator groups, not as characterizations of distinct races, but as a means of identifying factors affecting the health of the human population in all, or some, of its genomic diversity. The naming of groups—Yoruba, Han, Japanese—was portrayed as merely a means of indexing diversity, selecting from what were thought to be relatively stable population groups, geographically separated, where recent population mixing was considered rare. The HapMappers noted the "danger . . . that variants more common in certain populations will be mistakenly used to characterize entire populations"—they propose public education to counter such a danger.[21] But, of course, knowing that one set of samples is "Japanese" inevitably opens the possibility of using the data to investigate and interpret studies of disease frequencies and drug responses in people of Japanese descent. The bioethicists attached to the HapMap, perhaps not surprisingly, focused on conventional bioethical concerns: they seem to have been most worried that the consequences of the mapping of medically significant differences at the SNP level onto such groups would be increased discrimination or social stigma. They do not seem to have worried about the consequences for commercial investment for drug development, for public health policies in different regions, or for everyday clinical practice with different population groups and communities.

Many questioned the value of the HapMap Project, and its strategy in the search for the genomic bases of disease. However, the significance of research on the SNP variations relating to disease and treatability became even more pronounced with the publication of the first map of the human genome, and the recognition that far from the 100,000 genes that had been predicted, there were only about 30,000 coding sequences, and it was within such sequences that the genomics of differences must be located.[22] The focus now was on variations at the SNP level within this relatively small number of coding sequences, the functional properties of such variations, and the differing probabilities of functionally significant variations in different groups.

One key driver in the search for medically salient SNP variations came from the commercial aspirations of pharmaceutical companies and the biomedical industry, who saw opportunities for a genomic strategy for diagnosis, drug development, and marketing. Their hope was that the basis of susceptibilities to common complex disorders could be located at this SNP level, in the interaction of many diverse loci of small effect, and their expression in certain conditions. Biotech companies such as Celera Diagnostics and Genaissance invested heavily in attempts to develop genomic strategies for diagnosis and genomic-based drug development, and their publicity stressed the therapeutic advantages and commercial potentials of tests and drugs targeted at medically pertinent genomic differences. In the case of complex diseases, a raft of research projects studies the genomics of "stable" populations where particular conditions showed an increased prevalence. For example, in the search for genomic susceptibilities, a number of Western biomedical researchers went to China, and attempted to identify and genotype isolated populations with high incidences of particular diseases. In 1996 the Paris-based company, Genset, signed a letter of intent with the Chinese Academy of Medical Sciences, which operates twenty-five research institutes and all of China's hospitals. Genset, working with the French trading company Tang Frere International, was to employ researchers to diagnose diseases and collect DNA. According to one account:

> China is interesting to geneticists because its rural population had remained static for centuries, making each region different in its pattern of genes and diseases. This makes it easier to trace hereditary diseases back to a specific defective gene which may be unusually abundant where the disease is prevalent. "You can treat regional local populations almost like single families," explains [Genset President Pascal] Brandys. In addition to this, the diseases which affect China's population differ from those most prevalent in the West. This provides an opportunity to study new or relatively rare diseases such as

cancer of the gullet, which is much more common in China than in the West.[23]

But commercial considerations were not the only drivers for research that tried to identify genomic differences related to increased disease susceptibility in particular population groups. Medical geneticists concerned with clinical practice for such populations also considered that genetic information was important.[24] And, as we shall see later, pressure also came from the demands of patient and advocacy groups for research that could identify the genomic bases of the diseases that their families and communities suffered, and to develop effective screening tests and therapies. These multiple pressures embodied a view of the social significance of the links of genetics and population very different from that criticized by so many social scientists. This was the belief that to find a biological substrate for disease was to increase the possibilities of effective intervention in the name of health. It was in this field of probabilities, predictions and preventive interventions that new genomic conceptions of population differences took shape.

Ethnic Categorization in Health Research

Outside genetics, there had long been criticisms of racial and ethnic categorizations in health service and medical research. Critics argued that the categories used in such research overemphasized homogeneity within groups; produced misleading contrasts between groups; misunderstood the complexities affecting the ways individuals assigned themselves to such groups when asked; allowed simplistic, inconsistent, and misleading assignments to categories by researchers or clinicians; confused and confounded the associations between ethnicity, class, poverty, and lifestyle; conflated divisions variously based on nationality, skin color, country of origin, ancestry, descent, and self-identification; lacked any coherent conceptual grounding; and in general failed to capture the multiple and incommensurable dimensions of difference (Bhopal 2002, Bhopal and Donaldson 1998b, Bhopal and Rankin 1999, Bradby 1995, 1996, 2003, Dyson 1998, Fullilove 1998, Nazroo 1998, 2003). Where official data had been examined in detail, as in the case of the United States, researchers identified significant inconsistencies in racial and ethnic classification (Hahn 1992, Hahn et al. 1992). And while focus was mostly on the classification of what were, as of 2005 in the United Kingdom, referred to as "minority ethnic communities," some also pointed to the lack of specificity in categorizations of "white" and "Caucasian," and suggested that here, too, there was a lack of shared understanding about the meaning or use of such categories among the international research community (Bhopal and Donaldson 1998a). Some suggested guidelines for the use of such categories in biomedical publications (Kaplan and Bennett 2003). Others argued that any attempts at classification in terms of race or ethnicity reinforced the false belief that distinct racial groups existed, ignored the political, social, and cultural history and variability of classification schemes both outside and within biomedical research, and were unlikely to contribute to the understanding or amelioration of differences in health, illness, and treatment outcomes.

It is, therefore, not surprising that many of those involved in research in genomic medicine, especially in the United States and the United Kingdom, were uneasy with the forms of classification that were used to link issues of disease susceptibility or treatability to race or ethnicity. Many agreed with the social scientific critics that race was a biologically meaningless cultural construct; that within-group differences exceeded between-group differences; that genetic differences between racial or ethnic groups merely related to differences in skin pigmentation and physiognomy; and that race-based medical research and development might support racism (e.g. Cooper and Freeman 1999, Cooper et al. 2003, Kaufman and Cooper 2001, Schwartz 2001b). Others argued that there was no consistency in the use of the ethnic or racial categories, and that researchers themselves were confused or inconsistent about their applications.[25] Further, critics argued that the use of such classifications led to the downplaying of gene-gene and gene-environment interactions, accorded genomic knowledge far too much etiological significance, and exaggerated potential therapeutic benefits (Ellison and Rees Jones 2002). In 2002, a WHO report on Genomics and World Health, while highlighting the potential of genomics for health improvement in the poorest countries, nonetheless asserted, "The more that is learnt about the individual uniqueness and diversity of human beings, the less the concepts of "race" seem to have any meaning" (World Health Organization 2002: 70). Yet despite such criticisms, surveys of articles in leading genetics journals suggested that the use of race and ethnicity in genetic research was increasing (Ellison and Rees Jones 2002). For biomedical genomics at the start of the twenty-first century, population classification by race or ethnicity seems inescapable, elusive, and troubling.

In the United States in 1997, as a result of the OMB directive discussed earlier, it became a requirement of federally funded biomedical research to demonstrate that it took account of population diversity. From this point on, to encourage research in all sections of the U.S. population, and monitor for equity, the NIH required racial identification of participants in clinical trials.[26] Researchers were to use the standard definitions of ethnicity and race in the U.S. Census "to allow comparisons to other federal databases, especially the census and national health databases." The U.S. Census of 2000, like its predecessors from 1960 onwards, used self-definitions in gathering data, but it now separated the category Hispanic, treating it not as a race but as an ethnicity. In addition, it dealt with the issue of "mixed-ness" by allowing individuals the option to select more than one racial category. The schema is set out in the box below:

ETHNIC CATEGORIES

Hispanic or Latino A person of Cuban, Mexican, Puerto Rican, South or Central American, or other Spanish culture or origin, regardless of race. The term "Spanish origin" can also be used in addition to "Hispanic or Latino."

Not Hispanic or Latino

RACIAL CATEGORIES

American Indian or Alaska Native A person having origins in any of the original peoples of North, Central, or South America, and who maintains tribal affiliations or community attachment.

Asian A person having origins in any of the original peoples of the Far East, Southeast Asia, or the Indian subcontinent including, for example, Cambodia, China, India, Japan, Korea, Malaysia, Pakistan, the Philippine Islands, Thailand, and Vietnam. (Note: Individuals from the Philippine Islands have been recorded as Pacific Islanders in previous data collection strategies.)

Black or African American A person having origins in any of the black racial groups of Africa. Terms such as "Haitian" or "Negro" can be used in addition to "Black or African American."

Native Hawaiian or Other Pacific Islander A person having origins in any of the original peoples of Hawaii, Guam, Samoa, or other Pacific Islands.

White A person having origins in any of the original peoples of Europe, the Middle East, or North Africa.

As in 1977, the U.S. Census Bureau's statement about these categories in 2000 was carefully crafted: "the categories represent a social-political construct designed for collecting data on the race and ethnicity of broad population groups in this country, and are not anthropologically or scientifically based."[27] Nonetheless, many outside the United States will wonder at the coherence of this classification system—the sheer difficulty of thinking *that*.[28] And those of a critical mind will undoubtedly be struck by the reference back to "original peoples" and its resonances of racial classifications of the eighteenth and nineteenth centuries. Despite the acknowledgment that these categories have no basis in anthropology or science, the history of science demonstrates how often devices that begin as merely tools, very soon end up as theories (Gigerenzer 1991). And, as Ian Hacking has shown in other areas of the history of science, hypothetical entities have a tendency to become real when they are used to investigate something else (Hacking 1983). Inescapably, then, these census categories will come to organize the shaping of scientific truth, not only the techniques for selection of samples and collection of data but its analysis and the search for differences. In the very same moment as its contingent nature is recognized, population differentiation is mapped, named, and instrumentalized. Here, as elsewhere, classifications that may be arbitrary and contingent—"political constructs"—are made real in the very process of using them within a technology of investigation and analysis.

Biosociality

I have suggested that one driver of research on race, genomics, and health has come from patient and advocacy groups—that is to say, from the kinds of biosocial communities of active biological citizens that have key roles in the genomic medicine of the twenty-first century (Rabinow 1996a). These biosocial communities are a contemporary form of older collectivities defined by their members or by others in terms of a shared biology, such as race, family, or collective hereditary lineage, or a shared suffering, as in groups that have long sought to improve care for those with particular diseases. Contemporary "biosocialities" are similarly formed around beliefs in a shared disease heritage. They are frequently self-defined activist communities mobilized by the hope of a cure. That is to say, they are made up of those who affiliate themselves in various ways to a shared marker of identity, and who campaign energetically, for example by demanding resources for the biomedical research that might reveal the genomic bases of these diseases. These are sometimes actual groups or organizations meeting together at events, but sometimes they are more dispersed communities of identification, often virtual networks linked by email and the Internet.[29] And, in a number of cases, race or ethnicity is central to such forms of biosociality.

The most notable example is Howard University, which has already been mentioned. According to its website, Howard University "is a comprehensive, research-oriented, historically Black private university providing an educational experience of exceptional quality to students of high academic potential with particular emphasis upon the provision of educational opportunities to promising Black students."[30] In May 2003, Howard University announced that it had linked up with a commercial company, First Genetic Trust, on the Genomic Research in the African Diaspora (GRAD) Biobank project.[31] "The GRAD Biobank will be a highly secured comprehensive database of genetic information on 25,000 individuals carefully annotated with the individual's medical, environmental, lifestyle, and social context for translational genomic and clinical research on differential disease risk assessment, disease progression, and drug response in persons of African descent." Based on the hypotheses that "populations differ in the pattern of variation in DNA polymorphic markers . . . and population variation in DNA polymorphic markers correlates with disease susceptibility, progression and therapeutic response," its goal was "to provide the biomedical research community with the clinical resources to better understand the genetic, biological, and environmental basis for differential disease risk, disease progression and drug response in persons of African descent" and hence to "provide access to state of the art treatment and improve the wellbeing and quality of life for the community served by Howard University." Key, here, is not so much race, but the belief that a particular community has specific health needs that may have a genomic basis, and that research on that genomic basis is essential if these needs are to be met.

Howard University is not alone in its wish to use DNA sequences for the specific diseases of a "biosocial" community defined by race or ethnicity. For example, several organizations in the United States devote themselves to "Jewish Genetic Diseases"—notably the Center for Jewish Genetic Disorders at Mount Sinai Hospital, and the Chicago Center for Jewish Genetic Disorders. The Canavan Foundation advised that "1 in 9 Jews of eastern European descent is a carrier of one of the nine Jewish genetic diseases. Some of them, like Tay-Sachs and Canavan disease, are fatal but they CAN BE PREVENTED."[32] The means of prevention recommended is the screening of prospective parents, and prospective marriage partners, to see if they carry this disorder: at least one Rabbi in the United States refuses to officiate at a wedding ceremony unless the bride and groom have been tested. One particular organization, Dor Yeshorim, offers genetic testing to individuals for their carrier status for five "Jewish genetic diseases." Dor Yeshorim (the name means "the generation of the righteous") is an international genetic screening system used mainly by Orthodox Jews. The program was designed by Rabbi Joseph Eckstein in 1985, who lost four of his ten children to Tay-Sachs. Eckstein found all the available options either unappealing or irreconcilable with halachic

(Jewish) law. His solution was to try to eliminate the gene from the Jewish population entirely. Orthodox Jewish high school students who participate in the program have their blood drawn and then analyzed for carrier status for five disorders: Tay-Sachs disease, cystic fibrosis, Canavan disease, Fanconi Anemia, and Familial Dysautonomia. Instead of receiving direct results as to their carrier status, each person is given a six-digit identification number. Individuals are not told directly whether they are carriers, in order to avoid any possibility of stigmatization or discrimination, for if that information were released, carriers could potentially become unmarriageable within the community. Instead, couples considering engagement or marriage can call a hotline: if both are carriers, they will be deemed "incompatible." Not surprisingly, critics argue that this amounts to a eugenic program and remark on the irony that a Jewish organization is at the forefront of a "new eugenics."

Outside the United States, there are a number of other programs that seek to reduce the incidence of "genetic disorders" that are particularly prevalent in specific population groups. In China, until 2000, an explicit eugenic population policy was enshrined in law designed to prevent those with hereditary diseases from procreation, with a particular emphasis on mental disorders.[33] In most nations, however, such programs are not undertaken on eugenic grounds, that is to say, from the perspective of the quality of the population, but in attempts to reduce or eliminate severely disabling diseases. For example in Cyprus systematic programs of nationwide testing have been initiated, with the assent of the population, the church and the state, to identify and eliminate cystic fibrosis, once more by means of marriage counseling.[34] These examples suggest that the biopolitical implications of the framing of genomic knowledge in the categories of race or ethnicity are not inherent in the biotechnology itself: they take their character from national cultures and politics. And, in advanced liberal societies of the West, I suggest that the forms of collectivization being shaped in the links of ethnicity with medicine are not those of racial science but those of biosociality and active biological citizenship.[35]

Genomic Identity and Ethnicity

Race and ethnicity have long been central in practices of self-identification, frequently linked to lineage, constitution, blood, and biology. Hence they are subject to mutation through developments in genomics. Many social scientists are deeply suspicious of such mutations seeing them as an even more pernicious instance of the "geneticization of identity": "The individual affixed with a genetic label can be isolated from the context in which s/he became sick. . . . The individual, not society, is seen to require change; social problems improperly become individual pathologies" (Lippman 1992: 1472–73). In the context of race and ethnicity, critics suggest that this will lead to a resurgence of beliefs in biologically fixed, presocial racial identities and racial differences. Many would agree with Robert Simpson that "such thinking . . . invites soft practices of selection and exclusion which could easily pave the way for the hard eugenics which so badly scarred the twentieth century and might yet scar the twenty first" (Simpson 2000: 6). But a person's sense of his or her own identity has a complicated and variable relation to different body parts such as tissue, blood, and DNA (for an illuminating account, see Høyer 2002). And, in advanced liberal democracies at least, the biopolitics of identity is very different from that which characterized eugenics. The molecular rewriting of personhood in the age of genomics is linked to the development of novel "life strategies" for individuals and their families, involving choice, enterprise, self-actualization, and prudence in relation to ones genetic makeup. And these genetic practices of individuation provide new ways in which individuals are locating themselves within communities of obligation and self-identification delineated by race.

In 2003, in the United Kingdom, the BBC broadcast a documentary titled *Motherland—A Genetic Journey*. The makers of the program undertook a genomic analysis of DNA samples from individuals in Britain's African Caribbean community. The study showed that more than a quarter had a Y chromosome, inherited down the male line, that traced back to Europe, and thus claimed to demonstrate that, on average, 13 percent of any individual's ancestors were European.[36] It also used the Howard database to analyze the sequences in mitochondrial DNA, which is inherited down the female line. It matched these mitochondrial DNA markers with those collected by Howard from individuals in numerous villages in Africa, in order to identify the specific places from where the ancestors of these black Britons had been abducted. Far from being unique, this project drew its inspiration from a growing industry in genetic genealogy using mitochondrial or Y chromososomal sequences as markers to trace origins. In the United Kingdom, for example, Oxford Ancestors was established in 2001 to undertake such work.[37] In the subsequent two years alone, they performed over 10,000 analyses for those who wanted to trace first their paternal and later their maternal ancestry. While they assert, "[w]e are all human and we can prove it!" and that there is no connection between DNA sequences and race or ethnicity, they do classify their markers in terms such as "European" or "Native American."[38]

In the United States, firms such as Relative Genetics™ and Family Genetics™ form part of a rapidly developing commodification and geneticization of the search for identity in which "the globalized rhetoric of technoscience meets the intimacy of personal genealogies, identities and family relatedness" (Nash 2004: 2). As Catherine Nash has shown, the "genetic kinship" generated within this alliance of popular and scientific models of ancestry and identity is inescapably racialized as it tracks haplotypes, like surnames and family trees, to particular locales at home or abroad. Of course, there is much choice involved in the selection of the particular elements of the DNA sequence to define true identity—mitochondrial DNA to trace through the female line, Y chromosome markers that trace male ancestors, and so forth. And genomic identity fragments at the very moment it identifies, identifying by exclusion, splitting collective subjectivities identified by other means. Descent tracing can be deployed to support claims to the existence of a genetically pure, white British race, or to demonstrate the intermixing of slaves and slave owners, the black ancestors of white British families, or the common origin of all humans in Africa. The intermixing of race and genomics, that is to say, takes its character from the social relations, cultural values and political discourse within which it is deployed. As Nash argues, "Self-declared anti-racist, liberal scientists who repeat the argument that genetic research dispels the myth of pure, discrete "races," seem unwilling to acknowledge the implications of newly-constructed genetic kinship for existing social relations or for understanding the ways genetics is inevitably implicated in ideas of personhood, nationhood, cultural belonging, identity and community" (15).[39]

In advanced liberal democracies, the quest to discover the identify of "the real self" through the twin supports of the soma on the one hand and a community of allegiance on the other is not specific to race; it is an aspect of a wider shift in which somatic conceptions of individuality have come to the fore within a sociopolitical context where identification with the "natural" bonds of community has taken the place of the assertion of the rights and obligations of social citizens within a territorialized polis. These developments thus do not have much affinity with the race-based thinking of the twentieth century. But they do have much in common with the more general forms of biological citizenship that have been documented elsewhere.[40] But these issues take a rather different form in different social and cultural contexts.

This is particularly the case where individual identity—and sometimes individual rights to land or other benefits—is strongly tied to group membership and where genomic evidence threatens to reshape long held "traditional" views of the relations between, or differences between, particular communities. For example, in southern Africa there have long been contested relations between two very impoverished groups, the San and the Koi Koi. Despite physical and physiognomic similarities, each group has strong feelings about its own distinctiveness, its superiority, and the inferiority of the other; understandably they are reported to have responded with hostility to a recent request by a geneticist to use DNA evidence to see if they might actually have a common ancestry.[41] Dena Davis cites a number of other examples that suggest that genetic research might have the capacity to either undermine or corroborate a group's "creation story" or "communal narrative" (Davis 2003). Davis quotes a Lakota Sioux who expresses concerns that genetic research might undermine the view that the Lakota were created in the Black Hills at Wind Cave and were thus original inhabitants entitled to land, respect, and resources: such evidence might strengthen the arguments of those who consider the Lakota to descend from immigrants across the Bering Straits, and hence, in effect, to be like any other U. S. immigrant group. She discussed the case of the Lemba people, of southern Africa, where DNA sequencing and comparisons strengthened their own claims to be identified as Jews and weakened the case of those who disputed this. And she examines the case of Sally Hemmings, where DNA research on her descendants and those of Thomas Jefferson gave some (ambiguous) support to the view that this black slave had a sexual relation with the U.S. president, to the distress of not only of some of Jefferson's other descendents, but also to some of the descendents of Hemmings's son Eston, who had long taken on a white identity. The desire, and the fear, generated here, as Davis demonstrates, is not discrimination or eugenics, it is "damage to identity" (47). It is not clear that such identity damage can be countered by the usual ethical means of individual informed consent, for evidence from DNA taken from one or two willing individuals can be used to make claims about groups. Nor is community consultation an evident panacea, for in each case there are different "communities" with different investments, grounded in different narrative of lineage, similarity, and difference. Inescapably, that is to say, the generation and diffusion of genomic styles of thought in the delineation of populations will challenge and sometimes transform the very ways in which individuals and groups come to understand their affinities and distinctions. Rewritten at the genomic level, visualized through a molecular optic, bound up with novel self-technologies, identity and race are both transformed.

Pharmacogenomics and the Racialization of Pharmaceuticals

Pharmacogenomics is another driver in the direction of racial classification in medicine, and one that builds on and contributes to the increasing links of biosociality and identity with genomics. In pharmacogenomics, we are already seeing genetic segmentation of the treatment population according to race, together with arguments that genomic differences among racial and ethnic groups must be taken into account in health service policies and clinical practice (Burroughs et al. 2002, Cheng 2003, Kim et al. 2004, Munoz and Hilgenberg 2005). By the close of the twentieth century, almost all drug trials racially classified or selected their subjects, and included a pharmacogenomic element, collecting DNA samples that would allow sequence information to be correlated with efficacy, dosage, and adverse reactions. Many of those involved in pharmacogenetic drug development argued that genomic research confirmed longstanding findings of significant differences between population groups—such as those between Caucasians and Asians in the activity of key enzymes involved in drug metabolism—with implications for the efficacy and safety of treatments in the different populations. For example the enzyme Cytochrome P4502D6 (CYP2D6) catalyzes the oxidative metabolism of several clinically important classes of drugs. Many of these were already known to have lower metabolic clearance rates among Chinese, compared with Caucasians, and were prescribed at lower doses for Asian patients. Genomic analysis has demonstrated that a form of the enzyme with lower activity is more frequent in Chinese populations, and that this arises from genomic variations at the level of triplet repeats and single nucleotide substitutions. In an early example, Johanssen and his colleagues concluded that "important interethnic differences exist in the structure of the CYP2D locus, and they suggest that the frequent distribution of the C188 → T mutation among the CYP2D6Ch genes explains the lower capacity among Chinese to metabolize drugs that are substrates of CYP2D6, such as antidepressants and neuroleptic agents." (Johansson et al. 1994).[42] Many other researchers also pointed to "racial," "ethnic," or "interethnic" variations in the levels of activity of key drug metabolizing enzymes (e.g., Kalow 1982, 1989, 1991, Xie et al. 2001).[43]

In 2001 the *New England Journal of Medicine* carried two articles suggesting racial differences (in one case, differences between "black" and "white" patients) in response to drugs for heart failure (Exner et al. 2001b, Yancy et al. 2001). These papers generated considerable controversy in the pages of that journal (Bovet and Paccaud 2001, Exner et al. 2001a, Masoudi and Havranek 2001, Mcleod 2001, Ofili et al. 2001, Schwartz 2001a, 2001b). Some contributors argued that racial classification of this type was not, in itself, problematic. The data generated only indicated that certain population groups had a higher or lower proportion of those who would respond well, badly, or not at all to certain drugs, and this probabilistic information did not predict individual responses. Hence they believed that racial differences at the level of populations would be only the first indication of the need for further genomic testing before prescribing. But others argued that, for clinicians at least, race itself would become a proxy for these genomic variations in initial treatment decisions, with consequences that would inevitably be discriminatory. Indeed for pharmaceutical companies, race, by 2003, was already a proxy in the development and marketing of drugs, and number of drugs were on the market or under development for the treatment of conditions with high prevalence in specific minority groups in the United States, and marketed, not on the basis of a pharmacogenetic test, but on the basis of ethnic or racial identity, or even simply of skin color alone.

One example was Travatan, specifically marketed by Alcon for controlling elevated intraocular pressure in black patients.[44] A press release in December 2003 announced:

> American Legacy magazine and Alcon Laboratories will present the first annual TRAVATAN ® Eye Drops Project Focus Awards Presentation at the Forbes Galleries, 60 Fifth Avenue, New York, on Tuesday, January 6, 2004, 6–8 PM. The Awards will honor ophthalmologists for their work in treating glaucoma, and for their efforts to educate and promote awareness of glaucoma in the African-American community. Earlier this year, Alcon, the maker of TRAVATAN ® Ophthalmic Solution, launched the nationwide, multi-city urban outreach initiative, TRAVATAN ® Eye Drops Project Focus, to target at-risk members of the African-American community. Through a series of free glaucoma screenings, national and local advertising, distribution of educational material and community education seminars, the program encouraged African-Americans to become more conscious of eye care and their eye health. Partnering with Prevent Blindness America, the nation's leading volunteer eye health and safety organization, doctors and residents at major teaching hospitals, and local ophthalmologists, the TRAVATAN ® Eye Drops Project Focus screening teams have held free glaucoma screenings in 12 major cities. Screenings were conducted at large conventions and expos, as well as at local churches, community centers and major cultural events. Over 3,000 people participated in these screening activities with more than 15 percent of them actually being identified as high-risk prospects for glaucoma.[45]

The awards were hosted by Judge Mablean Ephriam, star of the popular syndicated television show, Divorce Court, who also collaborated with TRAVATAN® Eye Drops Project Focus in a campaign to advocate greater awareness of glaucoma risks among African Americans starting in 2004 with Glaucoma Awareness Month and that continues throughout the year. "I hadn't realized that African-Americans are six to eight times more likely to develop glaucoma than the general population," said Judge Ephriam. "When I learned that half of all African-American adults in the United States with glaucoma don't know they have it, I became very interested in helping people learn about glaucoma and its risk factors."[46] These methods for the marketing of Travatan are not unique to "race-based" pharmaceuticals. On the contrary, strategies for the commercial marketing of pharmaceuticals that were developed in the last decade of the twentieth century stressed the importance of developing disease awareness in a biosocial community, using disease awareness campaigns, setting up or funding disease support groups, linking up with celebrities, funding conferences, and other measures that would generate demand for the product. What one appears to be seeing in the case of Travatan, then, is merely the operation of this familiar strategy, however problematic it may be in other terms, rather than a set of practices specific to race. The "racialization" of certain conditions and their treatments may well be a consequence.

Another "racial" drug attracted more controversy, this time in relation to heart failure. It was widely claimed that, in the United States, blacks died from heart failure at a rate twice that of whites. This claim was key to the A-HeFT trial of treatments for African American Heart Failure, which was supported by the Association of Black Cardiologists. This trial led to the promotion of a particular drug, with the trade name of BiDil, which the manufacturers NitroMed claimed improved survival specifically in African Americans with heart failure. Many were critical of this development. Initially some argued that this "finding" had emerged simply by reanalyzing data from a trial that was not initially conceived in these terms. It was suggested that the researchers had tried out any number of ways of subdividing the trial population in the search for statistically significant differences, hoping to find anything that could be used to "save" a drug that did not show efficacy in the whole clinical trial population (Cooper et al. 2003). Others, suspicious of the very idea of race-targeted medicine, sought to undercut it by arguing that the figure on which it was based was flawed. Jonathan Kahn argued that the use of this erroneous figure was part of an attempt to deny that any differences that do exist in health between groups are socioeconomic and environmental. A focus on genetics, he argued, diverted attention from the social to the molecular and individual level, led to a misallocation of intellectual and material resources, and provided "fodder for those who argue that new links between race and biology must be recognized and acted upon" (Kahn 2003: 481). He documented the complex legal, commercial, and medical contingencies and negotiations though which the drug "became ethnic," suggesting, *inter alia*, that this was a strategic move by NitroMed to make themselves more attractive to investors, capitalizing by claiming to tailor therapies to the genetic profiles of ethically identified consumers, and enabling them to repatent the drug for African Americans in 2000, thus extending its patent life: "corporations," he suggested, "are using biology as a surrogate to get at race in drug marketing" (Kahn 2004: 28). Following a familiar line of critique, he suggested that to the extent that race becomes understood as a natural biological category, the door was opened for new forms of race-based discrimination, with the possibility of screening programs to identify and exclude individuals from particular forms of employment or insurance. From such a perspective, any association of race and biology, even in the new world of probabilities, and the new space of genomic activism and biological citizenship, is inescapably linked to a form of power that is negative, exclusionary, and malign.

However, the contemporary biopolitics of racialized medicine can be understood differently. As with Travatan, BiDil promotes itself by allying itself with the demands and concerns of a specific biosocial group, gaining the support of relevant community bodies and committed physicians, and claiming to address a neglected problem of an unempowered community. Indeed, subsequent trials of BiDil addressed the statistical uncertainties of the earlier research: the Phase III trial was stopped in July 2004 because of the assessment that the size and consistency of the benefits for African American patients made it unethical to deny the drugs to those randomized to the placebo arm. Once more, this development has generated controversy.[47] As the trial included only African Americans, should it then be licensed by the FDA only for this ethnic group? Is race, here, only an interim solution, until the underlying genetic sequences determining response are identified? Or is race as legitimate an initial indicator for clinicians, and for those seeking licensing of a drug, as age, gender, or body weight? A body of empirical research is growing on the complex relations between these developments in drug marketing and lay perceptions in the United States (Bates et al. 2004, Bevan et al. 2003, Condit et al. 2003). But, as already mentioned, the belief that drugs act differently in different population groups is already widely accepted in many countries in Asia. In China, for example, all drugs developed, trialed, and licensed on Caucasian populations seeking a license for importation have to undergo a further set of trials on Chinese populations to demonstrate safety and efficacy, no doubt also with an eye to the economic and other competitive consequences of such imports for the indigenous pharmaceutical industry (China Concept Consulting & the Information Centre of the State Drug Agency 2000). Of course, pharmaceutical companies, operating in market based economies, are concerned with market share and will use a variety of means to maximize this, including references to race or ethnicity. But the reflex suspicion of many critics does not seem to be helpful in understanding these new relations of genomics, identity, biosociality, and bioeconomics.

CHAPTER SIX

Conclusions: Race and Contemporary Biopower

Let me return to the special issue of *Nature Genetics* that I discussed earlier in this chapter. In the face of the controversies that I have examined, the U.S. Department of Energy, funder of the Human Genome Project, set up its own Web resources specifically devoted to minorities, race, and genomics.[48] And in 2004, together with the National Institutes of Health and Howard University, they sponsored the *Nature Genetics* supplement, "Genetics for the Human Race." This supplement originated from a May 2003 workshop held by the National Human Genome Center at Howard University in Washington D.C. The workshop was entitled "Human Genome Variation and 'Race,' " and both it and the special issue were proposed by researchers at Howard University. Represented at the workshop were most of the diverse actors that I have discussed in this chapter. Francis Collins, addressed directly the dispute between those who think race and ethnicity are significant in genomic biomedicine, and those who do not, and concluded:

> After reviewing these arguments and listening to the debate during the meeting at Howard University, one could conclude that both points are correct. The relationship between self-identified race or ethnicity and disease risk can be depicted as a series of surrogate relationships. . . . On the nongenetic side of this diagram, race carries with it certain social, cultural, educational and economic variables, all of which can influence disease risk. On the genetic side of the diagram, race is an imperfect surrogate for ancestral geographic origin, which in turn is a surrogate for genetic variation across an individual's genome. Likewise, genome-wide variation correlates, albeit with far-from-perfect accuracy, with variation at specific loci associated with disease. Those variants interact with multiple environmental variables, with the ultimate outcome being health or disease. . . . it is apparent why self-identified race or ethnicity might be correlated with health status, through genetic or nongenetic surrogate relationships or a combination of the two. It is also evident that a true understanding of disease risk requires us to go well beyond these weak and imperfect proxy relationships. And if we are not satisfied with the use of imperfect surrogates in trying to understand hereditary causes, then we should not be satisfied with them as measures of environmental causation either. (Collins 2004)

Perhaps we can agree, then, with the judicious assessment made by Charmaine Royal and Georgia Dunston, of the National Human Genome Center of Howard University: "there seems to be consensus that "race," whether imposed or self-identified, is a weak surrogate for various genetic and nongenetic factors in correlations with health status. We are at the beginning of a new era in molecular medicine. It remains to be determined how increasing knowledge of genetic variation in populations will change prevailing paradigms of human health and identity" (Royal and Dunston 2004: 57). In our present configuration of knowledge, power, and subjectivity, what is at stake in these arguments about human genome variations among populations is not the resurgence of racism, the specter of stigmatization, a revived biological reductionism, or the legitimation of discrimination: it is the changing ways in which we are coming to understand individual and collective human identities in the age of genomic medicine and the implications of these for how we, individually and collectively, govern our differences.

The new rationalities and technologies of power over life that have taken shape since the eighteenth century have not had domination and subordination as their universal principle, though they have certainly lent themselves to such repressive strategies. And while race has been a key element in the formation of modern powers over life, it has done so in a polyvalent manner, for "life as a political object" can and has been turned back against the controls exercised over it, in the name of claims to a "right" to life, to one's body, to health, and to the satisfaction of one's need for medical care (Foucault 1978: 145).

I think, therefore, that we need to locate the politics of racialized medicine within this contemporary complex of power, authority, and subjectivity. The disputes I have outlined exemplify clearly the centrality of the "making live" aspect of biopower to our present, its mobilization by nonstate authorities, its territorialization on collectivities other than "the national population," and its relation to the reshaping of individual identities in an age where health constitutes one key target of the ethical work humans do upon themselves. As we have seen in this research on population differences, the forces gathered around the question of health have pluralized, involving a range of powerful agencies within states, such as National Institutes of Health, and a range of transnational bodies such as the World Health Organization. Population differences have been marked out as potentially significant within the new circuits of bioeconomics that have taken shape, and have become strategically important within the large-scale capitalization of bioscience and mobilization of its elements into new exchange relations. The new molecular knowledges of human difference are being mapped out, developed, and exploited by a range of commercial enterprises, sometimes in alliance with states, sometimes autonomous from them, establishing constitutive links between human differentiation and biovalue. New kinds of biosocial associations and communities increasingly define their citizenship in terms of their rights (and obligations) to life, health, and cure and these active biological citizens demand that the particularities of their conditions be given weight in genomic biomedical research and the development of therapeutics. Around and within this assemblage at all levels, a host of bioethicists and critical social scientists—a whole "bioethical complex"—has taken shape encircling, publicizing, and intervening in the powers of doctors, researchers, corporations, and politicians and other authorities as never before. It is within this field of biosocialities and biological citizenship, of probabilities and uncertainties, and of the difficult ethical choices now facing individuals and communities as biology moves from fate to enter the field of the obligations of choice, that we need locate the molecular biopolitics of race in the age of genomic medicine and somatic identity.

Chapter 7
Neurochemical Selves

In 2001 the American Psychiatric Association chose "Mind Meets Brain" as the title for its annual meeting in New Orleans.[1] This was at the end of the decade designated "The Decade of the Brain" by the U.S. Congress and President George H. Bush on July 17, 1990. But it was not—or not just—the brain that was at stake here, it was "ourselves," or perhaps, our "selves," thus events included "Understanding Ourselves: the Science of Cognition" in 1999 and "Discovering Ourselves: The Science of Emotion" in 1998.[2] A few years earlier, some thought that the computer would be the paradigm of mental life: as the popular science writer, Matt Ridley, put it: "within the hardware of the brain there lurks a software called the mind" (on the back cover of Rita Carter's *Mapping the Mind*, 1998). But conceptions of the relation between mind and brain have progressed in a different way. They have been shaped by developments in brain imaging, in neuroscience, in psychopharmacology, and in behavioral genetics. Many have discussed the ways in which advances in the life sciences and biotechnology are increasing human beings' capacities to transform their life processes. But what happens when it is the self itself that is subject to transformation by biomedical technology. When cognition, emotion, volition, mood, and desire are themselves opened up to intervention?

Human beings, characteristically try to reform and improve themselves. Inescapably, at any historical moment, they do so in terms of knowledges and beliefs about the kinds of creatures that they are. Over the first sixty years or so of the twentieth century, human beings—at least in the advanced industrial and liberal democratic societies of the West—came to understand themselves as inhabited by a deep interior psychological space. They evaluated and acted upon themselves in terms of this belief. We need only to think of the rise of a psychological language of self description: the language of anxiety, depression, trauma, extroversion, and introversion. Or the use of psychological tests of intelligence and personality from vocational guidance to military promotion. Or the rise of psy technologies for marketing commodities. Or the proliferation of psychotherapies. But over the past half century, we human beings have become somatic individuals, people who increasingly come to understand ourselves, speak about ourselves, and act upon ourselves—and others—as beings shaped by our biology. And this somaticization is beginning to extend to the way in which we understand variations in our thoughts, wishes, emotions and behavior, that is to say, to our minds. While our desires, moods, and discontents might previously have been mapped onto a psychological space, they are now mapped upon the body itself, or one particular organ of the body—the brain. And this brain is itself understood in a particular register. In significant ways, I suggest, we have become "neurochemical selves."

Beyond Descartes

Over the last fifty years of the twentieth century, psychiatry gradually mapped out what it considered to be the neuronal and neurochemical bases of human mental life.[3] If one had to identify a turning point, perhaps it would be the decade from the mid-1980s to the mid-1990s. Let me take one text as an initial research site to illustrate what I mean. It is the second edition of *Biological Psychiatry*, by Michael R. Trimble, published in 1996 (Trimble 1996). The preface to the first edition, written in 1987, is included, and it makes interesting reading. In this first preface, which is five pages long, Trimble feels that he needs to justify the very existence and scientificity of biological psychiatry, which he does by referring to what he terms "the explosion of knowledge" in the previous thirty years. That is why he says he is going to start his book with a chapter setting out the "precursors" to today's biological psychiatry. I am reminded of a phrase of Hermann Ebbinghaus quoted by Edwin Boring in his attempt to establish the credentials of psychology as an experimental science: "psychology has a long past but only a short history" (Boring 1929: vii). Biological psychiatry too, it seems, has a long past—which establishes its respectability—but a short history—which establishes its scientificity. The long past runs from Hippocratic notions of disorders of the bile. It passes through the foundation of neurology by Thomas Willis in the seventeenth century: Willis thought that problems arose from problems with the blood or digestive system affecting the animal sprits that circulated through the cortex and out into the peripheral nerves. It flows through phrenology, brain localization, the German tradition of Griesinger and his successors, and the discovery of the origins of General Paralysis of the Insane (GPI) in the syphilis spirochete in the brain. This takes us to the mid-twentieth century, and the short history.

The short history begins with the shift from theories to hypotheses capable of testing by experiments, which allowed an accumulation of "findings." Progress was delayed, it seems, because of the persistent dominance of the neoromantic tide that had overtaken psychiatry at the start of the twentieth century and culminated in the psychoanalytic movement. But, as Trimble puts it, "psychological theories of pathogenesis have been overtaken by a wealth of neurochemical and neuropathological hypotheses and findings, especially with regard to the major psychoses." (Trimble 1996:x) Clinical observations on the effective treatment of patients with biological remedies—notably the results of giving chlorpromazine to psychotic patients—together with the psychopharmacological discoveries of the 1950s, "have given us not only a completely new image of the brain to work with, but also have allowed a more complete understanding of underlying functional and structural changes of the brain that accompany psychiatric illness" (x). "Functional" used to be a term that was used to describe a condition not related to a brain pathology—it was a disorder of psychological function rather than organic structure. But now, it seems, functional is to be used in a new sense; not as "an epithet for 'psychological' " but to designate a physiological disturbance. Trimble puts it bluntly: "with our present knowledge the distinction between "organic" and "functional" melts away, stripped of its Cartesian dualism" (x).

In 1987 Trimble had thought that the very idea of biological psychiatry might be regarded skeptically, or criticized or ignored because it rested on a complicated knowledge base, because of its bewildering pace of discovery, because of the residue of old critiques that the biological treatment of patients by doctors meant a lack of interest in the person who was the patient. But less than a decade later, the preface to his second edition contained no apologia and expressed no such concerns. Progress had been "truly remarkable," "awesome," and the ability of the new methods "to answer for us questions of principle relevance to the discipline would have left our forefathers breathless" (xv). As the present was reconfigured, its past was rewritten: "psychiatry is, and has always been, concerned with behavior in its widest sense, and has continually searched for knowledge of brain-behavior relationships and the somatic underpinnings of psychopathology" (19). And this claim about the somatic basis of psychopathology was supported by a whole apparatus of truth: Trimble's references, numbering around 1,200, go from Abdulla, Y. H., and Hamadah, K. (1970), 3,5 cyclic adenosine monophosphate in depression and mania, *Lancet*, I, 378–381 to Zipursky, R. B., Lim, K. O, Sullivan E. V., et al. (1992), Widespread cerebral grey matter volume deficits in schizophrenia, *Archives of General Psychiatry*, 49, 195–205 almost sixty pages later. This is a truth discourse with a vengeance, made possible by the growth of a significant truth community of researchers from different clinical and research institutions, with different specialities—medicine, computing, DNA—and with funding from a variety of public and private bodies. And the new truths of ourselves arise, not from philosophy, it seems, but from research: the research papers report the results of clinical trials, neurological research, and animal experimentation conducted in a range of spatial niches—usually laboratories, hospitals, or clinics. And the papers themselves are populated by some characteristic entities.

There are *brains* themselves. These psychiatric brains are imagined as "organs" like other organs of the body—with their regions and their constituents: the prefrontal cortex, the basal ganglia, the temporal lobes, the limbic system, the neurons in the cingulate cortex—and also some associated body systems such as cerebro-spinal fluid (CSF), or systems thought to behave in analogous ways to the brain, such as blood platelets. There are *brain chemicals*: 3,5 cyclic adenosine monophosphate, monoamine oxidase, acetyl choline, GABA, glycine, catecholamines (noradrenaline and dopamine), serotonin and its precursors and breakdown products, peptides (including endorphins, angiotensin, and the so-called releasing factors), together with their various precursors and breakdown products. Most of these are treated as known entities whose chemical composition can be written down or drawn in a diagram, although many of them were, not long ago, hypothetical entities whose existence at all, or whose existence in the brain, was in dispute. They have now become facts, things that can be invoked in accounts of other things. There are *brain functions* that often fuse cellular structures and molecular events: receptor sites, membrane potentials, ion channels, synaptic vesicles and their migration, docking and discharge, receptor regulation, receptor blockade, receptor binding, the metabolism of monoamines and glucose. At one point these too were hypothetical, tentative epistemic objects—now, however, they are accepted entities and processes. There are *drugs*, that is to say, fabricated chemical compounds whose molecular structure is known and that are believed to affect or mimic some or all of the above: tricyclic antidepressants, carbamazepine, fluphenazines, imipramine, mianserin, lithium, and all the rest of the contemporary psychiatric pharmacopoeia.

In addition, there are the *experimental model systems* within which the work of the researcher is conducted: sometimes these are human brains, sometimes cultures of cells in vitro, either nerve cells or other cells, such as blood platelets whose properties are taken to mimic those of neurons in particular respects such as receptor mechanisms, and sometimes animal models such as mice. There are the specific *investigative techniques*: biochemical assays of body fluids such as plasma, saliva, CSF and urine, EEG, studies of levels of binding of radioactive tracers, imaging techniques such as PET scans, which purport to show brain activity in terms of blood flow or energy utilization, plus the testing of patients on various tests and scales such as indices of depression or inventories of personality. There are *diagnoses*, apparently categorical ways of characterizing disorders of mood, emotion, cognition, and will, that enable the selection and differentiation of groups of subjects: depression, chronic schizophrenia, Alzheimer's disease, Parkinson's disease, and personality disorder. And thus they are also the *human subjects*. These subjects are sometimes included as participants ingesting the drugs, having their brains scanned, donating their tissues after death. They are sometimes invoked as potential beneficiaries of the results of the experiments or advances in knowledges some way down the line; they have an important place at the start of the research, as they are often referred to in order to obtain funding, and often appear again in the concluding passages of the research papers as justifications for the whole enterprise. Whatever their own experience of their troubles, these subjects need to be held constant, a feat achieved by identifying them with a standardized diagnosis in the Diagnostic and Statistical Manual of the American Psychiatric Association. Subjects thus embody or suffer from particular and discrete conditions, which account for behavioral symptoms and pathological conduct (low mood, excessive eating, panic, impulsive behavior, suicidal ideation or action) each of which is at least a candidate for a unique pathology.

Finally, in this assemblage, there are the general *truth technologies* that define and delimit how one can produce findings in psychiatry that can operate on the dimension of confirmation or disconfirmation, truth and error, that carry the effect of positivity. "Thinking in cases," to use John Forrester's term, was crucial for psychiatry as it became clinical in the nineteenth century (Forrester 1996). In the immediate postwar period, the early papers on candidate psychiatric drugs—for example in chlorpromazine and imipramine—deployed clinical observations of effects on patients. But now the efficacy of drugs was largely ascertained through the use of double-blind, randomized placebo controlled trials—or RCTs. As Healy has argued, the "various components of the randomized trial—randomization, placebo control and blind assessment—evolved separately and came together in the 1950s," greatly stimulated by the role of the U.S. Food and Drug Administration in adopting this as the standard for evaluating and licensing drugs (Healy 1997: 78). But the RCT for a psychiatric drug is only one potential terminal stage of a cascade of other truth technologies. There are those of animal experimentation. There is clinical research involving correlations of psychiatric observations and diagnoses with assays of body fluids, cultures of tissues, scans of brain functions, DNA sequencing. There is the combination of genealogical, clinical, experimental, and statistical techniques characteristic of research in behavioral genetics—twin studies, association studies, linkage studies, and the like. Battles over truth are often fought by means of conflicts over the relative veracity of the findings produced through each. But facts, observations, and explanations are only candidates for truth or falsity if their truthfulness is potentially ascertainable by one such authorized means.

The new style of thought in biological psychiatry not only establishes what counts as an explanation, it establishes what there is to explain. The deep psychological space that opened in the twentieth century has flattened out. In this new account of personhood, psychiatry no longer distinguishes between organic and functional disorders. It no longer concerns itself with the mind or the psyche. Mind is simply what the brain, does. And mental pathology is simply the behavioral consequence of an identifiable, and potentially correctable, error or anomaly in some of those elements now identified as aspects of that organic brain. This is a shift in human ontology—in the kinds of persons we take ourselves to be. It entails a new way of seeing, judging, and acting upon human normality and abnormality. It enables us to be governed in new ways. And it enables us to govern ourselves differently.

Visualizing the Abnormal Mind

If madness and sanity exist, how shall we know them? This was the question posed in David Rosenhan's classic paper, "On being sane in insane places" (Rosenhan 1973). How can the psychiatrist be sure that the person they encounter is suffering from a mental disorder, let alone what kind of disorder it was?

The reformed asylums of the nineteenth century answered this question through vision. They opened the inmates to a new form of visibility. There were many attempts to classify them through their appearance, notably that were compiled in the Atlas published by Esquirol in 1838 (Esquirol 1838) . Each diagnostic category was accompanied by a description of a typical case that focused on the physical appearance and bodily comportment of the afflicted individual. The description was accompanied by an account of an actual case. And an illustration—a line drawing—was supplied. As Sander Gilman has shown, in his exemplary analysis in *Seeing the Insane*, this style of representation linked the image of the mad person to their biography and made the varieties of insanity visible in terms of posture, gesture, skin color (Gilman 1982). For these new doctors of the mad, the visual image of the mad person was fused to the biography in the form of the case, and inscribed at the heart of psychiatric epistemology and diagnostic practice.

Clinical medicine was born in the nineteenth century. The gaze of the doctor plunged into the interior of the body. This clinical gaze linked symptoms visible on the surface of the body with the organic locations and lesions within. To diagnose an illness was to interpret symptoms in terms of the internal organic malfunctions that were their cause. In life, the gaze of clinical medicine was augmented by such devices as the stethoscope.[4] In death, it was confirmed by the dissection of the corpse. This new way of seeing was inscribed into the anatomical atlas—it was the body itself that had become ill.[5] As Michel Foucault put it: "The gaze plunges into the space that it has given itself the task of traversing. . . . In the anatomo-clinical experience, the medical eye must see the illness spread before it . . . as it penetrates into the body, as it advances into its bulk, as it circumvents or lifts its masses, as it descends into its depths. Disease is no longer a bundle of characters disseminated here and there over the surface of the body and linked together by statistically observable concomitances and successions . . . it is the body itself that has become ill" (Foucault 1973: 136).

The gaze of nineteenth-century mental medicine tried to acquire this depth but failed.[6] It remained focused on the surface of the body—posture, gaze, the color of the skin of the melancholic, the gestures of the maniac, the movements of the hysteric. It was instructed by images that remained more or less the same from the drawings in Esquirol's *Atlas* of 1838 to the carefully staged photographs that appeared four decades later in Bucknill and Tuke's *Manual of Psychological Medicine*. Thus the frontispiece to the third edition of Bucknill and Tuke's text, published in 1874, shows seven portraits (photographs taken at the Devon County Lunatic Asylum) illustrating the facial features, expression, and style of dress of those unfortunates suffering from Acute Mania, Acute Suicidal Melancholia, Secondary Dementia, Primary Dementia, General Paralysis, and Monomania of Pride (Bucknill and Tuke 1874). In each case, the visual image was central to the clinical practice of individuating the pathology.

In their different ways, the two figures of psychiatric modernity, Kraepelin and Freud, each mark a move away from the eye. Each opens the interior of the patient to medical knowledge by demoting observation in favor of interpretation. Kraepelin, working in Germany at the end of the nineteenth century, illustrated his textbook liberally with illustrations, but these were just that—illustrations. Their diagnostic role had been taken over by the case history: the chronology of symptomatology, etiology, and prognosis that was the diagnostic key (Kraepelin 1903). Freud famously turned away from the methods of his teacher Charcot, with his staged demonstrations of his patients, and the photographs he published in his *Iconographie* (Didi-Huberman 1982, Didi-Huberman and Charcot 2003). Freud referred to Charcot as a *visuel*—one who sees—but Freud was no *visual*—there are no photographs or images of patients in any of the twenty-four volumes that comprise Freud's collected psychological writings in English. In psychoanalysis, and in the whole array of psychotherapies that accompanied it, the eye gave way to the ear: it was the voice of the patient that was the royal road to the unconscious. Madness, as mental illness, neurosis, and psychosis, came to be located in a psychological space—the repository of biography and experience, the origin of thoughts, beliefs, moods, and desires. As David Armstrong has suggested, what was fabricated here was a new object—the mind:

> The mind was represented to the gaze in words. Whereas under the old regime the body of the patient had to be made legible to the physician's interrogation, under the new regime the body produced its own truth, which required not legibility but encouragement. The patient had to speak, to confess, to reveal; illness was transformed from what was visible to what was heard. (Armstrong 1983: 25)

A psy-shaped space opens up, and becomes the privileged object of the psychiatric gaze: the inner space of the individual. This was not the mind as it had been in the nineteenth century—"a space of rationality coterminous with the cerebral tissues" (26)—but a "moral" space between the organic brain on the one hand and the social space of conduct on the other, a space on which the sediments of familial and human relations were superimposed or inscribed, perhaps even those of collective existence in society. This space could not be seen, it could only be interpreted by analysts or imagined by poets or artists.

Across the first half of the twentieth century and into the 1960s, the interpretation of this psy-shaped space within the patient came to delimit the domain of psychiatry. Many of the troubles of psychiatry in the 1950s and 1960s flowed from its inability to demonstrate organic correlates of its diagnoses. This problem arose in criminal responsibility in the courts of law: psychiatrists often found that their own diagnoses of madness were unable to meet legal criteria of evidence and proof. It arose in cultural disputes, where critics argued that psychiatric diagnoses merely medicalized deviance and upheld the norms of a patriarchal social order. It arose in political controversies, where Western psychiatrists accused their brethren in the Soviet Union and elsewhere of political abuse of psychiatric judgments. At the most, in the antipsychiatry of the 1970s, psychiatric diagnoses were a "category mistake" illegitimately translating difference, disruption, and deviance into disease. Georges Canguilhem captured this dilemma in a prescient remark in his 1956 lecture at the Sorbonne (Canguilhem 1980). Like the Sorbonne itself, he suggested, the psy sciences stood between two paths. The uphill path led to the Pantheon where the wise and the great were commemorated. The downhill path led to the police station.

Of course, even nineteenth-century proponents of moral medicine like Esquirol thought madness was a disease of the brain. In Germany, Austria, and Switzerland in particular, neurologists tried to make the pathological interior of the subject of psychiatry visible. Like Broca, Fleschsig, and Wernicke, many psychiatrists pored over their microscopes. They scrutinized sections of brain from psychiatric patients, in a search for pathological lesions in the cerebral cortex, the frontal lobes, and the neurons. But despite dissecting innumerable brains of deceased asylum inmates, nineteenth-century psychiatrists were unable to find abnormalities in the brain that correlated with abnormalities of thought, conduct, mood, or will. These attempts to visualize psychiatric disorders in terms of lesions in the brains of dead patients—extracted, sliced, stained, and magnified—were much mocked by psychoanalysts in the early twentieth century. And they had little to offer psychiatrists: they had no therapeutic implications, and could only be used after the patient had died.

Developments in neuroanatomy proved more fruitful, usually extending experimental findings from animals to humans: in the process, the brain began to appear as an organ with localized regions for particular mental functions. As early as 1891, Burkhardt attempted to calm agitated and hallucinating patients by destroying the strip of the cerebral cortex between the areas of the brain thought to control sensory functions and those thought to control motor functions. In the 1930s Egas Moniz developed the technique of frontal lobotomy. Moniz was impressed by a paper by Jacobsen, who was investigating the effects of destruction of the prefrontal brain area of monkeys and chimpanzees on problem-solving. He mentioned in passing that the more excitable animals became much calmer following the surgery. Moniz decided to try it on humans and in 1936 reported the results for twenty patients on whom he had operated to destroy a portion of the frontal lobes, first by injection of alcohol, and later by cutting nerve fibers with a surgical leucotome (Moniz 1948, 1954, Valenstein 1986). These methods were taken up by Walter Freeman and James Watts in the United States, who invented a technique to enter the brain through the eye socket using an instrument modeled on the ice pick (Freeman and Watts 1942, 1950). By 1948, lobotomies had been performed on about 20,000 patients worldwide. In 1949 Moniz was awarded the Nobel Prize in physiology for this work. Of course this work is now, rightly, highly controversial—many if not most of those who received the treatment suffered irreparable loss of mental function. Nonetheless, these developments were linked to a new way of visualizing the brain as a differentiated organ traversed by localized neural pathways with specific mental functions amenable to localized intervention.

But could madness be seen in the living brain? Electron microscopes were invented in the 1930s: first transmission and then scanning. Brain tissue could now be imaged at resolutions 1,000 times greater than those possible with visual microscopy, but only postmortem. The properties of the skull ensured that the brain remained largely resistant to visualization with X Rays and other early techniques (Kevles 1997). However a series of inventions in the latter part of the twentieth century overcame this barrier: Computerized Tomography (CT) in the 1960s, Single Photon Emission Computer Tomography (SPECT), Positron Emission Tomography (PET), Magnetic Resonance Imaging (MRI) from the 1980s and then Functional Magnetic Resonance Imaging (fMRI). Most accounts that make use of these technologies write as if we can now visualize the interior of the living human brain and observe its activity in real time as it thinks, perceives, emotes, and desires—we can see "mind" in the activities of the living brain. Hence, it seems, we might be able to use these images of brain activity in different regions to make objective distinctions between normal and pathological functioning. Such claims led to huge investments in scanning apparatus in medical and research facilities, and to the increasingly routine use of such scans in medical procedures and research. Of course, the reality is much more complicated than its popular representation: brain scans produce digital data that is mapped pixel by pixel into a standard representation of brain space to produce these simulacra of the "real brain" (Beaulieu 2000, Dumit 1997, 2003). However the significance of this apparent verisimilitude of this mode of visualizing mind has not simply been rhetorical, or even clinical. It has been epistemological. The visualized living brain now appeared to be just one more organ of the body to be opened up to the eye of the doctor. A rapid tour around the World Wide Web at the start of the twenty-first century revealed what some thought they saw.

Take, for example, an organization called Brainviews, Ltd., which claims to provide state-of-the-art software for the study of brain and behavior and from whom you can buy a CD of an "animated brain" for under fifty dollars appropriate for "Medical, dental, nursing, neurobiology, biology, psychology, paramedical, physical, occupational and rehabilitation therapy, pharmaceutical and other medically related sales and training groups." The PET scans they presented appeared to show cycles of activity in the brain that were symptoms of bipolar illness—such illness now appeared to be incontrovertible, visible in the colors of the affected organ: "These brain scans show the metabolic activity that can be generated in the brain during hypomanic or manic phases of bipolar affective disorder. Red indicates the highest levels of metabolism followed by yellow, green and blue."[7] And images like this are not merely for education—they appear to have clinical utility. Thus, according to a psychiatrist working at the Amen Clinic for Behavioral Medicine: "SPECT can be helpful in the diagnosis and treatment in complex or resistant depressive disorders by differentiating it from other disorders, enhancing compliance by the patient being able to "see the changes in the brain," and by subtyping depression." SPECT can be helpful in the diagnosis and treatment in complex or resistant depressive disorders by differentiating it from other disorders, enhancing compliance by the patient being able to "see the changes in the brain," and by subtyping depression.[8]

The scans could thus be used to convince a reluctant patient that she was "really" ill—which must mean, it seems, that the organ of the brain itself must be abnormal:

> Leigh Anne came to see me fifteen months after the birth of her first child. . . . After diagnosing her with major depression I placed her on Prozac and began seeing her in psychotherapy. Her symptoms remitted after only several weeks. After several months Leigh Anne discontinued treatment. She associated taking Prozac with a course of action for "a depressed person." She did not want to see herself in that light or be stigmatized with that label. For several months after stopping she had no adverse reaction. Then the symptoms returned. . . . When she came to see me again Leigh Anne still didn't want to believe that anything was "wrong" with her, so she was still resistant to going back on medication. After I ordered a brain study to evaluate her deep limbic system, I was able to point out to her the marked increase in activity in that area of her brain. It provided me with the evidence needed to convince her to go back on Prozac for a while longer.[9]

Desire, too, was now a state of the brain with a visual form. For example, the craving of the cocaine addicts is visible in the brain: "Anna Rose Childress, Ph.D. . . . has used positron emission tomography (PET) scans to see the system of the brain that is involved when craving occurs in cocaine abusers. . . . She has found that particular areas of the limbic system—the part of the brain, including the hypothalamus and the amygdala, that is linked to emotion and motivation—"light up" during scans, showing these areas are being activated."[10] Interpretation, here, has fused with the image itself: the inescapably artful process of production of such pictures has disappeared, inscribed in the very architecture and algorithms of the computer programs that generate them.

These popular versions retell, in simplified terms, a way of seeing that is reshaping the psychiatric gaze, in studies of craving (Brody et al. 2002), depression (Beauregard et al. 1998), anxiety,[11] and much more, and is increasingly clinically deployed in differential diagnosis in psychiatry (see, e.g., Gupta et al. 2004, Hughes and John 1999, Krausz et al. 1996).[12] Perhaps the science journalist Rita Carter, writing in *Mapping the Mind* almost a decade ago, only slightly overstated the shift that was underway:

> The biological basis of mental illness is now demonstrable: no one can reasonably watch the frenzied, localized activity in the brain of a person driven by some obsession, or see the dull glow of a depressed brain, and still doubt that these are physical conditions rather than some ineffable sickness of the soul. Similarly, it is now possible to locate and observe the mechanics of rage, violence and misperception, and even to detect the physical signs of complex qualities of mind like kindness, humour, heartlessness, gregariousness, altruism, mother-love and self-awareness. (Carter 1998: 6)

When mind seems visible within the brain, the space between person and organs flattens out—mind is what brain does.

Molecularizing Psychiatric Diagnosis

This flattening of the distance between conduct and its organic basis, also characterizes the new diagnostic gaze of psychiatry. A key aspect of this is a finer and finer dissection of the varieties of mental disorder. The systematic classification of types of mental illness in the United States began, not as a clinical project, but as a statistical one. The 1840 census recorded the frequency of "idiocy / insanity." The 1880 census used seven categories—mania, melancholia, monomania, paresis, dementia, dipsomania, and epilepsy. The first official nosology, adopted in 1918, was an attempt to classify the large numbers of people confined in mental hospitals. The first *Diagnostic and Statistical Manual for Mental Disorders*, published in 1952, was prepared by a Committee on Nomenclature and Statistics of the American Psychiatric Association in the wake of psychiatry's wartime experience and conceived of mental disorders as reactions of the personality to psychological, social, and biological factors (American Psychiatric Association 1952). DSM II, published in 1968, was 134 pages and had 180 categories framed in the interpretative language of psychoanalysis (American Psychiatric Association 1968). DSM III, published in 1980, ran to almost 500 pages and is often seen as a response to the crisis in legitimacy of psychiatry over the 1970s (American Psychiatric Association 1980). The revised version of 1987 had 292 categories, each defined by a set of objective "visible" criteria (American Psychiatric Association 1987). Ideally, each of these categories was a distinct disorder, with a unique etiology and prognosis, amenable to a specific kind of treatment. DSM IV, published in 1994, which runs to 886 pages and classifies some 350 distinct disorders, from Acute Stress Disorder to Voyeurism (American Psychiatric Association 1994).

DSM IV cautions that individuals within any diagnostic group are heterogeneous: its categories are only intended as aids to clinical judgment. But it promotes an idea of specificity in diagnosis that is linked to a conception of specificity in underlying pathology. The broad categories of the start of the twentieth century—depression, schizophrenia, neurosis—are no longer adequate. Pathologies of mood, cognition, will, or affect are dissected at a different scale. The psychiatric gaze is no longer molar but molecular. And behind this molecular classification of disorders lies another image of the brain—that of contemporary neuroscience—and of therapeutic intervention—that of psychopharmacology. Initially it had been thought that although nerves themselves transmitted signals by chemical means, transmission between nerves, across the synapse, was electrical. However over the 1950s and 1960s it became accepted that the process was chemical and involved small molecules known as neurotransmitters, initially the monoamines dopamine, norepinephrine, epinephrine, acetylcholine, and serotonin. This new image of neurotransmission was intrinsically tied to research on psychiatric drugs.

The drugs administered to those thought mentally ill, up until the early 1950s, had been construed as rather general in their action, for example as sedatives or stimulants. But the "modern" psychiatric drugs that began to be developed from the 1950s onward—chlorpromazine for psychosis, reserpine for hypertension, the benzodiazepines for anxiety, imipramine for depression, and so forth—were all hypothesized to have specific effects on mood, thought, or conduct because of their specific action on the brain (Healy 1997, 2001). By 1955, technological developments such as the invention of the spectrophotofluorimeter enabled "the detection of variations in the levels of chemicals known to exist in the brain in minute amounts" and allowed Steve Brodie to correlate the administration of reserpine to rabbits with a fall in the levels of serotonin in their brain (2001: 106). As Healy puts it, "For the first time a bridge had been built between behavior and neurochemistry" (106). Across this bridge was to flow an accelerating stream of traffic between the clinic, the laboratory, and the factory. Each variety of disorder was soon assigned to an anomaly in a particular neurotransmitter system, and intensive research in the laboratories of universities and pharmaceutical companies sought to isolate the compounds whose specific molecular structure would enable them to target, modify, or rectify that anomaly. This research embodied and supported a dream of molecular specificity that, for fifty years, would weld psychiatric etiology to psychopharmacology, and underpin the creation and marketing of all modern psychiatric drugs claiming to be therapies for particular diagnoses.

These discoveries were implicated in that reshaping of the technologies of psychiatric truth exemplified in Trimble's textbook with which I began. A whole experimental system was brought into being for the development of drugs and the testing of hypotheses, involving animal models, notably rats, in-vitro systems, and complex new types of apparatus. New entities were brought into existence—receptor sites, membrane potentials, ion channels, synaptic vesicles and their migration, docking and discharge, receptor regulation, receptor blockade, receptor binding. These were first hypothetical, then demonstrated in the lab, then experimental entities in their own right, finally as realities that seem to be rendered visible by PET scans or other visualization techniques. New forces became involved, not just teams of research scientists, a whole host of new specialist journals, and the like, but also big grant-awarding agencies and, of course, the pharmaceutical companies. For these processes were not merely processes of discovery, but of intervention—the neurochemical brain becomes known in the very same process that creates interventions to manipulate its functioning.

By the start of the 1990s, these molecular brain processes had become accepted enough to be represented in conventionalized visual simulations. The best of these, such as *Essential Psychopharmacology* by Stephen Stahl, are replete with images simulating the neuronal processes underlying different pathologies (Stahl 1996). In Stahl's iconography, each neurotransmitter is allotted a distinct icon—norepinephrine, for example is a triangle—and each receptor is illustrated with a reciprocal icon—in this case a rectangle with a triangle shape cut out—into which the neurotransmitter fits like a key into a lock. Such images seem able to represent the normal state of a monoaminergic neuron as it releases its neurotransmitter—norepinephrine—at the "normal" rate. They seem to show the normal state of the functioning of the enzyme monoamine oxidase, which destroys norepinephrine, the "norepinephrine reuptake pump," which terminates the action of norepinephrine, and the norepinephrine receptors, which react to the release of norepinephrine. The magnification can be increased, for example, to illustrate the operation of an ion channel, and the ways in which it can be opened by an agonist or closed by an antagonist. These powerful and compelling simulations combine a matter-of-fact materiality, iconic character, and languid truthfulness.[13] They also diagram a molecular specificity of different types of disorder. Once imagined in this way, pathologies can be illustrated visually as variations form this "normal" state. Thus depression can be visualized as a depletion of neurotransmitters—a visual version of the famously oversimplified "serotonin hypothesis of depression" that understands depressed mood in terms of an abnormally low concentration of this neurotransmitter in the synapses in particular regions of the brain. Schizophrenia can be imagined in terms of the overactivity of dopamine neurons in the mesolimbic dopamine pathway, which may be illustrated with a moving image of a synapse with dopamine molecules jumping across it—clearly an excess of them in the "schizophrenic brain."

Or the analysis can be more complex but in the same thought style. For example, a specific variety of mental disorder might be explained in terms of an impairment in the ability of one particular subtype of dopamine receptor to modulate N-methyl-D-aspartate (NMDA) receptors. Thus, for example, Richard C. Deth, a professor of Pharmaceutical Science at Northeastern University, writing in *Scientific American* On-Line, provided this answer to the question, "What happens to the body and brain of individuals with schizophrenia?"

> Our research found that dopamine can stimulate the methylation of membrane phospholipids via the activation of the D4 dopamine receptor. Furthermore, we found that the D4 receptor is complexed to NMDA receptors, suggesting that the methylation of phospholipids could regulate NMDA receptor activity. Only in man and primates does the D4 receptor possesses a repeat structural feature, facilitating its complexation with synaptic NMDA receptors. This finding suggests that schizophrenia may result from an impairment in the ability of D4 dopamine receptors to modulate NMDA receptors at nerve synapses via phospholipid methylation. This modulation may be important for the normal attention and cognitive abilities of man.[14]

It is now in these terms that the action of psychiatric drugs is imagined—in terms of their capacity to block a particular reuptake pump, or binding onto particular types of receptors. And it is in these terms that they are invented, trialed, marketed, and prescribed. Prozac did not become an iconic drug because it was more effective than previous antidepressants. Its status was based on its claim to be the first drug whose molecule had been deliberately fabricated to disrupt one, and only one, aspect of a single neurotransmitter system. Fluoxetine hydrochloride claimed to increase the level of available serotonin in the synapses. It did this, it seemed, by disrupting the reuptake of serotonin because it locked into the precise sites on the neuronal membrane—the receptors—through which reuptake would otherwise have occurred. Hence all the hype about Prozac being a smart drug, a clean drug. It was its specificity that supposedly accounted for the fact that it did not produce all those unwanted effects associated with the earlier antipsychotics, derived by a combination of chance and clinical observation. The shuffling gait, dry mouth, tremors, and the like were produced by the "dirty" antidepressants because they hit a number of sites in a number of systems at the same time. But it appeared, that era was at an end—molecular disorders, from now on, would receive targeted molecular treatments.

By the start of the twenty-first century, it was in these terms that mental disorders and drug action were explained to potential patients. For example, a visit to the Prozac website in 2002 revealed the following:

> Depression is not fully understood, but a growing amount of evidence supports the view that people with depression have an imbalance of the brain's neurotransmitters, the chemicals that allow nerve cells in the brain to communicate with each other. Many scientists believe that an imbalance in serotonin, one of these neurotransmitters, may be an important factor in the development and severity of depression. . . . Prozac may help to correct this imbalance by increasing the brain's own supply of serotonin. . . . Some other antidepressant medicines appear to affect several neurotransmitters in addition to serotonin. Because Prozac selectively affects only serotonin, it may cause fewer side effects than other types of medications. . . . While Prozac cannot be said to "cure" depression, it does help to control the symptoms of depression, allowing many people with depression to feel better and return to normal functioning.[15]

By 2002, this was a wholly unremarkable text—unremarkable because it had become the routine popular account of depression and its treatment—distinctive only because of the animated illustrations of neurotransmission, with plenty of serotonin molecules jumping calmly across a synapse in "the normal brain"; a few slowly moving serotonin molecules in "the depressed brain"; and "the brain on Prozac," with the tablets blocking the reuptake sites, and hence restoring appropriate numbers of serotonin molecules to the synapse. The compelling character of the simulations disguised their wholly imaginary character, the reality effect inscribed a whole explanatory structure, diagnostic method, and treatment strategy within the images themselves.

Text itself can, of course, have the same function of truthfulness. Here is how the action of Paxil is described to physicians in the information issued by the manufacturers:

> The antidepressant action of paroxetine . . . its efficacy in the treatment of social anxiety disorder . . . obsessive compulsive disorder [OCD], and panic disorder [PD]) is presumed to be linked to potentiation of serotonergic activity in the central nervous system resulting from inhibition of neuronal reuptake of serotonin (5-hydroxy-trypt-

> amine, 5-HT). Studies of clinically relevant doses in humans have demonstrated that paroxetine blocks the uptake of serotonin into human platelets. In vitro studies in animals also suggest that paroxetine is a potent and highly selective inhibitor of neuronal serotonin reuptake and has only very weak effects on norepinephrine and dopamine neuronal reuptake. In vitro radioligand binding studies indicate that paroxetine has little affinity for muscarinic, alpha1-, alpha2-, beta-adrenergic-, dopamine (D2)-, 5-HT1-, 5-HT2- and histamine (H1)-receptors; antagonism of muscarinic, histaminergic and alpha1-adrenergic receptors has been associated with various anticholinergic, sedative and cardiovascular effects for other psychotropic drugs.[16]

We see here a whole style of molecular argumentation designed to emphasize the specificity of the neurochemical basis of the diagnosis and the mode of action of the drug. This style of thought is thus simultaneously pharmacological and commercial. This explanatory structure, with its rhetorics of specificity and selectivity, has been built into the process of fabrication of psychiatric drugs since the 1950s. It is embodied in the very idea that different drugs are antipsychotics, antidepressants, anxiolytics, and the like. Yet, as Joanna Moncrieff and David Cohen have argued, evidence for such specificity is actually extremely weak (Moncrieff and Cohen 2005). As histories of drug development show, these claims have been derived from the very ways the pharmacological research, the drug trials, the clinical observations and the licensing applications have been structured (Harris et al. 2003, Healy 1997, 2001). Indeed there is considerable evidence that so called "specific" drugs actually have a wide spectrum of action, and the claims for specificity often arise from the forms of evaluation used—for example, a focus on the reduction of a score on a scale supposed to measure depression or psychosis—or the dismissal of some phenomena as "side effects" (many of these studies are cited in Moncrieff and Cohen 2005).[17] These accounts of the molecular mechanisms underlying variations in mood, thought processes, and conduct bring us back to the question of the molecularization of diagnosis. But first, we need to explore one further dimension—that of genetics.

Decoding Fate

As is now well known, the draft sequence of the human genome published on February 11, 2001 did not produce a single master sequence, a genomic norm against which all variations could be seen as pathological mutations. Instead, there were millions of loci on the genome where individuals differed from one another by as little as a single base—an A substituted by a C, for example.[18] Every sequence identified as a "gene" now seemed to be marked by such variation—such Single Nucleotide Polymorphisms or SNPs. Only a few characteristics and disorders matched the earlier "Mendelian" ideas of genes as units of inheritance, single entities that exist in a small number of alleles, some of which are normal, others pathological. The consequences for psychiatry of this transformation in the truth discourses of biosciences have been profound.

In the nineteenth century, arguments about the inheritance of mental pathology all went along the same lines. An inherited predisposition or constitutional weakness was triggered into frank pathology by an exciting cause—loss of fortune, excessive consumption of alcohol, masturbation, and so forth. Mental disorders were placed among many manifestations of an inherited tainted constitution, whose passage down the generations could be visualized in a genealogical table whose blacked-out squares and circles—marking the affected men and women in the lineage—showed the defective germ plasm coursing down through the generations. Pedigree charts of varieties of madness, and their links with feeblemindedness, sexual immorality, criminality, and the whole catalogue of manifestations of degeneracy were common in eugenic advice up to the outbreak of World War II. Such images of inherited weakness led to guidance for those contemplating marriage, who were advised to check the family history of their prospective spouse for signs of these defects, and for those with such a history, who should think very carefully before procreating.

But in the contemporary style of biological thought, there is not a single inherited diathesis, or a constitution that can be healthy or degenerate. We all carry genomic vulnerabilities to different conditions, vulnerabilities that are small, discrete, molecular. There is still a logic of absolute genetic pathology—in disorders such as Huntington's—though even that fatal logic is not a logic of fatalism.[19] But such apparently implacable genetic pathologies are merely the extreme point of a rather different perception. This is the gaze of "susceptibility."[20] To say one is susceptible to a condition such as bipolar affective disorder or schizophrenia is not just to say, as in the past, that one is at risk because of a family history, or because epidemiological studies of populations place one in a risk category. Nor it is to say that one has "the gene for" the condition. The claim is now that mutations associated with increased susceptibility can be identified in precise loci in the base sequence of the genes that control the synthesis of the proteins involved in production and transportation of neurotransmitters, receptors, enzymes, cell membranes or ion channels regulating the activity of neurons.

Let me take an example more or less at random from a paper concerned with a subtype of schizophrenia that is associated with aggressive symptoms (Strous et al. 1997). The paper argues that this condition is linked to a variation in a single base in the gene for an enzyme called catechol-O-methyl-transferase (*COMT*). This gene has been located to Chromosome 22q11 and sequenced. The variation is a transition from G (guanine) → A (adenine) in the DNA sequence. It leads to valine being substituted for methionine in codon 158 of the enzyme. This enzyme has been shown to be involved in the breakdown of certain neurotransmitters. So an error in the makeup of the enzyme will disrupt this breakdown process, leading to more or less of the neurotransmitter in question being present in the synapses in particular regions of the brain. And the authors note other research that has claimed that impulsive behavior is correlated with anomalies in the synthesis or breakdown of these neurotransmitters. However bizarre and tendentious this argument, it is clear that our researchers are not imagining single genes and unitary pathologies. They are trying to discover variations in multiple loci in multiple gene systems resulting in continuous distributions of phenotypes and susceptibilities to particular disorders in specific environments.

In the thought styles of contemporary psychiatric and behavioral genomics, we are now in a postgenomic era of polygenetic susceptibilities (McGuffin et al. 2001, Plomin and McGuffin 2003). Thus, while the headline in *Scientific American* in March 2001 announced, "First gene for schizophrenia discovered," the actual claim was rather different—that a polymorphism in a base sequence on Chromosome 22 that coded for a protein called WKL1, thought to be involved in the production of ion channels that transport neurotransmitters into neurons, might increase the probability of an individual developing a particular subtype of schizophrenia:

> No single genetic mutation can ever account for the complex range of symptoms that arise in devastating neuropsychiatric disorders such as schizophrenia. But scientists from the Julius-Maximilians-University of Wuerzburg in Germany have zeroed in on one mutation of a gene on chromosome 22 that appears to play an important role in catatonic schizophrenia—a particularly severe form of the disease characterized by acute psychotic breaks and disturbed body movements. They will describe their finding, based on an analysis of a large pedigree, in an upcoming issue of *Molecular Psychiatry*. . . . Earlier linkage studies lead the team of geneticists, psychiatrists and neuroscientists to examine chromosome 22 more closely—and in particular, they focused on a gene encoding a protein called WKL1. This protein appears to share

> many features with ion channels, complexes that straddle a cell's membrane and help transport electric currents along neurons. (Mutations in one remotely related ion channel, the potassium channel KCNA1, cause a rare movement disorder called episodic ataxia.) Of significance, the researchers found the WKL1 gene transcript exclusively in brain tissue. Further study is needed to determine if the same WKL1 mutation occurs in other families with a history of schizophrenia, or in uninherited cases of the disease. Still, the scientists are hopeful that their discovery may help elucidate some of the biological mechanisms behind schizophrenia and ultimately one day lead to better treatment options.[21]

And biology is not destiny here—for the door apparently opens to the precision fabrication of therapeutic molecules that will target exactly this molecular anomaly, in exactly those individuals where genetic screening has shown this polymorphism to be present. Optimism abounded in the early years of the twenty-first century. Neuroscientists undertaking basic research claimed that "the discovery of novel neurotransmitters and venues of their activity afford multiple opportunities for therapeutic intervention" (Snyder and Ferris 2000: 1738) Others claimed that new understanding of the cellular neurobiology of depression would enable the development of new therapeutic agents that directly target molecules in these pathways (Manji et al. 2001a, 2001b) Psychiatric researchers expressed optimism about the implications for patients of their search for susceptibility genes and presymptomatic diagnostic tests (Plomin and McGuffin 2003), enthusiastically supported by those who have long argued that there is a genetic basis for mental disorder (Cloninger 2002), although many now sought to construct "complex" models of gene environment interaction (Caspi et al. 2002). Biotech companies, for instance Autogen and ChemGenex Pharmaceuticals based in Australia, launched a depression and anxiety gene discovery program, using animal models to try to find validated protein targets involved in the development of depression and anxiety, with an eye on the potential market for new medications.[22] This returns us to the issue of the reciprocal molecularization of treatment and classification.

In 2002 the American Psychiatric Association published a volume completing the initial phase of the planning for the next revision of the Diagnostic and Statistical Manual, DSM V, which began in 1999 (Kupfer et al. 2002). This intriguing text, which contains discussions of such issues as whether *disease, illness*, and *disorder* are scientific and biomedical concepts or sociopolitical terms that necessarily involve a value judgment, was made up of six principal chapters, each written as a "white paper" to stimulate discussion about the form that this revision should take. The second of these white papers, written by Dennis S. Charney;

David H. Barlow; Kelly Botteron; Jonathan D. Cohen; David Goldman; Raquel E. Gur; Keh-Ming Lin; Juan F. López; James H. Meador-Woodruff, Steven O. Moldin; Eric J. Nestler; Stanley J. Watson; and Steven J. Zalcman was entitled "Neuroscience Research Agenda to Guide Development of a Pathophysiologically Based Classification System" (Charney et al. 2002). These authors took a far more cautious view than some of the papers cited above. They pointed out that while there has been no shortage of neurobiological theories of causation for psychiatric disorders, psychiatry had thus far failed to identify a single neurobiological phenotypic marker or gene that is useful in making a diagnosis of a major psychiatric disorder or for predicting response to psychopharmacologic treatment. And, at the same time, they noted, DSM IV definitions are virtually devoid of biology, and are merely based on clusters of symptoms and characteristics of clinical course. They are admirably frank about the limits of what has actually been demonstrated, yet characteristically optimistic about the possibilities:

> Despite several decades of effort, no bona fide psychiatric disease gene has yet been identified with certainty, although the field is getting closer, and new advances in genetics (including the availability of the human genome sequence) portend rapid progress. Brain imaging studies in humans promise, for the first time, to provide detailed information about molecular and cellular substrates in the brain involved in a psychiatric disorder. Although currently available imaging techniques have thus far failed to provide diagnostic tests for psychotic, affective, or anxiety disorders, it is only a matter of time before such techniques have the spatial and temporal resolution and the chemical specificity to study relevant pathophysiological mechanisms. Finally, studies of brain samples obtained at autopsy should permit more detailed molecular analysis of the pathophysiology of psychiatric disorders. (34–35)

Hence their hope: to generate a research agenda that will, in the decades to come, lead to a diagnostic and classificatory system based directly on these molecular brain states:

> It is our goal to translate basic and clinical neuroscience research relating brain structure, brain function, and behavior into a classification of psychiatric disorders based on etiology and pathophysiology. It is possible, even likely, that such a classification will be radically different from the current DSM-IV approach. Prognostication is a risky business. However, we speculate that single genes will be discovered that map onto specific cognitive, emotional, and behavioral disturbances but will not correspond neatly to currently defined

> diagnostic entities. Rather, it will be discovered that specific combinations of genes will relate to constellations of abnormalities in many brain-based functions—including but not limited to the regulation of mood, anxiety, perception, learning, memory, aggression, eating, sleeping, and sexual function—that will coalesce to form disease states heretofore unrecognized. On the other hand, genes that confer resilience and protection will also be identified, and their interaction with disease-related genes will be clarified. The impact of environmental factors on gene expression and phenotype expression will be defined. The ability to discover intermediate phenotypes will be improved with advances in techniques such as neuroimaging. This will all lead to novel therapeutic targets of greater efficacy and specificity to disease states. Prediction of therapeutic response will be possible through genetic analysis and phenotype analysis. Disease prevention will become a realistic goal. (70–71)

To diagnose through the brain rather than through the visible symptoms, life changes, or course of the condition—this is the dream: a mode of diagnosis that, they admit, is very likely to radically revise existing classifications, linking disorders currently separated by superficial symptomatology, and dividing disorders illegitimately lumped together. In this style of thought, then, it is unthinkable that diagnosis—like treatment, and like the illness itself—should not be best when and if it can become a "brain" matter.

It is not surprising then, that commercial companies are already investing in a future in which diagnosis will indeed go straight to the brain. In the name of disease prevention, some hope, presymptomatic individuals will be tested for susceptibilities to particular psychiatric conditions, tested too for their likelihood to respond to certain drugs, administered the drugs on a preventive basis, and then brain scans will be used to see if those drugs are, in fact, appropriate for treating their individual condition.[23] Take, for example, the announcement by Aspect Medical Systems Inc. in May 2003, of the results of studies showing that EEG-based brain monitoring technology—which it claims is more effective than PET and MRI in capturing subtle changes in the brain over very short periods of time—predicts treatment response to antidepressant medications in depressed patients. The aim is to use brain monitoring technology to help clinicians more quickly identify optimal therapies for depressed patients:

> These milestones underscore the promise of Aspect's brain monitoring technology to help academic researchers, drug and device companies, and clinicians better utilize the biological markers to assess treatment response in neurologically-based diseases, such as depression and dementia," stated Philip Devlin, general manager of Aspect's Neurosci-

> ence division. "As a result of this research, we expect to have a greater understanding of how therapies benefit various patient populations, enabling the development of new, better targeted drugs and devices to treat the estimated nine million people worldwide who suffer from depression, and the millions more who suffer from the range of central nervous system disorders.[24]

Straight to the brain. And yet, there is something else, an assumption, a premise perhaps, in both this announcement and the arguments of those charged with revising DSM. Whether it is brain scans or genetic tests, all pathways through the brain seem to end in the use of psychopharmaceuticals. Let us return then to the drugs.

Consuming Psychopharmaceuticals

In the eugenic age, mental disorders were pathologies, a drain on a national economy. Today, they are vital opportunities for the creation of private profit and national economic growth. Indeed the profit to be made from promising effective treatment has become a prime motive in generating what counts for our knowledge of mental disorders. Over the last twenty years or so, in advanced industrial societies of Europe and North America, psychopharmacology has carved out a very significant market (Rose 2004).[25] Over the decade from 1990 to 2000, the psychiatric market increased in value by over 200 percent in South America, 137 percent in Pakistan, 50 percent in Japan, 126 percent in Europe, and a phenomenal 638 percent in the United States. At the end of the decade in the United States, sales of prescribed psychiatric drugs amounted to almost US$19 billion—almost 18 percent of a total pharmaceutical market of $107 billion, while the market in Japan, at $1.36 billion, amounted to less than 3 percent of a total pharmaceutical market of $49.1 billion. The growth of actual doses of psychiatric medication prescribed from 1990 to 2000 is less marked, with the United States showing an increase of 70.1 percent, Europe of 26.9 percent, Japan of 30.9 percent, South Africa of 13.1 percent, and Pakistan of 33.4 percent. In both the United Kingdom and the United States, one key growth area has been the new SSRI type antidepressants, increasing by around 200 percent over the decade, with a related, but smaller, decline in prescriptions for anxiolytics. A further feature, and one that has attracted much controversy, has been the rise in the prescription of psychostimulants, notably Ritalin (methylphenidate) and Adderall (dexamphetamine), for the treatment of Attention Deficit Hyperactivity Disorder in children. From the mid-1980s to the end of the twentieth century there was a remarkable growth in the diagnosis of this condition and in the use of these drugs. This was most marked in the United States, where prescribing rates rose eight-fold in the decade from 1990 to 2000. Similar, although less marked, rises can be seen in a number of other countries, notably Australia, New Zealand, Israel, and the United Kingdom. In 2004, for example, a National Health Service audit of children's health in Scotland showed that, between 1996 and 2003, the number of prescriptions of Ritalin for six to fourteen year olds went up from 69.1 to 603.2 per 10,000 population and it was estimated that about 5 percent of Scottish children "had ADHD" (National Health Service Quality Improvement Scotland 2004). Other studies showed rises in the use of these drugs in Israel, Canada, and elsewhere (e.g., Fogelman et al. 2003, Miller et al. 2001). And while for some this raises concerns about the rising levels of the disease, for others the concerns are the reverse: the rising medication of children.

Many earlier criticisms of the use of psychiatric drugs claimed that they were used as "chemical coshes" in control strategies seeking to pacify and normalize. But today, I suggest, such drugs do not so much seek to normalize a deviant but to correct anomalies, to adjust the individual and restore and maintain his or her capacity to enter the circuits of everyday life. Initially, of course, it was indeed the pacifying and disciplining effects of the drugs that were their main selling points. Thus an advertisement for Thorazine (the first neuroleptic, chlorpromazine), which appeared in *Archives of General Psychiatry* in 1962, shows a tiny patient lashing out at a huge and domineering eye under the caption, "When the patient lashes out against 'them' Thorazine quickly puts an end to his violent outburst," and an advertisement in the same journal a year later, for Dexamyl Spansules—a rather lethal-sounding combination of dexamphetamine and amobarbitol for the treatment of depression—features a calm and placid housewife in a neat apron vacuuming her house, presumably after the administration of the drug being promoted. As psychiatric systems in many countries across the 1980s became based on the use of drugs to maintain persons with mental disorders outside the asylum, drugs were largely marketed on the basis of their capacity to enable individuals to cope with life in the community. Thus the first images used in advertisements for SSRI drugs such as Luvox and Lustral stressed the way in which they might help patients cope with their depression, obsessions, intrusive thoughts, repetitive behavior, and the like and maintain a relatively normal existence. By the mid-1990s, however, advertisements for Prozac were full of smiling faces, pictures of the sun breaking through clouds and the like. What was marketed here was not coping, but hope.

Of course, psychiatric drugs are undoubtedly used, now as previously, in all manner of coercive situations and institutions to normalize conduct and to manage inmates. Similarly they are undoubtedly part of control strategies for patients "in the community." In these situations, those who take them often do so under explicit or implicit duress.[26] Many have leveled similar criticisms at the use of psychiatric drugs for childhood disorders. I have already noted the large increase in the use of drugs for the treatment of Attention Deficit Hyperactivity Disorder. This condition, and its predecessors of hyperactivity and attention deficit disorder, has long attracted criticism, with many arguing that it is a classic example of the labeling and medicalization of behavior that is problematic only to various authorities (Conrad 1976, Conrad and Potter 2000). However, I think that the rise of ADHD reveals a more complex process of the intertwined production of the disorder and its treatment. The drivers of this growth in the United States were multiple, including financial incentives provided to schools for children with disabilities that included ADHD, disease awareness campaigns and marketing campaigns directed at doctors undertaken by the pharmaceutical companies, not to mention the illicit trading of the drugs by the children to whom they were prescribed. But they also show the patterns of activism that we have seen elsewhere: pressures from parents themselves, identifying a disorder that had not been properly recognized and treated, claiming such neglect to be a scandal, forming pressure groups, allying with medical experts, and the demanding of attention to "their" problem (S. Rose 2005: 253–263). These drugs too, participate in a political economy of hope. This is nicely illustrated in a 1997 advertisement in the *American Journal of Psychiatry* for Adderall in the treatment of Attention Deficit Hyperactivity Disorder in children, which shows a smiling child reaching out to the future under the caption, "Soar confidently into summer and the new school year—try Adderall." And even here, as research has shown, parents, teachers, and even the children themselves speak frequently of the consequences of the drug not as imposing an external and alien constraint upon the child, but the reverse—as enabling the child to take control of him or herself, restoring the child to his or her true self again (Singh 2002, 2003, 2004).[27] Outside these practices of authoritative behavioral management, it is this conception of the role of drugs that is dominant. For those becoming neurochemical selves, these drugs promise to help the individual him or herself, in alliance with the doctor and the molecule, to discover the intervention that will address precisely a specific molecular anomaly at the root of something that personally troubles the individual concerned and disrupts his or her life, in order to restore the self to its life, and itself, again.

Consider, for example, fluoxetine hydrochloride. As we saw earlier, Prozac was initially promoted as a carefully sculpted molecule, fluoxetine hydrochloride, which had been specifically designed to target one specific condition—mild to moderate depression. But, very soon, it began to be promoted for a whole range of other disorders: anorexia, bulimia, obsessive compulsive disorder, panic disorder, and much more. And when Prozac itself went out of patent, Eli Lilly, the manufacturers, began marketing the same compound, fluoxetine hydrochloride, in different packaging for Premenstrual Dysphoric Disorder—a candidate diagnosis in DSM IV apparently marked by severe mood and bodily disorders around the time of menstruation and affecting 3 to 5 percent of menstruating women in the United States.

The form of this marketing is revealing. "Share Our Strength," says the Lilly website in 2002, drawing on several decades of feminist rhetoric—"you are not alone":

> Premenstrual Dysphoric Disorder, or PMDD, is a distinct medical condition that affects millions of women. It happens the week or two before your period, month after month. Its many symptoms markedly interfere with your daily activities and relationships. And left untreated it can worsen with age. But by understanding what may cause PMDD and what can be done to help relieve its symptoms, you may feel more in control. . . . While PMDD is not fully understood, many doctors believe it may be caused by an imbalance of a chemical in the body called serotonin. The normal cyclical changes in female hormones may interact with serotonin and other chemicals, and the changes may result in the mood and physical symptoms of PMDD. With PMDD, it may seem like you only suffer a few days a month, but over time those days add up. In fact, you can spend up to 25 percent of your childbearing years* dealing with these symptoms. The good news is your doctor can treat PMDD symptoms with an FDA-approved treatment called Sarafem.[28]

Jonathan Metzl has argued that advertisements for psychiatric drugs for women draw upon prevailing social and cultural anxieties about gender, for example, in the 1960s, showing images of visually threatening feminist women "tamed" by tranquilizers (Metzl 2003). In contrast, while Sarafem clearly embodies social, political, and ethical assumptions about what women want, it is not promoted to restore domesticity and subservience. On the contrary, it is tied to an ethic of self-control, lifestyle promotion, and self-realization, one in which women themselves, in collaboration with doctors and pharmaceutical companies, can retake charge of their own lives.

While depression was the exemplary diagnosis in the 1990s, perhaps now the best examples are the anxiety disorders—Social Anxiety Disorder, Panic Disorder, and Generalized Anxiety Disorder. Take Paxil, owned by GlaxoSmithKline, initially marketed as an SSRI antidepressant. Paxil came late to the market and its market share was not spectacular. But in

2001 it was relicensed as the first SSRI specifically approved for the treatment of Generalized Anxiety Disorder, a condition that, according to Glaxo Smith Kline's direct-to-consumer advertising on U.S. television, affected "many millions of Americans." As recently as 1987, the section on prevalence of this disorder (coded 300.02) in DSM III-R said, "When other disorders that could account for the anxiety symptoms are ruled out [they previously stipulated that the disorder should not be diagnosed if the worry and anxiety occurs during a mood disorder or a psychotic disorder, for example], the disorder is not commonly diagnosed in clinical samples" (American Psychiatric Association 1987: 252). By the publication of DSM IV, in 1994, the same section read, "In a community sample, the lifelong prevalence rate for Generalized Anxiety Disorder was approximately 3 percent, and the lifetime prevalence rate was 5 percent. In anxiety disorder clinics, approximately 12 percent of the individuals present with Generalized Anxiety Disorder" (American Psychiatric Association 1994: 434). In this move, GAD was simultaneously reframed so the diagnosis could coexist *with* mood disorders, and could be separated out *from* the general class of mood disorders. By April 2001, when the U.S. Food and Drug Administration (FDA) approved Paxil for the treatment of GAD, it was widely being claimed that GAD affected "more than 10 million Americans, 60 percent of whom are women."[29]

Licensing allows marketing for the licensed indication and GlaxoSmithKline immediately engaged in a marketing campaign in the United States. What was characteristic about this campaign is that it marketed, not so much the drug, Paxil, as the disease itself. In part this was by means of "direct-to-consumer advertising" suggesting to individuals that their worry and anxiety at home and at work might not be because they are just worriers but because they are suffering from a treatable condition. "Paxil ® . . . Your life is waiting," the TV advertisements and the Paxil website announced.[30] And the narrative in these representations was instructive: "*I'm always thinking something terrible is going to happen, I can't handle it,*" says the first female character; "*You know, your worst fears, the what ifs . . . I can't control it, and I'm always worrying about everything,*" says a second young woman; "*It's like a tape in your mind, it just goes over and over. . . . I just always thought I was a worrier,*" says a third; "*It's like I never get a chance to relax. At work I'm tense about stuff at home. At home I'm tense about stuff at work,*" says a worried-looking young man. And then, the reassuring voice of the female narrator: "*If you are one of the millions of people who live with uncontrollable worry, anxiety, and several of these symptoms. . . .* (symptoms roll across the screen: worry . . . anxiety . . . muscle tension . . . fatigue . . . irritability . . . restlessness . . . sleep disturbance . . . lack of concentration . . .) *you could be suffering from Generalized Anxiety Disorder and a chemical imbalance could be to blame. Paxil works to correct this imbalance to relieve anxiety.*" (Now we see images of previous characters, now happy and playing with children, washing cars, and so forth, while the narrator rushes through a list of side effects: "*Prescription Paxil is not for everyone. . . . Tell your doctor what medicines you are taking. . . . Side effects may include decreased appetite, dry mouth, sweating, nausea, constipation, sexual side effects, tremor or sleepiness. Paxil is not habit forming.*" And now we return to the first female character, who addresses us directly to announce, "*I'm not bogged down by worry anymore, I feel like me again, I feel like myself. . .*" The drug thus does not promise to create a false self, on the contrary, it is through the drug that the self is restored to itself. If there is one theme or promise that runs though all these promotional materials it is this: with this drug, I can get my real self back, I can feel like myself, I can feel like me again.

The United States is one of the few countries that allow such direct-to-consumer advertising of prescription drugs—which has grown into a US$2.5 billion a year industry since drug advertising legislation was relaxed in 1997—but it is not the only country where "disease mongering" has become a key marketing tactic.[31] As Ray Moynihan and others have pointed out, this process involves alliances between drug companies anxious to market a product for a particular condition, biosocial groups organized by and for those who suffer from a condition thought to be of that type, and doctors eager to diagnose underdiagnosed problems (Moynihan et al. 2002). Disease awareness campaigns, directly or indirectly funded by the pharmaceutical company that has the patent for the treatment, point to the misery caused by the apparent symptoms of this undiagnosed or untreated condition, and interpret available data so as to maximize beliefs about prevalence. They aim to draw the attention of lay persons and medical practitioners to the existence of the disease and the availability of treatment, shaping their fears and anxieties into a clinical form. These often involve the use of public relations firms to recruit the media, supplying experts who will give their opinions to substantiate the stories, and providing victims who will tell their stories. Thus Roche promoted its antidepressant Auroxix (moclobemide) for the treatment of social phobia in Australia in 1997 by using a public relations company to place stories in the press, an alliance with a patients' group called the Obsessive Compulsive and Anxiety Disorders Federation of Victoria, funding a large conference on social phobia, and promoting maximal estimates of prevalence. These are not covert tactics, as a quick glance at the "Practical Guides" published on the Web by the magazine *Pharmaceutical Marketing* will show.[32]

Yet ethical arguments can intervene upon bioeconomics in instructive ways in our era of shareholder value, where negative publicity and distrust can rapidly, sometimes irreparably, damage the brand image of a drug. By 2005, in response to widespread criticism that their direct-to-consumer and other advertisements exaggerated the benefits of their products and downplayed the risks, many companies shifted their marketing strategies. But this "ethical" shift was actually to increase investment in the education of their potential consumers. Thus, for example, in a move widely regarded as an attempt to regain public trust after adverse publicity, Wyeth (mission statement: "leading the way to a healthier world") launched a telephone support program, The Dialogues: Time to Talk, in which nurses contact Effexor patients, invite them to join the program and talk through problems of depression and seek to increase medicine compliance. According to Wyeth's press release of August 16, 2005:

> Over the course of several months, patients who enroll receive education materials that have been developed in consultation with physicians and patients to address common therapeutic needs of patients taking EFFEXOR XR. Patients enroll via telephone or Internet and receive a Welcome Kit containing a Dialogues: Time to Talk membership card and health-related reminder items. Also included are a Dialogues: Time to Talk magazine that discusses treatment-related issues, brochures that answer questions about depression and possible side effects of their medication, and a helpful checklist of key points to cover at the patient's next meeting with the physician.[33]

In an analogous move at the same time, Pfizer announced "fundamental changes" in its direct-to-consumer advertising campaigns, planning to include "consumer friendly" risk-benefit summaries, and to invest an amount similar to that it spends on "branded ads" on "disease awareness with advertisements that do not mention a product, such as the recent 'Why Live With Depression' campaign that featured [*Sopranos*] actress Lorraine Bracco." Karen Katen, Vice-Chairman and president of human health at Pfizer put it thus: "DTC advertising is demonstrably helpful to patients, but it should be refined to be even more helpful. . . . We're announcing changes to our DTC advertising to strengthen its educational benefits—and to motivate patients to take earlier action and work with their healthcare providers to take more control over their health."[34] And we might note Pfizer's mission statement, carried on all of its press releases: "We dedicate ourselves to humanity's quest for longer, healthier, happier lives, through innovations in pharmaceutical, consumer and animal health products." Bioeconomics and ethics are intrinsically intertwined in the pharmaceutical biopolitics of the neurochemical self.

Neurochemical Citizenship

Elsewhere I have stressed the emergence of novel patterns of biological activism around genetic and somatic understandings of selfhood.[35] However, in relation to psychiatry and psychopharmaceuticals, such activism takes a different form. Some psychiatrists, even those previously associated with "genetic reductionism," argue that "far from increasing the stigma, advances in genomics have the opposite effect," and point to Ronald Reagan's announcement that he was suffering from Alzheimer's disease as a sign that stigma has been reduced by "progress that has been made in understanding its basis at a molecular level. We predict that this is the start of a trend and that identifying genes involved in behavioral disorders will do much to improve public perception and tolerance" (McGuffin et al. 2001: 249). A number of organizations campaigning around mental illness, notably the National Alliance for the Mentally Ill (NAMI) in the United States, and SANE in the United Kingdom, do embrace a conception of mental disorders as diseases with a genetic component and argue that this recognition will reduce stigma associated with such conditions and lead to effective treatment.[36] Some prominent individuals have written "pedagogic" accounts of their own experiences of depression that narrate a journey from hostility to an illness model of their condition, through to reluctant acceptance of drug treatment, recovery, and conversion to biomedical and genetic ideas of the origins of mental disorder (Solomon 2001, Styron 1990). Certainly in some ways, once located within the brain, mental disorders can sometimes, and somehow, escape from the stigma of madness, and become simply diseases like any others.

Thus it is not surprising that for some conditions, largely those affecting children, one can see patterns of biological activism by parents that have similarities to those found in relation to nonpsychiatric conditions. However there is one key difference: while in relation to conditions such as the dystrophies, Huntington's Disease, epidermolysis bullosa, and the like, few contest a genetic etiology, genetic etiology is precisely what is contested in the biopolitics of mental disorders. Thus one of the key characteristics of parent activism in these areas is to dispute suggestions that the conditions of their children have anything to do with social conditions or parental management. Take, for example, Attention Deficit Hyperactivity Disorder. I have already mentioned the activist groups of parents in the United Kingdom campaigning for the recognition of ADHD and the increased availability of psychostimulant treatments for this condition. In the United States, such parents groups are more powerful and active in relation to ADHD and its status as an illness with genetic causes and the need for basic genomic and biomedical research. CHADD (Children and

Adults with Attention Deficit / Hyperactivity Disorder), whose advisory board contains psychiatrists, psychologists, and those working on psychopharmacology, for example, say that "Although precise causes have not yet been identified, there is little question that heredity makes the largest contribution to the expression of the disorder in the population. Research does not support the popularly held views that AD / HD arises from excessive sugar intake, food additives, excessive viewing of television, poor child management by parents, or social and environmental factors such as poverty or family chaos."[37] Their website contains details of brain abnormalities that researchers have linked to ADHD, and links to other sites explaining the genetics of mental disorders, and giving advice on the use of pharmaceutical treatments of the condition.

Another area where one sees such patterns is that of autism spectrum disorder. Chloe Silverman has charted these developments, examining the genetic initiatives of two U.S.-based parent advocacy organizations, NAAR (National Alliance for Autism Research) and CAN (Cure Autism Now), each of which, in 1997, established and financially supported programs in autism genetics, although using different strategies and enrolling the scientists and researchers in different ways, and with different approaches to the bioeconomic quandaries that such work entailed (Silverman 2003, 2004). But in autism too, the biogenetic model is contested: there are groups such as DAN! (Defeat Autism Now!) that certainly support biomedical research into autism but resist current arguments from genetics and neurochemistry, and there are groups that reject a disease model completely.[38] And, more generally, few of the organizations that are members of the U.S.-based Genetic Alliance are advocating for genetic research on psychiatric conditions.[39] Information distributed by the U.K.-based MIND and Mental Health Foundation does include genetic causes in their accounts of "why people get" conditions from ADHD to schizophrenia, but they are not driven by active biological or genetic citizens seeking research to identify the genetic basis of these conditions.[40] Many campaigning organizations for those with particular conditions such as depression and manic depression do publish updates on current genetic research on their condition, and often carry requests from researchers for individuals to participate in clinical trials.[41] However the organizations themselves focus mainly on more traditional issues such as improving services, giving advice on employment and self-help, aiding recovery, and combating stigma.[42] In relation to neurochemical selves, biological citizenship often takes a different form.

That is not to say that those so diagnosed do not seek to become skilled, prudent, and active, and to take their own share of the responsibility for managing their mental health. In this area too, such persons are increasingly demanding control over the medical practices that subject them, seeking multiple forms of expert and nonexpert advice in devising their life strategies, and demanding that medics act as the servants and not the masters of this process. But the biopolitics of psychiatry is heavily overlain by the long history of critical, radical and anti psychiatry, which has almost always been premised on the belief that explanations of mental health problems in terms of brain disorders, and accounts that give a significant role for genetics in the etiology of such problems, are not only epistemologically and ontologically suspect (as in critiques of the biomedical model as essentialist, reductionist, individualist, and so forth) but tied to depersonalizing, uncaring, degrading, and repressive strategies for governing those with mental health problems.[43] Indeed the activism of those diagnosed with mental health problems, treated with drugs, and probed for genetic association, usually takes the form of a vociferous denunciation of the biomedical model, its reductionism, its reliance on chemical treatments, and the excesses allegedly committed in its name.[44] Campaigning mental health organizations that support biomedical research and are in turn supported by pharmaceutical companies are vilified, with the critics arguing that these are largely made up, not of those with mental health problems themselves, but of family members wishing to divest themselves of any responsibility for the genesis of the condition in their own conduct toward the affected individual by locating it squarely in biology. And contemporary inheritors of the mantle of antipsychiatry have launched a multipronged attack on the validity of research upholding the "biomedical model" of mental health problems, on the efficacy of psychiatric drugs, and on the integrity of the pharmaceutical companies that develop, test, and market them. In a key moment, Lauren Mosher, one of the key figures in the development of community-based and nonpharmaceutical treatments for those with mental health problems, and at that time working with the National Institute of Mental Health, resigned from the American Psychiatric Association, in a letter that was widely circulated on the Internet and elsewhere:

> The major reason for this action is my belief that I am actually resigning from the American Psychopharmacological Association. Luckily, the organization's true identity requires no change in the acronym. . . . At this point in history, in my view, psychiatry has been almost completely bought out by the drug companies. The APA could not continue without the pharmaceutical company support of meetings, symposia, workshops, journal advertising, grand rounds luncheons, unrestricted educational grants etc. etc. Psychiatrists have become the minions of drug company promotion. . . . No longer do we seek to understand whole persons in their social contexts—rather we are there to realign our patients' neurotransmitters.[45]

Mosher criticized the APA for entering into "an unholy alliance with NAMI," thus supporting parents who wish to control their mad / bad offspring, condoning the use of drugs known to have toxic effects, and fabricating DSM IV, a political document that serves to make psychiatrists money while promoting categories that have no scientific validity.[46] It is, argued Mosher, the true consumer groups that the APA should align itself with, "i.e. the ex-patients, psychiatric survivors, etc."

Many of those critical of psychiatric services drew upon the arguments of David Healy, that the pharmaceutical companies concealed negative trial results in obtaining licenses for some of the most significant of the SSRI group of psychiatric drugs, ignored or buried evidence that in a significant minority of cases such drugs produced severe adverse effects that increased the risks of suicide and homicide, and that, in key instances, there was an unholy alliance between the pharmaceutical industry, university departments dependent on grants, and supposedly independent psychiatric researchers who actually had financial interests in the compounds they were evaluating (Healy 2002, 2004, and Healy et al. 1999).[47] Other activists argued that the pharma companies downplayed the difficulties of withdrawal from these drugs and hence are guilty of creating large-scale dependency (Medewar 1997). Still other critics drew on research that suggested that the companies grossly overstated the efficacy of drugs that, in their own trials were often no more effective than placebo (Moncrieff and Kirsch 2005). And, by 2004, these assaults had achieved some success on one front at least—destabilizing the claims of efficacy and safety of SSRI psychopharmaceuticals. In December 2003, the Chairman of the Committee on Safety of Medicines in the United Kingdom, Professor Gordon Duff, advised that most of the antidepressant drugs in this group should not be used to treat depression in children and adolescents under eighteen, mirroring concerns that were also being expressed by the U.S. Food and Drug Administration (Ramchandani 2004). In December 2003 the U.K. Medicines and Healthcare Products Regulatory Authority banned the use of most SSRI and SNRI antidepressants in children, although it made an exception for Prozac, but in April 2005 the European Medicines Agency recommended that doctors be issued with strong warnings against the use with children of all SSRI and SNRI products—including Lilly's Prozac, Cymbalta and Strattera, GlaxoSmithKline's Seroxat, Lundbeck's Celexa and Lexapro, Pfizer's Zoloft, Wyeth's Effexor, and Akso Nobel's Remeron—on the grounds that they increased the frequency of suicide-related behavior and hostility. The debate over the use of these drugs for adults remained unsettled. But these developments, linked as they are to a much more general activism of suspicion and critique in relation to "big pharma," and the growing industry of claims for damage by recipients of such drugs and their families, gives a particularly agonistic character to the contemporary biopolitics of neurochemical selves.[48]

Neuropolitics

By the 1990s a fundamental shift had occurred in psychiatric thought and practice. No matter that there was little firm evidence to link variations in neurotransmitter functioning to symptoms of depression or any other mental disorder in the living brains of unmedicated patients, although many researchers are seeking such evidence and occasional papers announce that it has been found. And no matter that most of the new smart drugs are no more effective than their dirty predecessors for moderate or severe depression: they are favored because they are claimed to be safer, and to have fewer "unwanted effects." A way of thinking has taken shape, and a growing proportion of psychiatrists find it difficult to think otherwise. In this way of thinking, all explanations of mental pathology must "pass through" the brain and its neurochemistry—neurons, synapses, membranes, receptors, ion channels, neurotransmitters, enzymes, etcetera. Diagnosis is now thought to be most accurate when it can link symptoms to anomalies in one or more of these elements. And the fabrication and action of psychiatric drugs is conceived in these terms. Not that biographical effects are ruled out, but biography—family stress, sexual abuse—has effects through its impact on this brain. Environment plays its part, but unemployment, poverty, and the like have their effects only through impacting upon this brain. And experiences play their part—substance abuse or trauma for example—but once again, through their impact on this neurochemical brain. A few decades ago, such claims would have seemed extraordinarily bold; for many medicopsychiatric researchers and practitioners, they now seem "only common sense."

And, in the same movement, psychiatry became transformed by its capitalization. One of the criticisms of the private madhouses before the spread of public asylums was that they were generating what was termed "a trade in lunacy" in which profit was to be made by incarceration—leading to all manner of corruption (Jones 1972). No one made enormous sums out of public psychiatry in the nineteenth century, or indeed up until the middle of the twentieth. One of the eugenic arguments in Nazi Germany was that the care of the psychiatric ill was an enormous drain on the public purse (Burleigh 1994). Of course, as we all know, in the second half of the twentieth century, psychotherapy and counseling became big business. But psychiatry itself—in the mental hospitals, the clinics, the GP's surgeries, and the private psychiatric consulting room—also became a huge and profitable market for the pharmaceutical industry. Only the large pharmaceutical companies can now afford the risk capital involved in developing, trialing, and licensing of a new psychiatric drug. And because contemporary psychiatry is so much the outcome of developments in psychopharmacology, this means that these commercial decisions are actually shaping the patterns of psychiatric thought at a very fundamental level. The factories of the pharmaceutical companies are the key laboratories for psychiatric innovation, and the psychiatric laboratory has, in a very real sense, become part of the psychopharmacological factory. Many of these large multinational conglomerates make a considerable proportion of their income from the marketing of psychiatric drugs, and their success, or failure, in attracting market share is key to maintaining the shareholder value of the company. Paul Rabinow's assessment of the new life sciences is especially apt for psychiatry—the quest for truth is no longer sufficient to mobilize the production of psychiatric knowledge—health—or rather, the profit to be made from promising health—has become the prime motive force in generating what counts for our knowledge of mental ill health (Rabinow 1996a).

Of course, to identify this new medicoindustrial complex and to point to its power is not in itself to critique it. In a situation where only investment of capital on a large scale is capable of producing new therapeutic agents, such linkages of health and profitability might well be the inescapable condition for the creation of effective drugs. In the early decades of the discovery and use of such drugs, in the 1950s and 1960s, those working in drug companies felt themselves to be in a virtuous alliance with those concerned with public health, jointly identifying untreated psychiatric conditions and developing and marketing effective treatments.[49] By the twenty-first century, however, the situation is more complicated. Those concerned with public health repeatedly draw the attention of policymakers to the burden of undiagnosed psychiatric disorder, especially depression and anxiety disorders, and the need for increased recognition and treatment (European Commission Health and Consumer Protection Directorate-General 2005, World Health Organization 2004). At the same time, medical and pharmaceutical information companies repeatedly draw the attention of the pharmaceutical companies to the potential market in treating these conditions. Thus, to take only one example, a 2004 "Stakeholder Insight" report by Datamonitor entitled "Anxiety Disorders—More Than Just a Comorbidity" promotes itself as follows: "Although a fifth of the total population across the seven major markets suffers from an anxiety disorder only a quarter of these individuals are diagnosed and therefore treated. As a result, drug manufacturers are failing to maximize revenues" and gives readers three reasons to purchase the report (that costs a mere US$15,200):

- Assess the size of the drug-treated population by gaining an understanding of the prevalence of anxiety with comorbid psychiatric disorders;
- Target physicians more effectively, through an understanding of prescribing behavior;
- Identify commercial opportunities by benchmarking current products and assessing unmet needs and future market potential.[50]

This is only one of dozens of similar reports on unexploited sectors of the psychiatric pharmaceutical market: once more, allying the pharmaceutical industry with the simultaneous promotion of public health and private shareholder value.

However, the consequences of many of the developments we have charted here cannot be reduced to a debate about either commercial interests or about efficacy, as if illness and treatment were independent of one another. The most widely prescribed of the new generation of psychiatric drugs treat conditions whose borders are fuzzy, whose coherence and very existence as illness or disorders are matters of dispute, and which are not so much intended to "cure" a specific illness—a transformation from a pathological to a normal state—as to modify the ways in which vicissitudes in the life of the recipient are experienced, lived, and understood. It is undoubtedly the case that psychiatric drugs today are conceived, designed, and disseminated in the search for biovalue. But they are entangled with certain conceptions of what humans are or should be—that is to say, specific norms, values, and judgments internalized in the very idea of these drugs. An ethics is engineered into the molecular makeup of these drugs, and the drugs themselves embody and incite particular forms of life in which the "real me" is both "natural" and to be produced. Hence the significance of the emergence of treatments for mental ill health lies not only in their specific effects, but also in the way in which they reshape the ways in which both experts and lay people see, interpret, speak about, and understand their world.

The psy discourses that took shape across the twentieth century brought into existence a whole new way of relating to ourselves—in terms of neuroses, trauma, unconscious desires, repression, and, of course, the theme of the centrality of sexuality to our psychic life. To say we have become "neurochemical selves" is not to say that this way of relating to ourselves has now displaced or replaced all others: different practices and locales embody and enjoin different senses of selfhood, and the idea that each culture or historical period is characterized by a single way of understanding and relating to ourselves is clearly mistaken.[51] But I suggest that a neurochemical sense of ourselves is increasingly being layered onto other, older senses of the self, and invoked in particular settings and encounters with significant consequences. Individuals themselves and their authorities—general practitioners, nurses, teachers, parents—are beginning to recode variations in moods, emotions, desires, and thoughts in terms of the functioning of their brain chemicals, and to act upon themselves in the light of this belief. To grasp the world in this way is to imagine the disorder as residing within the individual brain and its processes, and to see psychiatric drugs as a first line intervention, not merely for symptom relief but for ways of modulating and managing these neurochemical anomalies. It is, of course, important to criticize the use of these drugs as agents of control, to point to their false promises, adverse effects, and bioeconomic rationale. But it is also important to note this wider shift in which such drugs are becoming central to the ways in which our conduct is governed, by others, and by ourselves.

As is well known, Gilles Deleuze (Deleuze 1995) suggested that contemporary societies are no longer disciplinary, in the sense identified by Foucault; they are societies of control. Where discipline sought to fabricate individuals whose capacities and forms of conduct were indelibly and permanently inscribed into the soul—in home, school or factory—today control is continuous and integral to all activities and practices of existence. Deleuze himself makes reference to "the extraordinary pharmaceutical productions, the molecular engineering, the genetic manipulations [that are] are slated to enter the new process" (178, translation modified). We do not have to accept the whole of his dystopian picture to recognize something of our own form of life in it. In the field of health, the active and responsible citizen must engage in a constant monitoring of health, a constant work of modulation, adjustment, improvement in response to the changing requirements of the practices of his or her mode of everyday life. Similarly, the new psychiatric and pharmaceutical technologies for the government of the soul oblige the individual to engage in constant risk management, to monitor and evaluate mood, emotion, and cognition according to a finer and more continuous process of self-scrutiny. The person, educated by disease awareness campaigns, understanding him- or herself at least in part in neurochemical terms, in conscientious alliance with health care professionals, and by means of niche-marketed pharmaceuticals, is to take control of these modulations in the name of maximizing his or her potential, recovering his or her self, shaping the self in fashioning a life. The forms being taken by contemporary neurochemical selfhood, the blurring of the boundaries between treatment, recovery, manipulation, and enhancement, are intimately entwined with the obligations of these new forms of life. They are intrinsic to the continuous task of monitoring, managing, and modulating our capacities that is the life's work of the contemporary biological citizen.

Chapter 8
The Biology of Control

> *The fable of intelligible freedom.* The history of those feelings by virtue of which we consider a person responsible, the so-called moral feelings, is divided into the following main phases. At first we call particular acts good or evil without any consideration of their motives, but simply on the basis of their beneficial or harmful consequences. Soon, however, we forget the origin of these terms and imagine that the quality "good" or "evil" is inherent in the actions themselves, without consideration of the consequences . . . that is, we take the effect to be the cause. Then we assign the goodness or evil to the motives, and regard the acts themselves as morally ambiguous. We even go further and cease to give the particular motive the predicate good or evil, but give it rather to the whole nature of the man; the motive grows out of him as a plant grows out of the earth. So we make man responsible in turn for the effects of his actions, then for his actions, then for his motives and finally for his nature. Ultimately we discover that his nature cannot be responsible either, in that it is itself an inevitable consequence, an outgrowth of the elements and influences of past and present things; that is, man cannot be made responsible for anything, neither for his nature, not his motives, nor his actions, nor the effects of his actions. And thus we come to understand that the history of moral feelings is the history of an error, an error called "responsibility," which in turn rests on an error called "freedom of the will."
>
> —F. Nietzsche [1878] 1944, *Human, All too Human*, Aphorism 39

In our biologized culture, not merely the sicknesses of human beings, but also their personalities, capacities, passions, and the forces that mobilize them—their "identities" themselves—appear to be explicable, potentially at least, in biological terms.[1] In this chapter, I want to consider one key site for this biologization of the human soul—that of crime. In the closing decades of the twentieth century, a new biological criminology began to take shape. Newspaper reports, television documentaries, films, and novels have popularized this with claims about "the gene" for crime and speculative drama documentaries about a future where the criminal mind can be detected with brain scans, and managed with drugs. However, most of these "new" biological criminologists rejected earlier criminological assertions that crime was committed by those of a particular, defective, biological type or that there was a "gene for crime." They focused not on "crime" in general, but on violent, aggressive, and antisocial behavior, and suggested that they were able to account for such behavior through the explanatory regimes of molecular genetics, neurochemistry, and neurobiology, supported by evidence from family histories or twin studies, and by direct indicators of abnormality from EEGs, CAT scans, PET scans, MRIs, and DNA sequencing.

In the 1990s, as such arguments developed momentum, many sociologists and cultural critics regarded them with alarm, believing that they embodied genetic essentialism or neurogenetic determinism. Nelkin and Lindee, for example, on the basis of analysis of popular accounts of the relations between genetics and crime, claimed that they manifested a "belief in genetics destiny [which] implies that flaws and failings are inscribed in an unchangeable text—the DNA—that will persist in creating criminals even under the most ideal social circumstances," and suggested that this would deflect attention from the need for social reform, education, and rehabilitation in tackling the violence of American society in favor of biological techniques for the control of criminal behavior (Nelkin and Lindee 1995: 96, cf. S. Rose 1995). From another direction, the new biological criminology was one of the nodal points for the social anxieties around which new discourse of neuroethics took shape at the start of the twenty-first century. Those concerned with the ethical implications of neuroscience suggested that evidence for the biological bases of criminal behavior would strike at the heart of the doctrine of free will and personal responsibility that lies at the heart of most legal systems. They suggested that we might be moving to a situation in which biological evidence from brain scans or genetics would enable prospective criminals to be identified and confined in advance of committing any crime. They were concerned that the novel technique of "brain fingerprinting" would invade the "neural privacy" of the individual, providing incontrovertible evidence from brain scans as to whether a suspect or a witness was lying in testimony (Gazzaniga 2005, Illes 2005; for general overviews of these issues, see Marcus and Charles A. Dana Foundation 2002).

In this chapter, I examine the extent to which the evidence supports the concerns of the critical sociologists and troubled neuroethicists. I analyze arguments about the biological bases of violent crime as they operate in the courts and in the criminal justice system, and as they have been developed by biological criminologists themselves. I suggest that some important changes are taking place in control strategies, but these cannot be understood either in terms of the jurisprudential binary of free will versus determinism or through the sociological binary of biological reductionism versus social causation. Biological styles of thought are entering the criminal justice system and jurisprudential debate, but in an uneven and contested manner, and they are unlikely to overturn or radically transform legal reasoning or the premises of the criminal justice system.

Far from biological explanations of conduct mitigating responsibility in the criminal justice system, as suggested by the neuroethicists, the resurgence of such explanations has gone hand in hand with a renewed emphasis on the moral culpability of all offenders, irrespective of biological, psychological, or social dispositions, and a move away from logics of reformation to those of social protection that is linked with some rather general shifts in the government of conduct in advanced liberal societies.[2] Rather than seeing the reactivation of the eugenic strategies of the first half of the twentieth century that sought to eliminate members of subpopulations whose tainted constitutions predisposed them to immorality and criminality, as suggested by the sociologists, these new biological conceptions of the origins of pathological conduct are bound up with a new "public health" conception of crime control. In these strategies, sociopolitical interventions are legitimated not in the language of law and rights, but in terms of the priority of protecting "normal people" against risks that threaten their security and contentment. Biological factors are merely one set of factors among others predisposing individuals to antisocial conduct, and "therapeutic interventions" are proposed for the good of both the individual and society. This requires the preemptive identification and management of "risky individuals," and risk-generating environments. These seek to reduce the riskiness of individuals thought to be potential offenders where possible, and, where not, seek their indefinite containment in the name of public safety.

A New Eugenics?

The criminological enterprise that flowered throughout Europe and North America in the last decades of the nineteenth century was founded on the belief that the criminal was a certain type of person, that the propensity for crime was inscribed in the identity of the individual in the form of a specific and identifiable aberration or abnormality that could be identified by physical signs: the visible marks of criminality.[3] The idea that criminality was written in the body in the form of an inherited, unalterable, degenerate physical and moral constitution, threatening the character of the race and demanding control by sequestration, sterilization, or elimination, was one element—though not the key one—in the strategies of eugenics that proliferated in the first four decades of the twentieth century. In eugenics, the criminal was linked with the feebleminded, the insane, the tubercular, the alcoholic, the prostitute, the habitual gambler into a single degenerate and heritable identity, and in many countries the powers of the state and law were used to control these threats in the name of racial health.[4]

These brief reminders of history are relevant here, because critics of the new biological criminology tend to suggest that it is part of a new eugenics (Duster 1990, Horgan 1993) or, at the very least, show a renewed conviction that defective identities are indelibly inscribed within the corporeal body and implacably override cultural, social, or psychological forces. Such critics suggest that this new determinism operates at the level of the genome itself—an inherited pathological identity that can be *made visible* by tests for genetic abnormality in fetuses, by gene sequencing, brain scanning, investigations of brain biochemistry, and the like.[5] Thus Dreyfuss and Nelkin argue that "genetic essentialism" is on the rise in the criminal justice system, challenging key legal precepts, altering the perception of the person by positing that "personal traits are predictable and permanent, determined at conception, "hard wired" into the human constitution . . . [and if known, would] largely explain past performance and could predict future behavior" (Dreyfuss and Nelkin 1992: 320–321).

Those working in the field take a different view. They suggest that such simplistic assertions are not part of the discourse of their science: they are generated by an unholy interaction between sensationalist popular media and paranoid politically motivated groups, although naïve researchers sometimes play into their hands. For example, a 1995 editorial in *Psychiatric Genetics*, after reporting a range of steady advances in the search for the genetic bases of schizophrenia, bipolar illness, Alzheimer's, panic disorder, Tourette's, alcoholism, and autism, and advances in identifying the complex molecular mechanisms involved, expressed the following concern:

> In the world of scientific politics, our field continues to struggle against demons within and without. At times we in psychiatric genetics may have been guilty of optimistic reports based on preliminary findings: unfortunately this has provided fuel for our detractors. Any

> premature consideration of complex social issues in reductionist terms is also seized upon by the media. Our statements are then held up for public ridicule by scientists and media polemicists who are intrinsically opposed to human behavioral genetics in any form. . . . Psychiatric genetics and behavioral genetics are pigeonholed with eugenics and are vilified as bad science and bad policy. It is asserted that we are naively mechanistic in our view of human behavior. Assumptions that genes influence behavior and behavioral disorders are viewed as absurd. We must work hard in this climate to give a balanced view of our science. . . . (Editorial 1995: 4)

The writer was particularly concerned about an article in *Scientific American* in 1993 entitled "Eugenics Revisited" (Horgan 1993). This reviewed the arguments advanced by behavioral geneticists in relation to homosexuality, intelligence, alcoholism, schizophrenia, manic depression, and crime. It found severe flaws in all the major studies: the much publicized claims by Thomas J. Bouchard and his group in Minnesota that their studies of identical twins reared apart had discovered a high genetic component in "traits" ranging from intelligence to taste in clothes; Robert Plomin's "fishing expeditions" to catch the specific genes that differentiate school children on tests of intelligence; claims by Kenneth Blum and his group at the University of Texas Health Science Center to have discovered a genetic marker for alcoholism (reported on the front page of the *New York Times* in 1990); and by Dean Hamer and others to have discovered the genetic basis of homosexuality (Blum et al. 1990, Bouchard et al. 1990, Hamer and Copeland 1994, Thompson et al. 1993). Janice Egeland's group at the University of Miami School of Medicine had withdrawn their well-publicized claim to have linked manic depression in an Amish population to a genetic marker on gene 11 (Kelsoe et al. 1989, Kidd et al. 1987).[6] Miron Baron and his group at Columbia University had withdrawn their claim to have linked a marker on the X chromosome to manic depression in three Israeli families (Baron 1989, Baron et al. 1987). Even the enthusiastic Hugh Gurling at University College London Medical School had admitted that his claim to have found linkage in Icelandic and British families between genetic markers on Chromosome 5 and schizophrenia was probably based on a false positive, since subsequent studies showed a much reduced linkage (Gurling 1990, Sherrington et al. 1988). Glenn Walters's reanalysis of thirty-eight studies of families, twins, and adoptees from the 1930s to the present found that statistical evidence for their claims to have demonstrated a genetic component to crime was very weak: the better the study, the smaller the relationship (Walters 1992). Horgan concluded that most such research was scientifically worthless: even if it was possible to identify a genetic component in some behavior, this would produce only a slightly elevated risk of the disorder or conduct in question, and the discovery would be more likely to lead to discrimination than to therapeutic benefits.

In the United States, disputes over research on the biological basis of antisocial conduct came to a head over a National Institute of Health-funded conference entitled "Genetic Factors and Crime," that was planned for the University of Maryland in 1992.[7] NIH withdrew the grant after much controversy, on the basis that the prospectus gave "the distinct impression that there is a genetic basis for criminal behavior, a theory that has never been scientifically validated" (quoted in Marshall 1993: 23). The grant was eventually reinstated, and the conference took place in September 1995 accompanied by protests and placards reading "Jobs not Prozac" and "This Conference Predisposes Me to Disruptive Behavior Disorder."[8] One of the principal accusations was that because African Americans were overrepresented among those convicted of crime in the United States, any attempt to explore genetic factors in criminal conduct implied that this overrepresentation arose from their biology and thus was "inherently racist."

The Cases

These themes in the public discourse of science and culture are significant in their own right. But to understand the micropolitics of contemporary control practices, we need to examine the ways in which these new biological games of truth and identity are being played out. One key site for such an investigation is the courtroom.[9] Over the twentieth century, despite the ambitions of psychology, psychiatry, and criminology, programs of criminological positivism made few inroads into the courtroom and the trial process. Criteria for attributing criminal responsibility are diverse and complex, and vary from jurisdiction to jurisdiction and in relation to the conceptions of personhood ascribed to different kinds of legal subject. Thus, juvenile offenders were comprehensively psychologized throughout the English-speaking world and in most European countries in the first seven decades of the twentieth century; female offenders in this period have tended to be seen as the more or less passive victims of psychological or other forces sweeping them into pathological conduct.[10] But, in the main, the criminal courts have remained rather hostile to psychological and psychiatric assaults on the doctrines of free will, rationality, and responsibility when determining guilt. Only after the trial and verdict, in the determination of the sentence, in the interventions of probation officers, social workers, and psychiatrists, and in the psychological technologies of reformation utilized in the prisons, has the focus shifted from "what have you done?" to "who are you and why do you act as you do?" What, then, has been the impact of the new biological criminology? Let us consider a few cases.

First, there were the XYY cases in the late 1960s and the 1970s (cf. Denno 1988, Denno 1996). These were cases where gross chromosomal abnormalities—the possession of an extra Y chromosome (the sex chromosome whose presence together with one X chromosome determines male-ness)—were used in the defense of individuals charged with violent crime. Research findings in the 1960s claimed that there was a disproportionate number of XYY males in maximum security institutions in the United States and other countries, and that such individuals showed immaturity, defective development, or inadequate control of aggressive instincts and emotional responses. By the mid-1970s, the XYY defense had been mounted in five cases in the American courts. But in no case did it succeed: the reasoning usually being that an insanity defense should only be accepted if an etiological relationship can be established between the defendant's mental capacity and the genetic syndrome, which has "so affected the thought processes as to interfere substantially with the defendant's cognitive capacity or with his ability to understand or appreciate the basic moral code of his society," and that "presently available medical evidence is unable to establish a reasonably certain causal connection between the XYY defect and criminal conduct" (quoted from Denno 1988: 21). XYY defenses ceased after later research revealed major flaws in the studies they had relied upon.[11]

In the "premenstrual syndrome" cases in the United Kingdom in the 1980s, the legal reasoning was similar (discussed in Allen 1984). A successful defense of PMS required more than general arguments about the correlation of PMS with mood disorders and so forth. There had to be convincing evidence that there was a clear causal link between the mental condition of this particular individual in the premenstrual period and the act in question. In two well-publicized cases, evidence was presented, and accepted, that the crimes had been committed while the defendants were suffering from severe symptoms of premenstrual tension. The courts heard arguments made by Dr. Katharina Dalton that this biomedical condition was causally responsible for the criminal behavior of the young women involved. Ms. English killed her lover by running him down with a car; she was found not guilty of murder but guilty of manslaughter due to diminished responsibility and was conditionally discharged without punishment. Ms. Smith was found guilty and sentenced to probation for threatening to kill someone, one year after a previous diminished responsibility conviction for having stabbed a colleague to death at work. She appealed her conviction for the second offense on the grounds that she should not have been found guilty because there was no moral fault. The

Appeal Court judges rejected the appeal, and also rejected the idea of a "special defense" of premenstrual tension that would absolve women from criminal responsibility. But they commended the trial judge, who had instructed the jury that Ms. Smith was "morally guiltless" and that they "should proceed on the assumption that her behavior was attributable to the fact that she had insufficient of this hormone . . . she knew what she was doing but she could not control herself . . . she had lost her moral safeguards" (quoted in Allen 1984: 26–27). While this seemed to open the door to a limited biologization of pathological identity, the PMS defense is now used rarely if ever.[12] The argument has followed a different path, one that is instructive: it is to emphasize the need for the early detection and treatment of those suffering from PMS before any criminal behavior has resulted (Fishbein 1992).

Biological defenses have been pursued in other ways. Perhaps most significant have been the cases in the United States where evidence from brain scans has been introduced into court in order to support a defense of biological impairment.[13] In the John Hinckley case, where the defendant had attempted to assassinate the U.S. president, the defense claimed that computerized axial tomography (CAT) scans of the brain provided organic evidence that Hinckley was schizophrenic.[14] Hinckley's acquittal on the grounds of "Not Guilty By Reason of Insanity" (NGRI) gave added impetus to the U.S. campaign to reform the insanity and diminished capacity defenses.[15] These were severely limited in some thirty-nine states, transformed into "Guilty But Mentally Ill" in eight others (a verdict that allows any sentence up to and including death), and abolished entirely in Illinois and Idaho (Moran 1991). Nonetheless, evidence from the new technologies for visualizing the brain found its way into the American courts in the 1980s. In particular, the results of positron emission tomography (PET) were thought to be relevant to sentencing decisions, on the presumption that biological evidence might be relevant to the most appropriate disposal (Anderson 1992). By 1992, for the first time in the United States, a court allowed an expert to draw upon evidence from a PET scan in determining the defendant's sanity, although in the end the matter was resolved by lowering the charge from murder to manslaughter and avoiding a trial.[16] It is worth considering some later cases where brain scans were invoked in some detail.[17]

Kip Kinkel was accused of murdering both his parents in 1998 and then returning to the school in Springfield, Oregon from which he had been expelled for carrying a gun to class, and opening fire in the cafeteria, killing one student and wounding eight others, one of whom later died.[18] Kinkel abandoned the option of pleading not guilty by reason of insanity just before jury selection, but under Oregon law the judge holds a hearing before sentencing, at which both defense and prosecution present evidence in relation to the severity of the sentence. Most experts called by the defense to support the claim that Kinkel was mentally ill did so on the basis of clinical interviews and psychological tests, although a private investigator was used to imply that there was a genetic element involved, testifying about her discovery of multiple cases of mental illness in Kinkel's extended family, including schizophrenia. The defense also called Dr. Richard J. Konkol, a pediatric neurologist who showed several images from a SPECT scan apparently revealing areas of decreased blood flow in the prefrontal cortex, the temporal occipital and parietal lobes; he suggested that this was consistent with other research evidence that shows a correlation between deficiencies in gray matter in these regions and the onset of childhood schizophrenia. The judge prefaced his final ruling with reference to a 1996 change to the Oregon State constitution that shifted the focus of criminal punishment from "the principle of reformation" to "the protection of society," and argued that the protection of society should indeed be given most importance in his ruling in this case. He sentenced Kinkel to 25 years for the murders, and a further 40 months for each attempted murder, making a total of more than 111 years in prison without parole. Jesse Barton, Kinkel's lawyer, filed a petition for review with the state's Supreme Court, arguing lower courts misinterpreted Oregon's sentencing guidelines in putting emphasis on the "protection of society" principle, yet ignoring "reformation." He argued not only that the sentence violated constitutional rights against cruel and unusual punishment, but also that "Record evidence shows that owing to a genetic predisposition, and therefore through no conscious fault of his own, defendant suffers from a mental illness resulting in his committing his crimes of conviction." The National Alliance for the Mentally Ill, who have issued their own position paper arguing against the criminalization of people with mental illness, also filed a petition for review of Kinkel's sentence.[19] But the Court of Appeals called Kinkel's crimes among "the most horrific in Oregon's history" and concluded that the "protection of society" consideration held greater weight than the state's three other sentencing guidelines: the appeal was dismissed.[20]

This was just one of a number of high-profile murder cases in the United States, in which defense lawyers sought permission to introduce brain scans as evidence of mental illness in support of an insanity plea. Thus, in mid-2002, in the trial of Cary Stayner for the horrific sexual assault and murder of three tourists in Yosemite, defense lawyers supported their plea of not guilty by reason of insanity with evidence from a host of psychiatrists over a two week period, which included reference to brain scans and genetics.[21] Experts sharply disagreed, however, over the significance of Stayner's brain scans. Dr. Joseph Wu, for the defense, saw abnormalities in the images that could account for the defendant's violent tendencies, while Dr. Alan Waxman, called by the prosecution, saw nothing of the sort: on August 26, 2005, the jury took less than five hours to find Stayner guilty of three counts of first degree murder.

Genetics has also been called upon by defense lawyers in the U.S. courts in support of pleas of insanity or mitigation of sentence. I have already mentioned the way in which Kinkel's defense team invoked genetics in their arguments. Stayner's lawyers also used genetic arguments, calling on Dr. Allison McInnes, assistant professor of psychiatry and human genetics at the Mount Sinai School of Medicine in New York, who showed the jury a genealogical table of Stayner's family history, with putative sufferers from mental disorders marked in different colors. *Fresno Bee* reporter Cyndee Fontana described the testimony thus:

> The story of Cary Stayner's family tree rose in bursts of bright color from a white horizontal chart. . . . Yellow for psychosis. Green for obsessive-compulsive disorder. Red for substance abuse. Purple for pedophilia. Even more colors for more mental diseases ranging through four generations down to Stayner himself—the fruit of a family gene pool marked by psychiatric disorders. "So many different mental disorders," said Dr. McInnes as she led a jury through the branches of men and women, the flashes of color, that yielded the man convicted last week of murdering three tourists in February 1999.[22]

The jury, as we have seen, was not convinced. One of the first cases where evidence from contemporary molecular genetics was invoked, in addition to family history, was that of Stephen Mobley (cf. Denno 1996). Mobley was convicted of shooting to death the manager of a Domino's Pizza store in the back of the neck, in February 1991. Mobley, who was twenty-five at the time of the offense, had a lengthy history of treatment for "inability to control his impulses or to internalize any kind of value system." In the trial, his attorneys did not try to introduce genetic evidence as a defense, but as possible mitigation in relation to the sentence. The genetic evidence was based on a family history that was claimed to show four generations of violence, aggression, and behavior disorder in uncles, aunts, and grandparents. The lawyers argued that this was relevant because of a study by Han Brunner (Brunner et al. 1993) of a family history of violence in the Netherlands. This study seemed to identify a syndrome in which borderline mental retardation was linked to abnormal behavior, including violence and aggression: genetic linkage studies showed this syndrome to be associated with a point mutation in a gene regulating the production of an enzyme—monoamine oxidase A (MAOA)—linked to changes in levels of various neurotransmitters.[23] The Superior Court of Hall County denied the request, arguing that Mobley could not be compared with the Netherlands family because he was not borderline mentally retarded, and noting that Brunner and his colleagues had acknowledged that the inhibition of MAOA had not been reported to cause aggressive behavior in adult humans. A jury found Mobley guilty in February 1994 and he was sentenced to death. After a stay of execution granted in 2002 by the Federal Appeals Court in Atlanta, and several more appeals on various grounds, Mobley was executed in February 2005.

Present evidence thus suggests that biological and genetic defenses have largely failed to displace older conceptions of responsibility within the practice of the criminal law, at least in the United States where one might expect such developments to be furthest advanced.[24] There has been rather more success in pleas for mitigation of sentence, but this has long been an aspect of the trial process more open to psychological, psychiatric, and social expertise. Biological arguments seem to enter the courtroom not because legal personhood has become biological, but because defense lawyers, especially in the United States, will utilize anything they can in defense of their clients. Indeed legal philosophers who have considered the claims of contemporary biological criminology find no good reasons why neurogenetic discoveries in relation to criminality *should* alter prevailing legal conceptions of freedom, responsibility, or desert. Genetic accounts of the relation between "criminogenic genes" and conduct are no more "determinist" than those that point to the effects of background, environment, or biography. The argument from genetics is that the ability to control oneself or act differently was reduced by an inherited genotype. The argument from psychology and sociology is that the capacity for self-control was reduced by one's upbringing or environment. Judges and juries are likely to be as resistant to the exculpatory claims of the former as they have been to the latter.

It might be thought that the objectivity and demonstrability of biological accounts could give them a greater impact. Arguments of psychologists, psychiatrists, and social workers usually fail to satisfy the courts, and are often demolished by lawyers and ridiculed in the media. The courtroom display of a physical inscription—a PET scan or a DNA profile—has a greater rhetorical force. But as with DNA "fingerprinting," where brain scans have been used in controversial cases such as that of Cary Stayner cited above, the effects of the trial process are to expose the messy complexity that lies behind such sanguine assertions of truth (for discussion in relation to DNA fingerprinting, see Alldridge 1992, 1994, Lynch 1998, M'charek 2005).[25] Opposing teams of lawyers hire biological experts to attack each other's truth claims, with consequences that would delight contemporary sociologists of scientific knowledge: the most robust scientific claims are shown to be contestable products of dubious technical procedures, debatable selection of comparator populations, questionable leaps of causality, and loosely controlled discretionary judgment. There is no reason to suppose that brain imaging, and neurochemical and genetic arguments will not be subjected to the same destabilization if they begin to enter the criminal trial process on a regular basis.

As long-standing debates about the role of other forms of knowledge in the legal process have demonstrated—notably that from psychology—when the judiciary defend the nongenetic, nonpsychiatric fictions of free will, autonomy of choice, and personal responsibility, this is not because legal discourse considers this a scientific account of the determinants of human conduct. Rather, legal arguments deem it necessary to proceed as if human beings had free will, for reasons to do with prevailing notions of moral and political order.[26] Indeed, as the judge's considerations in the Kinkel case in Oregon exemplify, the trend of legal thought in the United States seems to be increasingly toward the protection of society rather than the mitigation of responsibility. In this context, the argument from biology is likely to have its most significant impact, not through the maneuverings of defense lawyers but in the determination of the sentence. For if antisocial conduct is indelibly inscribed in the body of the offender, it seems that it is not mitigation of punishment that is required, but the long-term pacification of the irredeemable individual in the name of public protection, even if this means the rejection of many rule of law considerations, such as those concerning the proportionality of crime and punishment.

I have already mentioned the trend of many states to reform their insanity and homicide laws to allow persons suffering from mental disorder to face all the penalties the law allows for the sane.[27] In Texas, for example, an article of the code of criminal procedure required juries assessing the appropriateness of the death penalty to consider, among other things, "whether there is a probability that the defendant would commit criminal acts of violence that would constitute a continuing threat to society": mitigating evidence such as genetic predisposition is thus a double-edged sword, which may diminish blameworthiness for the crime at the same time as it indicates the probability that the criminal may be dangerous in the future and is beyond redemption, hence justifying the death penalty.[28] To this one can add the rise of demands for preventive detention for "psychopaths," "pedophiles," and other "monstrous individuals" thought to be constitutionally incorrigible and a permanent threat to "the public."[29] Ours is an age where political rationalities increasingly stress the moral obligations of individuals for their own action and their own welfare, and their obligations to their families and communities. In this political and moral context, it is not surprising that courts and legislatures, faced with what is perceived as a threatening epidemic of violent crime and by repeated panics about such monstrous individuals as serial killers, psychopaths and sexual predators, increasingly insist that moral culpability, especially in relation to violent or antisocial conduct, should not be mitigated by any social, familial, medical, or biological factors.

We may, as Nietzsche predicted in 1878, have come to recognize that "freedom of the will is an error." But we cannot, it seems, abandon the idea of responsibility. On the contrary, within the criminal justice systems of our contemporary cultures of individual accountability, we reconceptualize offenders as creatures inescapably required to bear full responsibility for the outcomes of their actions, and deem these actions to be moral choices whatever their material causes.

The Experts

"Neuroethicists" in the early years of the twenty-first century have worried about the implications of genetics and brain scans for ideas of criminal responsibility. As far as the brain is concerned, many papers routinely cite the case of Phineas Gage, apparently a respectable individual before his personality changed in 1848 when a tamping iron was driven through the frontal lobes of his brain by an explosion on the railway line he was helping construct: following the accident, Gage was impulsive and lacked inhibitions and so forth. Gage is just one of a series of subjects of neuropsychology called upon to support an argument that our growing knowledge of the ways in which aggressive or criminal behavior can be exacerbated by brain damage, brain anomalies, or neurochemical changes will trouble the traditional doctrine of free will. This has led to many speculative articles, and much convoluted reasoning about the relations between bran and mind, determinism and freedom. However most of those who consider this issue seriously are in agreement with Michael Gazzaniga: neuroscience "will never find the brain correlate of responsibility, because that is something we ascribe to humans—to people—not to brains. It is a moral value we demand of our fellow, rule-following human beings" (Gazzaniga 2005: 101). The attribution of responsibility, to put it simply, is a matter of the organization of our modes of government, not a reflex of our knowledge of the brain. This, indeed, is what emerges from the review of cases; it supports the conclusion drawn by Stephen Morse—that existing legal principles are perfectly capable of handing any arguments and evidence from neuroscience, at least in the foreseeable future (Morse 2004).[30] And, while a few tendentious enthusiasts make wild claims about the implications of the new biology of conduct, and some draw eugenicist and racist conclusions,[31] this also seems to be the view of the majority of contributors to the scientific discourse of biological criminology, whether they are discussing the implications of genetics, neurochemistry, or brain scans.

Not, of course, that all biomedical researchers took this view in the early 1990s. In a comprehensive review in 1996, Evan Balaban produced a critical review of animal and human research on the biology of aggression published in the previous ten years (Balaban et al. 1996). He pointed to the media attention given in the early 1990s to claims of having discovered the genetic or neurochemical basis of aggression in mouse models, and noted that there had been a significant shift in the thinking of some researchers from the consensus among those studying aggressive behavior in the 1980s. While in the 1980s, he suggests, most leading researchers in the field argued that any attempt to find biological correlates of aggression were based on simplistic, distorted, or erroneous interpretations of the available evidence this situation had changed by the 1990s. He quotes from papers arguing that abnormalities in the serotonin mechanism are associated with impulsive, aggressive behavior, and suggestions that some will use genetic findings that link aggression to alleles of specific genes in mice to suggest implications for understanding aggression in humans (he cites Coccaro 1992, Golden et al. 1991, Tecott and Barondes 1996, Virkkunen et al. 1994). But, he argues, a meta-analysis shows that there are no differences in mean levels of the serotonin metabolite in CSF that is cited as evidence in such arguments (5-H1AA), and that there is no demonstrated association between low 5-H1AA levels and impulsive aggression or criminality, and suggests that the specific focus on serotonin is misleading, since variations in serotonin levels have been implicated in such a wide variety of behaviors. Criticizing the breathtaking oversimplifications in thinking and interpretation of results that characterize such claims, he takes particular exception to Hen's wish to identify "mean genes" (Hen 1996) which, among much else, ignores "pleiotropic" changes—that is to say multiple changes arising from the same alteration in the genome—associated with the "knocking out" of certain genes in mice and rats. "At the present time," Balaban concludes, "it is a scientific non-sequitur to speak of mean genes or a specific neurochemistry, of aggression" (Balaban et al. 1996: 35).

Clearly, for Balaban and those who think like him, developments in the biology of aggression should have no impact on the criminal justice system. But this is also the position of the majority of those who do think the science supports a link between genetics, neurochemistry, and aggression or impulsivity. Consider an article from 1995 in the *Journal of Forensic Psychiatry*, written by Heikki Vartiainen, the leader of a Finnish team of researchers who had studied different aspects of the neurochemistry of aggression (Vartiainen 1995).[32] It was entitled "Free will and 5-hydroxytriptamine." 5-hydroxytriptamine (known as 5-HT) is the neurotransmitter serotonin. The paper reports abnormalities in serotonin levels in the cerebrospinal fluid (CSF) of people who have taken their own lives and in offenders who have carried out violent crimes, some of whom show excessive sensitivity to alcohol. It links this to reports of a hereditary factor in some alcoholic and violent offenders, to brain scans showing abnormalities in individuals with neuropsychometric deficits, and to arguments that impulsive aggressive individuals have decreases in certain aspects of the serotonin uptake mechanism in the medial prefrontal cortex. Not all biological factors are hereditary, but nonetheless, "The relationship between a low serotonin turnover and impulsive, aggressive behavior seems to be obvious. . . . A display of uncontrolled and uncharacteristic anger following minimal provocation can be biologically explained—a decrease in brain 5-HT manifesting itself as aggressive behavior" (7).

Neurogenetic determinism? Perhaps. Yet on the basis of his findings, Vartiainen does not propose revising criteria for judgments of guilt in criminal trials. He argues that "since all behavior is biologically based, attributing causation to a given type of conduct as biological and calling it therefore an illness, tells us nothing about the social, moral, or legal implications which that behavior ought to have" (8). His view is that sentencers should not be concerned with whether a biological condition weakens legal responsibility, but with the protection of society and the reduction of the likelihood of recurrence of violent acts. In arguing in this way, he is typical of contemporary biocriminology. Biological accounts of propensities to antisocial behavior deploy conceptions of moral responsibility and individual culpability that are unmitigated by evidence of biological causation, and in which social classifications such as illness have no legitimate relevance (cf. Dinwiddie 1996). The jurisprudence that is called for is one of public health. Offenders should be held morally accountable for the consequences of their conduct irrespective of heredity, neurophysiology, or neurochemistry; thus they should be liable for whatever is the most appropriate sentence in the light of the need for public protection.[33] Tests may help us identify people vulnerable to biologically based diseases and their consequences; where we can help them, therapy might be appropriate, but when past conduct shows incorrigibility or expert evidence suggests untreatability, then the need for public protection should determine the appropriate disposal.

I will return to these themes later in this paper. But let me stick with the researchers for a bit longer. Most biological criminologists writing at the end of the twentieth century hedged their analyses with qualifications. They were quick to acknowledge that crime as such does not exist; that lawbreaking acts are heterogeneous; that crime is culturally and historically variable; that infraction of law is common; that those arrested, charged and convicted are not representative of those who break the law but a skewed sample produced through all sorts of social processes.[34] In this new positivism, conduct was never simply "caused" by biology and significant biological malfunctions may themselves be the result of environmental assaults, etcetera. As far as genetics is concerned, David Wasserman was confident in asserting, in 1995:

> No mainstream researchers believe that there are single genes that cause violent or anti-social conduct; all regard behavioral phenotypes like criminal behavior as arising from a complex interaction of many genes and environmental factors. None believe that genetic influence makes criminal behavior less mutable, and many suspect that the most effective ways of countering genetic influence will involve social and economic reforms. Finally, few of these researchers advocate, or believe their findings would support, mandatory screening, involuntary medication, or harsher sentences. (Wasserman 1995: 15)

Indeed even Han Brunner, whose work was cited in the Mobley case, was stating, by 1996, that his research gave no support for the notion of an "aggression gene," despite having been interpreted in this way by the popular press: "the notion of an "aggression gene" does not make sense, because it belies the fact that behavior should and does arise at the highest level of cortical organization, where individual genes are only distantly reflected in the anatomical structure, as well as in the various neurophysiological and biochemical functions of the brain . . . although a multitude of genes must be involved in shaping brain function, none of these genes by itself encodes behavior" (Brunner 1996: 16).

Or take Adrian Raine, whose studies demonstrating brain abnormalities in individuals accused of murder and pleading Not Guilty by Reason of Insanity (NGRI) have been cited by defense lawyers seeking to mitigate the responsibility of their clients. He also makes regular appearances in the media. For example, discussing his appearance in a speculative BBC television documentary called "If . . . We Could Stop the Violence," broadcast in 2004, Raine set out his controversial views. He asserts that brain imaging studies demonstrate that murderers, psychopaths, and individuals with aggressive antisocial personalities have poorer functioning in the prefrontal cortex—that part of the brain involved in regulating and controlling emotion and behavior. He claims that they also show that the brains of criminals are physically different from noncriminals, revealing an 11 percent reduction in the volume of gray matter (neurons) in the prefrontal cortex. Further, he argues, evidences shows that genetic processes account for 50 percent of antisocial and criminal behavior and also shape antisocial behavior in children. He concludes that, "One of the reasons why we have repeatedly failed to stop crime is because we have systematically ignored the biological and genetic contributions to crime causation. We instead need to focus efforts on new interventions that will improve brain structure and function." Intriguingly, these include some rather uncontroversial suggestions—better childhood nutrition, perhaps giving prisoners diets rich in fish oil, and early intervention such as prenatal and perinatal health care—although he does argue for the use of medication for severely aggressive children.[35]

Discussing potential future developments, Raine's speculations join those of popular journalism and science fiction: he imagines a future in which new drugs might correct the neurotransmitter brain abnormalities that cause violence, and where reparative brain surgery might be carried out on prisoners to correct the faulty neural circuits that give rise to violence. And yet, when discussing these issues in his academic papers, his views on the question of criminal responsibility echo those we have already cited. Thus, in 1997, considering the implications of his brain scans on violent offenders, he argued that "the neural processes underlying violence are complex and cannot be simplistically reduced to single brain mechanisms causing violence in a direct causal fashion. Instead, violent behavior probably involves disruption of a network of multiply interacting brain mechanisms that predispose to violence in the presence of other social, environmental, and psychological predispositions" (Raine et al. 1997: 503). Raine and his team stress that the findings cannot be generalized to other violent offenders, and assert that "these data do not demonstrate that murderers pleading NGRI are not responsible for their actions, nor do they demonstrate that PET can be used as a diagnostic technique [and] our findings cannot speak to the issue of the cause (genetic or environmental) of the brain dysfunction, nor do they establish causal direction" (505).

Evan Balaban, reflecting on the presentations at the Maryland conference, concluded, however, that such qualifications made by those studying the biology of aggression were inadequate: the repeated demand for more research on the significance of genes and biology in the origins of violent crime itself contributes to the illusion that a potential magic bullet may be discovered that will make the streets safe (Balaban 1996). Carey and Gottesman, in the same symposium, disagree. They assert that "it ain't all genetic and it never was all genetic," and that "phrases like 'nature versus nurture,' 'the aggression gene,' and 'my genes made me do it,' belong to the history of science, science fiction and social satire, not to serious behavioral genetics" (Carey and Gottesman 1996: 89). But they were confident that the current generation of molecular genetic research will find polymorphisms associated with various aspects of antisocial behavior—not a "crime gene" but a number of loci of small effect that together influence temperament, motivation, and cognition and which influence the probability that, in certain environmental and biographical conditions, an individual will engage in an antisocial act.

Diane Fishbein, of the U.S. Department of Justice, was even more hopeful about the implications of a biologization of criminal identity. After outlining a program of research tasks to assess the relevance and significance of genetic findings for crime and violence prevention, she concluded, "Studies suggest that a subgroup of our population suffers from genetic vulnerabilities that overwhelm most environments" (Fishbein 1996: 93). But this was a cause for optimism, for "genetic traits are not immutable, they are alterable in a social environment. . . . Not only do these individuals stand to benefit greatly from the research, but the public may eventually give way to more tolerance of behavioral aberrations, understanding that behavior is not entirely volitional at all times in all individuals . . . there is little evidence that present tactics are effective; thus we need to move forward into an era of early intervention and compassionate treatment that genetic research may advance" (93). We can see something taking shape here that is far more significant than the debate over free will and responsibility: the emergence of a new human kind—the person "genetically at risk" of antisocial conduct.[36] Genetically at risk individuals are those thought to have a susceptibility to develop a condition on the basis of DNA testing or family history; hence they may be treated in the present as if they are highly likely to be affected in the future in a manner that would warrant intervention, even where they show no present signs of the problem in question, and even though the certainty, nature, timing, and severity of any difficulty cannot be predicted. It is here that the new biological criminology intersects with more general changes in strategies of crime control—the new centrality accorded in many jurisdictions, notably in the United States and United Kingdom, Australia, and New Zealand, to social and public protection.

Crime Prevention, Public Health, and Social Protection

There is a control strategy here, but it is neither a new eugenics nor a genetic determinism, at least in the senses that such terms are generally understood: the belief that the nature and life-course of an individual is predetermined by a fixed and unalterable inherited constitution. Contemporary biocriminology does not suggest that biology is destiny. It operates in the same thought style as the rest of contemporary molecular biology and molecular neuroscience, involving the logic of susceptibility, prediction, and prevention. And it is not concerned with threats posed to the national gene pool through the rate of reproduction of defective stock. The problem space in which it operates is very different, shaped by an apparent "epidemic" of antisocial, aggressive, and violent conduct that is thought to arise from a diminution of self-control, reasonableness, maturity, judgment, tact, and reasoning. To control these anticitizens, who seem to lack all the self-governing capacities that are at the heart of civilized moral agency in an advanced liberal society, a two-pronged strategy is taking shape. On the one hand, one must understand the conditions that lead to such antisocial conduct in order to identify the individuals with these propensities, and to intervene upon them to reduce the risk that they pose to their families and communities. On the other hand, one must prioritize the protection of the public from the threats to physical and mental health that such individuals and their actions represent.[37]

Within this problem space, a new research program on the biology, neurology, neurochemistry, and genetics of crime has taken shape, seeking to locate biological processes, genetic markers, and risk factors for aggressive and antisocial conduct, and to develop techniques for identifying risky individuals. Through adoption studies, hormonal research, neurophysiological studies, and studies of intellectual ability, attention deficit disorder, and minimal brain dysfunction, researchers are searching for links between specific biological abnormalities and the propensity to commit violent crime, with a view to early identification, preventive intervention, and effective treatment. It is in these terms that a new molecular biopolitics of control is taking shape.

Of course, there is nothing new in the belief that research into the backgrounds and characteristics of current offenders will enable one to develop instruments that will objectively identify "presymptomatic offenders" who are constitutionally "predisposed" to crime or "at risk" of offending, the and hence can legitimately be the targets of preventive intervention (e.g., Glueck and Glueck 1930, 1934, Glueck et al. 1943). Sociological criminology was founded in a rejection of such beliefs, dating in its modern form to Edwin Sutherland's attack on biological explanations of crime (Sutherland 1931). From the end of World War II through to the late 1970s, such arguments were largely expelled from the truth discourse of criminology—they appeared inextricably associated with scientific racism. When Wilson and Herrnstein published *Crime and Human Nature* (Wilson and Herrnstein 1985)—arguing that human rationality was subject to biological constraints, including genetic predispositions to impulsivity, aggressiveness, and low intelligence, and marshaling a range of empirical evidence to support their claim that these were associated with criminal conduct—their argument was harshly criticized, associated with the work of other biological reductionists, and dismissed as politically motivated (Cohen 1987, Gibbs 1985). But in the subsequent years, there were a multitude of proposals for "integrated" approaches to criminal behavior in which biological factors formed one key dimension. These arguments are still contested by most sociologists, who link them to sexism, racism and fascism. But they are achieving the status of truth, for example finding their way into introductory textbooks of criminology where explanations of violence involving biochemical, genetic, and neurophysiological factors are increasingly presented as based on sound empirical evidence (Wright and Miller 1998).

The arguments, now, operate within the contemporary style of genomic thought—not in terms of monogenetic determinism but polygenetic susceptibilities. That is to say, the search is for variations in the human genome at the SNP level which, in combinations, may modulate responses to environmental stimuli or increase the liability of the development of particular behavioral characteristics. A cascade of papers were publicized in the closing years of the twentieth century and the early years of the twenty-first century claiming to have discovered susceptibility loci relating to depression, anxiety disorders, and the disorders of children such as ADHD and conduct disorder.[38] And, of course, many such papers focused on aggression, impulsivity, and other undesirable forms of behavior. Take, for example, a paper published in 2000, entitled "A regulatory polymorphism of the monoamine oxidase-A gene may be associated with variability in aggression, impulsivity, and central nervous system serotonergic responsivity" (Manuck et al. 2000). The authors write:

> This study presents preliminary evidence of an association between polymorphic variation in the gene for monoamine oxidase-A (MAOA) and interindividual variability in aggressiveness, impulsivity and central nervous system (CNS) serotonergic responsivity. An apparently functional 30-bp VNTR in the promoter region of the X-chromosomal MAOA gene (MAOA-uVNTR), as well as a dinucleotide repeat in intron 2 (MAOA-CAn), was genotyped in a community sample of 110 men. All participants had completed standard interview and questionnaire measures of impulsivity, hostility and lifetime aggression history; in a majority of subjects (n=75), central serotonergic activity was also assessed by neuro-psychopharmacologic challenge (prolactin response to fenfluramine hydrochloride). The four repeat variants of the MAOA-uVNTR polymorphism were grouped for analysis (alleles '1+4' vs. '2+3') based on prior evidence of enhanced transcriptional activity in MAOA promoter constructs with alleles 2 and 3 (repeats of intermediate length). Men in the 1 / 4 allele group scored significantly lower on a composite measure of dispositional aggressiveness and impulsivity (P<0.015) and showed more pronounced CNS serotonergic responsivity (P<0.02) than men in the 2 / 3 allele group. These associations were also significant on comparison of the more prevalent one and three alleles alone (encompassing 93 percent of subjects). Although in linkage disequilibrium with the MAOA-uVNTR polymorphism, MAOA-CAn repeat length

> variation did not vary significantly with respect to behavior or fenfluramine challenge in this sample. We conclude that the MAOA-uVNTR regulatory polymorphism may contribute, in part, to individual differences in both CNS serotonergic responsivity and personality traits germane to impulse control and antagonistic behavior. (9)

Such arguments may still appear to be framed within a causal logic that postulates genes as units with given effects, that are prior to, and independent of, environmental inputs. The same is true of much research on animal models. Thus, for example, in 2003, Evan Deneris and his colleagues at Case Western Reserve University, working with mice, reported the discovery of the Pet-1gene—only active in serotonin neurones—which when knocked out produced elevated aggression and anxiety in adults compared to wild type controls (Hendricks et al. 2003). The Case Western press release, ignoring the complexities, was headed "Researchers Discover Anxiety And Aggression Gene In Mice; Opens New Door To Study Of Mood Disorders In Humans." It asserted:

> Serotonin is a chemical that acts as a messenger or neurotransmitter allowing neurons to communicate with one another in the brain and spinal cord. It is important for ensuring an appropriate level of anxiety and aggression. Defective serotonin neurons have been linked to excessive anxiety, impulsive violence, and depression in humans. . . . Antidepressant drugs such as Prozac and Zoloft work by increasing serotonin activity and are highly effective at treating many of these disorders.

Deneris himself commented "The behavior of Pet-1 knockout mice is strikingly reminiscent of some human psychiatric disorders that are characterized by heightened anxiety and violence."[39] But more sophisticated researchers, including those working with animal models, suggest that genotypes, for example those affecting the levels of activity of enzymes metabolizing neurotransmitters, interact with environmental insults, thus, for example, modulating the effects of childhood mistreatment.[40] The reference paper here for humans is usually that by Caspi and his colleagues, which was based on research with a large cohort study, and claimed that a "functional polymorphism in the promoter region of the serotonin transporter (5-HTT) gene was found to moderate the influence of stressful life events on depression. Individuals with one or two copies of the short allele of the 5-HTT promoter polymorphism exhibited more depressive symptoms, diagnosable depression, and suicidality in relation to stressful life events than individuals homozygous for the long allele" (Caspi et al. 2002: 386).

Considering the implications of this kind of research on behavioral genomics for violence prevention in 2003, Katherine Morley and Wayne Hall of the Australian Institute of Criminology, helpfully summarize the candidate gene variants that have been nominated by different groups as having a potential bearing on an "individual's liability to develop antisocial behavioral characteristics"—variants in the genes for elements of the serotonergic system linked to impulsivity; those for elements in the dopaminergic system linked to ADHD; those for elements in the noradrenergic system linked to ADHD, impulsivity, and hostility; and those linked to the activity of enzymes involved in the metabolism of neurotransmitters linked to ADHD, impulsivity, aggression, conduct disorder, and criminal conviction—nonetheless stressing that "an individual will only have a significantly increased risk of engaging in antisocial behavior if they carry a large number of variant genes" (Morley and Hall 2003: 4).[41] And they conclude that while "Genetic research is beginning to identify genetic variants that may have some bearing on an individual's liability to develop antisocial behavioural characteristics," this was not a matter of single genes—instead, the issue was relocated in the contemporary style of thought about "susceptibilities." Indeed their report was entitled "is there a genetic susceptibility to engage in criminal acts," and they argued:

> This review of genetic research on antisocial behavior has summarised growing evidence for a genetic contribution to antisocial behavior but it has also indicated that it is highly unlikely that variants of single genes will be found that significantly increase the risk of engaging in violent behavior. Instead it is much more likely that a large number of genetic variants will be identified that, in the presence of the necessary environmental factors, will increase the likelihood that some individuals will develop behavioural traits that will make them more likely to engage in criminal activities. (4–5)

Even in the early 1990s, explanations of violence taking roughly this form were entering into strategies of control. At that time, the U.S. National Institute of Mental Health launched its National Violence Initiative, under its director Frederick K. Goodwin. In this initiative, psychiatrists would seek to identify children likely to develop criminal behavior, and would try to develop intervention strategies. The *Chicago Tribune* reported in 1993 that this program raised the hope "that violent behavior can eventually be curbed by manipulating the chemical and genetic keys to aggression . . . anti-violence medications conceivable could be given, perhaps forcibly, to people with abnormal levels" (quoted in Citizens Commission on Human Rights 1996).[42] The official report from this initiative, issued in 1993 and 1994 in four volumes, called for more research on biological and genetic factors in violent crime and on new pharmaceuticals that reduce violent behavior (Reiss and Roth 1993, Reiss et al. 1994). The veteran campaigner against psychiatric violence, Peter Breggin, obtained leaked copies of many of the planning documents. He claimed that the proposal for the initiative pointed to an "emerging scientific capacity to identify the individual determinants of behavior—at the biochemical, psychological and social / environmental levels," and that "solutions must reflect increasing scientific and clinical capacities to target the individual determinants of violence" and linked these specifically to genetic and neurochemical risk factors. It proposed establishing research centers for "the testing of a variety of interventions aimed at the individual, family and community," and the summary of the proposal stated that "minority populations are disproportionately affected" (Breggin 1995–96, Breggin and Breggin 1994).

In 1992 Breggin launched a media campaign to publicize and protest the initiative—a campaign aimed particularly at mobilizing African American activists. It was this campaign that enmeshed the Maryland Conference on Genetics and Crime, which Breggin saw as an elaboration of the rationale for the NIMH Violence initiative. Certainly by 1992 the U.S. federal government, in partnership with the MacArthur Foundation, was sponsoring a large-scale initiative entitled the "Program on Human Development and Criminal Behavior," to the tune of some US$12 million per year. This was based on the view that "advances in the fields of behavior genetics, neurobiology, and molecular biology are renewing the hope that the biological determinants of delinquent and criminal behavior may yet be discovered" (Earl 1991 quoted in Breggin 1995–96). Hence the project aimed at screening children for biological, psychological, and social factors that may play a role in criminal behavior, and proposed to follow subjects over an eight-year period, with a view to ultimately identifying biological and biochemical markers for predicting criminality. While this umbrella program was withdrawn as a result of the controversy surrounding the violence initiative, individual projects from the program continue to be sponsored by the federal government.

Breggin sees this as a racist program of surreptitious governmental social control. He certainly overstates his case. Franklin Zimring, who served on the National Academy of Science Panel on the Understanding and Control of Violent Behavior "doubt[s] that genetics will ever play a major role in violence prevention in the United States. . . . The prediction of violence even in previously violent adults is an error-prone exercise. The selection of children at high risk for serious violence as adults is pure science fiction" (Zimring 1996: 106). However, Diane Fishbein argued, "Once prevalence rates are known for genetically influenced forms of psychopathology in relevant populations, we can better determine how substantially a prevention strategy that incorporates genetic findings may influence the problem of antisocial conduct" (Fishbein 1996: 91). At a minimum, she believed, the evidence "suggests the need for early identification and intervention" (91). As Daniel Wasserman pointed out, biological criminologists hope that neurogenetic research into antisocial behavior, while it will not discover "causes," might identify markers and genes associated with that behavior (Wasserman 1996: 108). Programs of screening could then be established to detect individuals carrying these markers; preemptive intervention might be planned to treat the condition or ameliorate the risk posed by the affected individual. And there are many who believe on this basis that biological expertise could be part of the risk prevention strategies undertaken by a variety of agencies of social control (Fishbein 2000).

Understood in these terms, it is clear that these genetically and biologically inspired initiatives are only one element within a complex of programs that address the issue of crime control in terms of risk management, located within strategies for the promotion of public health. Indeed, most violence prevention initiatives being developed across the United States are conceived in these terms.[43] To combat the phenomenon of crime understood as a kind of "epidemic," a whole variety of tactics is required. These include preventive intervention before criminal behavior reaches a serious level, and attempts to identify, treat, or sequester risky individuals. But they also include attempts to strengthen "immunity" and "resistance" through support to communities and families, through the work of the churches and voluntary organizations, and through a range of more familiar schemes for moral and environmental regeneration.[44] As the 1995 NIMH Program Announcement for research on violence and traumatic stress put it, "the effects of violence and trauma constitute a major public health problem for all Americans . . . interpersonal violence has in recent years come to be widely viewed as a serious public health problem."[45] Within this conception of violence prevention as public health, biological factors are now thought of as one set of risk factors for perpetration of violence, interacting with intrapersonal, familial, peer, community, and cultural factors, and with other traumas or toxins such as experience of violence, alcohol, or drugs. The early detection and treatment aspirations of biological criminology are only one of a range of tactics within this widespread reshaping of control mechanisms in which the work of many professionals, from genetic researchers through psychiatrists, police, and social workers, has come to be understood in terms of the identification, assessment, communication, and management of risk.[46]

Tactics include: minimizing risk in populations as a whole; identifying and targeting high risk zones—which may be particular geographical spaces or particular groups, communities, or subpopulations; seeking to identify the presymptomatic individual at risk through the analysis of combinations of factors statistically and clinically linked to the problematic conduct or pathology in question. And, post hoc, risk is to be reduced by subjecting problematic or offending individuals to risk assessments, entering them on risk registers, deciding on their treatment in relation to risk levels, subjecting them through risk monitoring, reforming them through intervention programs designed to build the capacities and competencies necessary for them to monitor and control their own risk or, if they are thought incorrigibly risky, incapacitating them by permanent incarceration as in the "three strikes and you're out" policies, and proposals for preventive detention of sex offenders, those still deemed "risky" at the end of a prison sentence, or those diagnosed with particular "personality disorders."

The New Biology of Control

In the biology of control of the first half of this century, to explain individual human characteristics through inheritance was to claim, first, that they were the property of particular subpopulations and, secondly, that they were unalterable. The new biology of control differs on both these dimensions. There are, undoubtedly, behavioral geneticists who still think in terms of population groups, especially enthusiasts for evolutionary biology and evolutionary psychology, and even evolutionary sociology.[47] Latter-day eugenicists such as Charles Murray and Richard Herrnstein, not to mention the more outlandish characters funded by explicitly racist organizations, clearly fall into this camp. And it is certainly the case that some violence studies, especially in the United States, have focused on racial groups, either as an explicit choice or as a consequence of selecting subjects whose family members were known to the criminal justice system—a practice that will include large numbers of African Americans and Latinos in any American city, given the over-representation of such individuals in the prisons and on probation for reasons that have little to do with biology. But such practices are contested within the science itself, where most behavioral geneticists regard it as quite mistaken to seek group dispositions for antisocial conduct. Risk, here, is understood in clinical terms: while individuals in certain groups may carry an "elevated risk" for specific conditions—as in the case of sickle-cell anemia—the practices that follow are not concerned with the control of such population groups en masse, but with the identification of specific individuals where a biological or familial predisposition may, in certain developmental or social circumstances, lead them to violent or antisocial conduct. The aim is either to restore such individuals to a condition where they can exercise adequate controls over their own will—by therapies that may be biological, psychological, or even entail changing the environment that might excite or provoke expression of these predispositions—or else to sequester them. The calls for confinement may evoke echoes of earlier demands for the segregation of those whose constitution and reproduction rendered them a threat to the health of the race, and the consequences of confinement may be equally unpalatable for those concerned. But the rationale is different. Actual or potential offenders are to be confined, not as members of a defective subpopulation or a degenerate race whose reproduction is to be curtailed, but as intractable individuals unable to govern themselves according to the civilized norms of a liberal society of freedom.

Unlike the eugenics of the first half of the twentieth century, to place something on the side of nature is no longer to place it on the side of the unalterable. Even in the hypothetical case of a gene being discovered that in a particular form, did predispose to antisocial, violent, or aggressive behavior, this would not be taken as an unalterable mark of fate, justifying forcible sequestration, sterilization, or euthanasia. In the current overstated rhetoric of molecular biology and neurogenetics, once one has identified the genetic basis for an undesirable characteristic, and once one has identified individuals genetically at risk, interventions to reduce that risk can then begin: psychopharmacology, gene therapy, environmental control, skills in life management, cognitive restructuring. Within conceptions of crime control as public health, new control possibilities open up for the utilization of such risk-minimization techniques in connection with biological conceptions of the bases of violent or antisocial behavior.

Full-scale screening of the inhabitants in the inner cities might seem too controversial to contemplate in most jurisdictions. However, in 2004, President George W. Bush's New Freedom Commission on Mental Health proposed a program of widespread screening for undiagnosed psychiatric disorders, coupled with adoption of the methodology of the Texas Medication Algorithm Project. TMAP, initiated in 1995, involved the development of an "algorithm"—actually a flow chart to guide practitioners through a treatment plan from first diagnosis—involving increasing uses of psychopharmaceuticals depending on response, up to and including Electro Convulsive Therapy (ECT). This proposal was widely criticized, not least on the grounds of the apparent links between the politicians proposing it and the pharmaceutical companies who partly funded it and stood to benefit from it. Yet the President's Commission proposed coupling this treatment program with a proposal for comprehensive mental health screening for "consumers of all ages," including preschool children: "Each year," wrote the Commission, "young children are expelled from preschools and childcare facilities for severely disruptive behaviors and emotional disorders." Schools are in a "key position" to screen the

52 million students and 6 million adults who work at the schools (Lenzer 2004: 1458).

There is, as yet, no evidence that such screening makes use of genetic profiling or neurochemical assessments. Nevertheless, there is an increasing likelihood that there will be proposals for genetic screening of disruptive school children, with preemptive treatment a condition of continued schooling: indeed at least one group has suggested that their multigene test could have such diagnostic applications (Comings et al. 2000). Similarly, one can foresee postconviction screening of petty criminals, with genetic testing and compliance with treatment made a condition of probation or parole. Or one can envisage scenarios in which genetic screening is a condition for employment, or genetic therapy is offered to disruptive or delinquent employees as an alternative to termination.[48] There are suggestive precedents. Many psychiatric medications, for example antabuse for alcoholism and lithium for manic depression, were introduced in this way. Unlike the negative eugenics of an earlier era, the contemporary biologization of risky identities in the name of public health offers biological criminologists a role as therapeutic professionals—therapeutic for individuals and for society itself.

A Critical Biopolitics of Control?

Contemporary biological criminology gains its salience within a specific problem space. On the one hand, this is the apparent "epidemic" of crimes involving brutality, aggression, impulsivity, antisocial conduct, or self-glorification through violence. These crimes are viewed in highly moral terms: they are acts that seem to show wanton disregard for the moral constraints on the conduct of free individuals in a liberal society. They are not pathologies of a population group, but of individuals who reject the bonds of moral community and who violate the norms of freedom and self-control that lie at the heart of the moral order of an "advanced" liberal society. Extraordinary moral panics have surrounded these "evil" persons, apparently beyond any hope of reform and unwilling or incapable of restraining their perverse desires. They have led to widespread demands for the identification and preventative detention of such monstrous individuals. Proponents of claims about the inherent biological propensity of some persons to incorrigible, antisocial, aggressive, and violent behavior not only support these specific proposals, they also imply that the web of preventive detention in the name of the protection of the public may need to be spread wider, in order to embrace all those persons whose very makeup renders them dangerous to others. It appears that a new branch must be added to the conventional apparatus of the criminal justice system whose role will be the permanent sequestration of those whose biology places them permanently beyond the reach of treatment or reform.

On the other hand, biological criminology is likely to play a more modest role, as an element in the more general rise of public health strategies of crime control, focusing on the identification of, and preventive intervention upon, actual or potentially antisocial individuals, using screening programs in childhood or adolescence, in school or in the juvenile courts, to identify such individuals at an early stage, and enable them to be treated. Practices for the identification, calculation, and management of biological risk factors will take their place among a host of others in an expanded role for control professionals in a political and public sphere suffused by insecurity, characterized by the political dominance of the precautionary principle, where risk thinking dominates much of mental health practice, and where psychiatric professionals have already been given the obligation of governing, and being governed, in the name of risk.

The combination of the idea of susceptibility, the emergent technologies of screening, and the promise of preventive medical intercession with drugs is potent. Especially in a world in which the preventive prescription of psychiatric medication has become routine. Even more when it is embedded in a political and public sphere suffused by the dread of insecurity. Critical analysis needs to move beyond the traditional dichotomies—free will versus determinism, society versus biology—for these cannot help us understand the relations of power, knowledge, ethics, and subjectification that are taking shape within these new practices of control. Instead, perhaps, a critical biopolitics of control needs to ask what are the benefits, what are the dangers, what are gains, and to whom, and what are the costs, and to whom, of strategies of control that seek to identify and govern biologically risky individuals in the name of public protection?

Afterword

Somatic Ethics and the Spirit of Biocapital

The events traced in this book are not episodes of a single narrative—there is and will be no single point of culmination or transformation. If mutations are occurring in the relation between life and politics, we are neither at the beginning nor at the end, but in the middle. No doubt many of the hopes embodied in these practices will be dashed, most fears will prove unfounded, many impediments and complications will block or divert "implementation," and some quite unexpected or unanticipated things will occur. When innovations do reach the consulting room or the clinic, the procedures and interventions that now seem daringly radical will soon appear normal and become routinized. Reputations and fortunes will be made and lost, but many bioeconomic predictions will turn out to have been wildly optimistic, many strategies for discovery and commercialization will fail, and biotechnological change will be gradual rather than revolutionary, incremental rather than epochal. Most of the alterations that do occur will be at the practical and technical level, in a multitude of small modifications in clinical and therapeutic procedures. And these changes will soon become so integral to our ways of seeing, thinking, and acting that it will be hard to recognize their novelty.

In writing the essays in this book, I have therefore tried to avoid breathless epochalization, and to remember that we are not blessed or cursed by being at some turning point. We do not stand at a unique moment in the unfolding of a single history, but in the midst of multiple histories. But I have tried to demonstrate that our future, like our present, is emerging from the intersection of a number of contingent pathways that, as they intertwine, might make some differences to how we live, and to who we think we are. I hope that I have managed to show that the speculation about our move into a "posthuman future" is overblown, as are many of the worries associated with it. But that nonetheless, in all manner of small ways, I think that things will not be quite the same as they were.

Writing "in the middle," then, I have tried to describe some of the mutations that are making a difference that can be diagnosed. In the molecular biopolitics of our present, many aspects of our human vitality have already become technical, opened up to manipulation and modification in the operating theater, the clinic, the schoolroom, the military, and in everyday life. We do often still speak of many aspects of our vitality as given by our nature, think of certain aspects of our bodies and our minds as natural, and define ourselves, at different times as healthy or unhealthy, well or sick, not least because we inhabit so many practices—schooling, employment, insurance—that require us to do so. For these and other reasons, it is quite possible for many still to imagine human life as an essential endowment—human nature—and the human being as a living creature with natural organic norms. Yet the very borders between life and death, borders that are still so final, have become so open to negotiation and dispute. As, indeed, is the liveliness of all those entities such as tissues and ova, hovering between life and death, oscillating between vitality in a test tube or vat and information in a database or biobank. And in so many of our everyday and our medical practices, human bodily and mental capacities are not taken as given, biology is no longer destiny, judgments are no longer organized in terms of a clear binary of normality and pathology, and the familiar distinction between illness and health is blurring. It is becoming increasingly difficult to pretend that there is a line of differentiation between interventions targeting susceptibility to illness or frailty on the one hand, and interventions aimed at the enhancement of capacities on the other. In the world of risk, susceptibility, prudence, and foresight, we see new practices and styles of judgment where identification of biological—genetic, neuronal, etc.—risk can switch the affected individual, or potential individual, onto circuits of compulsory treatment, constraint, and even exclusion. And, in the case of an ovum, sperm, or fetus, such a diagnosis can lead to an irreversible diversion from the path of potential life into the realm of nonlife. But we also see another dream, another hope—from doctors, geneticists, biotech companies, and many "afflicted" individuals and their families—that presymptomatic diagnosis might be followed by technical intervention at the biological level to repair or even improve an organism, and hence a life, that would otherwise be painful, short, or suboptimal. The political vocation of the life sciences today is tied to the belief that in most, maybe all cases, if not now then in the future, the risky, damaged, defective, or afflicted individual, once identified and assessed, may be treated or transformed by medical intervention at the molecular level. This does not so much imply that we now think of the body as a machine, but rather that humans have become even more biological, at the same time as the vitality of the body has become increasingly open to machination.

In and through such developments, human beings in contemporary Western culture are increasingly coming to understand themselves in somatic terms: corporeality has become one of the most important sites for ethical judgments and techniques. In this regime, each session of genetic counseling, each act of amniocentesis, each prescription of an antidepressant is predicated on the possibility, at least, of a judgment about the relative and comparative quality of life of differently composed human beings and of different ways of being human. As biomedical technique has extended choice to the very fabric of vital existence, we are faced with the inescapable task of deliberating about the worth of different human lives—with controversies over such decisions, with conflicts over who should make such decisions and who should not, and hence with a novel kind of politics of life itself. This politics is not one in which our authorities claim—or are given—the right, the power, or the obligation to make such judgments in the name of the quality of the population or the health of the gene pool. On the one hand, in the new forms of pastoral power that are taking shape in and around our genetics and our biology, questions about the value of life itself infuse the everyday judgments, vocabularies, techniques, and actions of all those professionals of vitality—doctors, genetic counselors, research scientists, biotechnology executives, drug company employees, and more—and entangle them all in ethics and ethopolitics. And, on the other hand, the politics of life itself poses these questions to each of us in our own lives, in those of our families, and in the new associations that link us to others with whom we share aspects of our biological identity. Our biological life itself has entered the domain of decision and choice; these questions of judgment have become inescapable. This is what it means to live in an age of biological citizenship, of "somatic ethics," and of vital politics.

Of course, many profound inequalities of life and health will persist, or will be tackled in other ways. Heart disease, obesity, stroke, diabetes, and other common complex disorders will prove resistant to genomic explanations or interventions and will demand investment in the more mundane yet more effective measures of the medicine of public health. Millions of lives will continue to curtailed by factors that have little need of fancy genomic medicine or neurogenetics—by poverty, by the lack of good food, pure water, or decent sanitation, by the price of drugs, by

AIDS, malaria, and much more. There is a politics of life here too, which I have not discussed in the preceding chapters, a politics of NGOs, of philanthropy, of the innovative endeavors of some biomedical organizations to address the health problems of the poor, and of a growing cosmopolitan consciousness. Yet here too, a somatic ethics is taking shape: the sense that all human beings on this planet are, after all, biological creatures, and that each such creature exercises a demand on each other simply by being a creature of this sort. Perhaps this "biological reductionism" should not be a cause for critique but the grounds for a certain optimism.

■ ■ ■

Max Weber famously argued that there was an "elective affinity" between a certain religious ethic of worldly asceticism that he saw in Calvinism and the early emergence of capitalism in Europe and North America (Weber 1930). This thesis has, of course, been the subject of extensive debate, interpretation, and empirical refutation. But it was grounded in Weber's more profound insight that central to the ways in which human beings conduct their lives in different times and places was what he termed a "soteriology," a way of making sense of one's suffering, of finding the reasons for it, and thinking of the means by which one might be delivered from it. At least in part it is in relation to such a soteriology, in its various historically and geographically specific forms, that human beings identify and interpret their unease, and conduct a life that enables them to accept or overcome it. Today, this provokes the following question: Is there some particular affinity between the "economies of vitality" that I have delineated in the preceding chapters, and the contemporary "somatic ethics" that I have also described? Is there a relation between the birth of the bioeconomy and the emergence of the living biological body as a key site for the government of individuals, as the contemporary locus for so much of our unease and discontents, as the site of hope and potential overcoming? What are the links between the contemporary salience of biocapital, and all the novel forms of ethical work that human beings do to themselves in the name of health, longevity, and their vital existence?

We need, in thinking about these questions, to distinguish the sense of an ethic that is invoked here from that entailed in the idea of bioethics. In an earlier chapter I asked why it was that bioethics had achieved such salience today, and provided some preliminary and provisional answers. Bioethics certainly can operate as a legitimation device within the regulatory technologies of government as they deal with these highly controversial issues of life and its management (Salter and Jones 2002, 2005). Bioethics provides the essential ethical guarantees that enable elements—tissues, cells, eggs, sperm, embryos, body parts—to move legitimately around the circuits of biocapital so that they can be combined and recombined in settings from laboratory to clinic (Franklin 2003). Bioethics can serve to insulate researchers from criticism, and from the detailed examination of the nature and consequences of their activities, by routinizing and bureaucratizing the processes whereby they can obtain "ethical clearance" for what they do. Bioethics often seems to arise from an alliance, perhaps an unhealthy one, between those who want or need an ethical warrant for their commercial or scientific activities—whether they be pharmaceutical companies or those whose careers depend upon research with human subjects—and those—philosophers, theologians, ethicists, and others—who see here a potential locus for grants, recognition, a professional vocation, and a public role. And, as some critics claim, there are certainly moments when bioethics, and the clean bill of health it can offer, seems to be "for sale": when bioethicists, in taking subsidies for their educational activities, accepting grants, and acting as consultants to biotechnology and pharmaceutical companies, may have betrayed the trust vested in them, legitimating the unacceptable, at the cost of human lives (Elliot 2004).

Such critical analyses of the sociopolitical role of bioethics are instructive, although they need to be nuanced in relation to empirical studies of the actual role of different aspects of the discourses, practices, forms of expertise, and strategic engagement of bioethics in different places and practices (Lopez 2004). But alongside the urge to critique, perhaps we need to attend to what it is that this demand for bioethics manifests. Perhaps, at the simplest, we need to distinguish between two general senses in which the biological and the ethical are intertwined.

On the one hand stand those practices and ways of thinking that might more accurately be termed "biomorality," whose aim is to develop principles, and promulgate codes and rules as to how research or clinical work in biomedicine might be conducted. At a time when the somatic, the bodily, the "bio" have become so central to our forms of life, we should not be surprised that one response is, indeed, to seek to discipline these difficulties, to find some algorithms to adjudicate about them, to standardize procedures for the potentially conflictful decisions concerning them. In this way, these problematic issues are transformed into technical questions: Have the proper procedures been followed? Have the proper permissions been obtained? Is confidentiality assured? Has informed consent been obtained? Bioethics, here, like accountancy, legal regulation, audit and the like, has indeed become an essential part of the machinery for governing the bioeconomy, for facilitating the circuits of biological material required for the generation of biocapital, and for the government of all those practices in which life itself is the object, target, and stake.

But we also need to consider another sense in which we can think of bioethics. This other sense concerns the ethical considerations deemed relevant by participants—not just patients and their families, but also researchers, clinicians, regulators, and even those working in the world of commerce—in their actual conduct of themselves and their lives in relation to the dilemmas that they face and the judgments and decisions that they must make. We can learn much about this ethic from all the detailed ethnographies of biosocial communities that I have drawn on in previous chapters. These demonstrate the ways in which those who I have termed "biological citizens" are having to reformulate their own answers to Kant's three famous questions—What can I know? What must I do? What may I hope?—in the age of the molecular biopolitics of life itself. But I think we can also see something similar in studies of the ethos of the authorities and professionals enmeshed in contemporary vital politics, in those working in and for commercial biotechnology and pharmaceutical companies and perhaps even in those stock market actors and investors whose concerns seem purely financial. While they may have their own share of cynicism, pragmatism, ambition, greed, and rivalry, they are also inescapably searching for, assembling, and inventing ways in which they might evaluate, adjudicate, and ethically justify the decisions they must make when human vitality is at stake.

It is this sense that is closest to the notion of somatic ethics that I have developed in this book. Ethics, here understood as a way of understanding, fashioning, and managing ourselves in the everyday conduct of our lives.[1] If our ethics has become, in key respects, somatic, this is in part because it is our "soma"—or corporeal existence—that is given salience and problematized—to some extent at least, our genome, our neurotransmitters—our "biology." Second, it is also because the authorities that now articulate the rules for living, whose injunctions shape our relations with ourselves, now include not merely doctors and health promoters, but so many other "somatic" experts, genetic counselors, advice and support groups, projects for the public understanding of genetics, and, of course, bioethicists. Third, it is because the forms of knowledge that are shaping our understandings of ourselves are themselves increasingly "biological"—medical, of course, but also coming more directly from genomics and neuroscience, in their popular presentations, their scientific elaborations, and in the hybrid forms they take within lay discourses of everyday life. And, fourth, it is because our expectations—the ways in which we are shaping our hopes for salvation, for the future for ourselves—are themselves shaped by considerations about the maintenance of health and the prolongation of earthly existence, and about the future beyond that. This future is now less seen as one in which we will ourselves achieve immortality in a promised land, than as one in which we will live forever through our offspring. It thus seems to require us to commit our own energies to the future biological health of those who will carry us forward into the future—to our children and their own healthy lives. The management of health and vitality, once derided as obsessive or narcissistic self-absorption, has now achieved unparalleled ethical salience in the conduct of the lives of so many.

This, then, is the somatic ethical economy that perhaps has an elective affinity with a certain form of capital, biocapital, and with the capitalization of life itself. Of course, we do not have to decide between a materialistic or a spiritualistic interpretation of these developments.[2] Somatic ethics and biocapital have been locked together since birth. For only where life itself has achieved such ethical importance, where the technologies for maintaining and improving it can represent themselves as more than merely the corrupt pursuit of profit and personal gain, can place themselves in the service of health and life, would it be possible for biocapital to achieve such a hold on our economies of hope, of imagination, and of profit. In this sense, I suggest, somatic ethics is intrinsically linked to the "spirit of biocapital."

Coda

This spiritualization of the flesh, this sensualization of ethics would, no doubt, be greeted with a certain disdain by the critical intellectuals writing at the turn of a previous century. They wanted to set us humans the assignment of inventing a less mundane and animalistic, a more elevated and spiritual ethic. But, in the words that Weber used as he concluded his famous essay, "this brings us to the world of judgments of value and faith, with which this purely historical discussion need not be burdened. The next task would be rather to show the significance [of this development] for the content of practical social ethics, thus for the types of organization and the functions of social groups from the conventicler [the irregular assembly of believers] to the State" (Weber 1930: 182).

The essays in this book have merely attempted the beginning of this task. Among the many criticisms that might be made of them is, no doubt, their absence of the familiar tropes of social critique. For all the talk of biopower, who has the power, who wields it to what ends, who gains and who loses? For all this talk of biopolitics, where is the analyses of hidden interests, of the role of class, the new forms of inequality and exploitation that are arising? There is, it might be said, too much description, too little analysis, too little criticism. Where so many judge, however, I have tried to avoid judgment, merely to sketch out a preliminary cartography of an emergent form of life and the possible futures it embodies. And in doing so, not to judge, but I hope, to help make judgment possible. To open the possibility that, in part through thought itself, we might be able to intervene in that present, and so to shape something of the future we might inhabit.

My attitude to criticism arises from the sense that the social and human sciences might learn something from the new epistemologies of life that are taking shape in the biosciences. Sociology and biology were born as disciplines within the same historical and geographical space in the nineteenth century. They shared an epistemology of depth. Historians of the social sciences have long emphasized the significance of biological models and metaphors—of biological styles of thought—in the explanatory systems that took shape in the second half of the nineteenth century and the first half of the twentieth: organisms, functions, systems, and all the rest of it. Perhaps, then, it is not too much to suggest that as the styles of thinking in biology mutate, so then should the styles of thought in those disciplines seeking to understand their social organization and consequences. The critical social sciences also need to understand that the most profound thought is that which remains on the surface.

Notes

Introduction

1. The assumed reference of this "we" was to the inhabitants of the advanced liberal nations of the first world, although sometimes it seemed to encompass the whole of humanity. It does not. A tiny proportion of biomedical resources are directed to the health problems of the majority of the world's population. Médecins Sans Frontières reported in 2004:

> Ten years ago, the world spent US$30 billion on health research of which under 10 percent was spent on 90 percent of the world's health problems—a disparity known as the "10/90 gap." Today global spending on health research has more than tripled to under US$106 billion, yet the amount allocated to the R&D for drugs to treat 90 percent of the global disease burden has risen by a mere US$0.3–0.5 billion to around US$3.5 billion, mainly due to contributions from private foundations, governments, and charities. Thus, the 10/90 gap doesn't just persist, in percentage terms it shows alarming growth over the last decade.
>
> (*http://www.msf.org/content/page.cfm?articleid=3534F412–8F82–4E5E-B4459FE9B5C666AF* accessed 15 January 2005).

Of 1,393 new chemical entities brought to market between 1975 and 1999, only 16 were for tropical diseases and tuberculosis. There was a 13-fold greater chance of a drug being brought to market for central-nervous-system disorders or cancer than for a neglected disease (Trouiller et al. 2002).

2. This idea of an "emergent form of life" is, of course, not original. As I say in chapter 3, I adopted it from the title of a symposium organized by Stefan Beck and Michi Knecht at Blankensee in Germany in 2003. It was also used by Michael

Fischer as the title of a recent collection of his anthropological essays (Fischer 2003). I discuss my use of it in chapter 3.

3. As the millennium approached, Sarah Franklin and I independently started to work with this idea of "life itself"—an idea that we both adapted in different ways from themes in the writings of Michel Foucault. Her paper (Franklin 2000) was written about the same time as my own (Rose 2001) and I benefited from reading a draft while finalizing mine.

Chapter 1 Biopolitics in the Twenty-First Century

1. Nor did the 1980s and 1990s see the "death of the clinic," as Donna Haraway once suggested in "A Cyborg Manifesto": "The clinic's methods required bodies and works; we have texts and surfaces. Our dominations don't work by medicalization and normalization anymore; they work by networking, communications redesign, stress management. Normalization gives way to automation, utter redundancy" (Haraway 1991).

2. This is not the place for a full account of all these developments, but Richard Horton, editor of the very influential medical journal, *The Lancet*, provides a very interesting overview of some of the key issues (Horton 2004).

3. To take just one example, Mintel reported, in 2003, that sales of self-testing equipment such as blood pressure monitors and blood glucose testers had increased dramatically, that almost 60 percent of Britons had at least one piece of diagnostic medical equipment, and that sales of such equipment had increased by 32 percent over five years and were predicted to continue growing (reported in *The Guardian*, October 29, 2003).

4. From the 1960s to the 1980s, this threat to the authority of the clinician was responded to by a number of attempts to develop a philosophy of medical practice from the standpoint of the clinician at the bedside, and to give a philosophical and conceptual justification for the priority of clinical judgment (see, for example, Engelhardt and Towers 1979, Feinstein 1967). I am grateful to Uffe Juul Jensen for putting the issue so clearly in a seminar presentation entitled " 'We really need a whole new philosophy of medical knowledge'—and more" given to the BIOS Centre in November 2005. He considers that his own book (Jensen 1987) was also part of this movement to defend the primacy of clinical practice.

5. I use the term "molar" here in the sense that the Oxford English Dictionary defines as "Of or relating to mass; acting on or by means of large masses of matter. Often contrasted with molecular."

6. This, of course, is the "clinical gaze" whose archaeology is traced by Michel Foucault in *Birth of the Clinic* (1973); a gaze beautifully illustrated in the catalogue of the *Spectacular Bodies Exhibition* mounted in London in 2000 (Kemp and Wallace 2000).

7. For an introduction to PXE, see *http://www.geneclinics.org/profiles/pxe* (accessed Jan. 12, 2004).

8. For an analysis of the invention of just one of the technologies, the polymerizing chain reaction, see Rabinow 1996b.

9. Others have noted this shift and analyzed it in various ways, some drawing from literary or cultural studies, and focusing on shifting metaphors and tropes in the language and writing of prominent biologists (see, for example, Doyle 1997).

10. A useful chronology of blood transfusion can be found at *http://www.bloodbook.com/trans-history.html*. See also, Starr 2002.

11. For an analysis of the early stages of this process see Scott 1981. For an acute analysis of cultural differences in organ donation, see Lock 2002. For critical analyses of the commodification of organs, and the organ trade, see Scheper, Hughes, and Waquant 2002.

12. On the general question of "tissue economies," see Waldby and Mitchell (2006).

13. See her on-line publication Hannah Landecker, no date, "Living differently in time: plasticity, temporality and cellular biotechnologies," at *http://culturemachine.tees.ac.uk/Articles/landecker.htm* (accessed December 1, 2005). Thanks to one of the reviewers of this manuscript for directing me to this article.

14. I discuss this in more detail in chapter 3.

15. For information on Steve Mann, see *http://about.eyetap.org/cyborgs/* (accessed Jan. 12, 2004)

16. For information about Kevin Warwick, see *http://www.kevinwarwick.com/* (accessed Jan. 12, 2004)

17. Thanks to Btihaj Ajana for suggesting this formulation—becoming more biological—which also has resonance with Sarah Franklin's remarks on "mo biology" in her inaugural lecture at the LSE on November 24, 2005. On the issue of cyborgs, one reviewer of this manuscript suggested that cyborgism was not so much the amalgam of the organic and inorganic, of human and artefact, but a certain conceptualization of systems of communication and control. This is certainly one interpretation. However Ian Hacking, in a different conception of the cyborg, points out that the word was first used in 1960 by two multitalented individuals. The first was Manfred Clynes, who had carried out extensive work on computer-assisted biofeedback control in the laboratory of Nathan Kline at Rockland State Hospital, before he developed the technology for the CAT scan. (Hacking 1998) Kline, the coinventor of the term, was a clinical psychiatrist who Hacking tells us was also, at one time, personal psychiatric consultant to Papa Doc Duvalier. He also played an important role in the development of psychopharmacology (Healy 1997). Kline and Clynes's cyborg was invented at the behest of NASA to free humans in space from having to continually check and adjust to the environment: "The point was to supplement a human being, to make it possible to exist *qua* man, as man "*not changing his nature, his human nature* that evolved here" (Hacking op. cit.: 209; he is quoting Clynes from p. 47 of Gray et al. 1995). For fuller discussions of divergent understandings of cyborgism, the reader is referred to Gray's later book (Gray 2000).

18. Many thanks to Ian Hacking for letting me see a text of this seminar presentation, entitled "The Cartesian Vision Fulfilled: Analog Bodies and Digital Minds," given in a series on "digital scholarship, digital culture" at Kings College, London in February 2004. Despite the collective and distributed nature of cognition and our distrust of the author function, misunderstandings of his argument are mine alone.

19. Face transplants became a reality in December 2005, not long after Hacking made this argument.

20. I discuss this in chapter 8 below.

21. See chapter 3 below.

22. Tim Tully's work with his company Helicon, on the development of a memory enhancing compound based on his work with fruit flies, which I discuss in chapter 3 of this volume, received extensive publicity in April 2002, when it was dubbed Viagra for the brain (see *http://www.forbes.com/global/2002/0204/060_print.html*) (accessed Jan. 12, 2004).

23. See chapter 7 below, and Rose 2004.

24. Adriana Petryna has also used this term, but in a more restrictive sense (Petryna 2002); see the discussion in chapter 3 below.

25. For Chromosome 18, see *http://www.chromosome18.org/index.htm* (accessed Jan. 12, 2004); for Genetic Alliance, see http://www.geneticalliance.org/.

26. I have termed this the birth of the "neurochemical self." See chapter 7 below, and for earlier versions Rose 2003b, 2004.

27. I am leaving to one side the areas of crime control and psychiatry where these famous ethical principles have only limited purchase (see chapter 8 below).

28. The burgeoning of "bioethics" has, of course, attracted the attention of sociologists, anthropologist, and historians, and I discuss their role at several points in this book. However, it may be helpful to provide some basic information on this strange phenomenon at this point. While some consider bioethics to date back to ancient times and embrace all reflections on ethical issues concerning life, death, and medicine, Roger Cooter tells us that the term "bioethics" appeared in print only in 1970 in an article by Van Renselaer Potter and was taken up independently by R. Sargent Shriver and André Hellgers when they named the Joseph and Rose Kennedy Institute for the Study of Human Reproduction and Bioethics in Georgetown, Washington D.C., in 1971. The former "saw it as a new discipline combining science and philosophy," while "the Georgetown philosophers and theologians regarded it as a branch of applied ethics." Controversies over medical experiments and other criticisms of medical practice and research in the 1960s and 1970s "facilitated the empowerment of "bioethicists" to advise on the ethical limits of medicine and biotechnology" (Cooter 2004, c.f. Potter 1970; see also Jonsen 1998).

Outside the United States, certainly before the 1990s, such reflections were either the province of ethicists or more usually of medical lawyers, as, for example, in Ian Kennedy's 1980 series of "Reith Lectures" in the United Kingdom criticizing the disguised moralism underpinning many medical decisions. For Kennedy, bioethics is a rather unappealing term used in the United States for what he prefers to see as a mixture of ethics, law, philosophy, sociology, and politics as they relate to medicine (Kennedy 1981: vii).

Reflecting on the criticisms of its funding base, its restricted notion of ethics and its rejection of social scientific methods of analysis, Cooter remarks that "bioethics seems destined to a short lifespan. . . . To many, its embrace of everything bearing on human life renders it, paradoxically, bankrupt" but suggests that "it signposts the emergence of a set of tensions and realignments within the social relations of late-20th-century medicine" (Cooter 2004: 1749).

29. For a compendium on the "official history" of medical ethics in the United States and the role of the American Medical Association, see Baker et al. 1999.

30. For a critical analysis of the involvement of bioethicists with the U.S. pharmaceutical industry and biotechnology corporations, see Elliot 2001.

31. Many of these questions are discussed in a special issue of *The Tocqueville Review* 23:2 (2003). Carl Elliot is, in my view, one of the most insightful commentators on the dilemmas of bioethics (see in particular, Elliot 1999).

32. ELSI = Ethical, Legal, and Social Implications. The Human Genome Project institutionalized this trend, seeking to deflect social and political criticism by setting aside a proportion of its funds for work on ELSI issues. The European Commission Framework projects—a series of large calls for research across Europe that have had a large bioscience and genomics element—also now include funding for ELSI research. The Wellcome Trust has a biomedical ethics program also set up to fund such research.

33. Once more it is necessary to stress that there is nothing novel in itself in close relations between industrial corporations and the development of scientific research, outside and inside universities. Indeed the image of scientific knowledge as developing within the sequestered space of the university laboratory, funded by public money, detached from commercial imperatives, mobilized only by Mertonian norms of disinterestedness and the like applies, if at all, only to a small number of disciplines in an exceptional period of fifty years or so in the mid-twentieth century. The novelty, if it is to be found, lies in the particular configuration taking shape around the life sciences that I describe below.

34. I have argued elsewhere that images of the development of scientific disciplines that trace a vector running from the laboratory to society described in the language of "application" are misleading, especially in those domains of knowledge that have what Michel Foucault termed a "low epistemological threshold." The psy sciences, for example, were "disciplined" around their fields of application—in industry, the schoolroom, the military, the courtroom—and only later established in the university (Rose 1985). In particular, the impact of military priorities and funding should not be underestimated, in even the most apparently theoretical of disciplines such as example mathematics.

35. The collection of papers edited by Sarah Franklin and Margaret Lock began to develop the idea of biocapital in intriguing ways, pointing to the new hybrids of knowledge, technology, and life involved in patenting, sequencing, mapping, marketing, purifying, branding, marketing, and publicizing new life forms (Franklin and Lock 2003b): these studies contributed greatly to my own, less ethnographic approach to these issues. As I write, Sarah Franklin's own development of these ideas, in her study of cloning, is in the press, and I thank her for letting me read some of the manuscript in advance of publication (Franklin, forthcoming 2006).

36. As Franklin points out, in Volume 3 of *Capital,* Marx discusses the capitalization of cattle and sheep breeding in his account of the process in which capital became an independent and dominant force in agriculture; Franklin suggests that, in many ways, the cloning of Dolly the sheep—made possible only through the investment of venture capital and with the aim of creating transgenic "bioreactor" sheep that could create marketable enzymes for treating human diseases in their

milk—binds the oldest definitions of capital as "stock" to the newest forms that this takes in contemporary biocapital. Human aspirations thus become literally "embodied" in the vital living existence and capacities of capitalizable entities. This, then, can serve as an exemplar for many other instances of the capitalization of "stock" or "livestock" in the bioeconomy, for example in the capitalization of stem cells (Franklin, forthcoming 2006).

37. The conferences of BioCapital Europe, organized by Fortis Bank and Life Sciences Partners, enable biotech companies to present themselves to an invited audience of venture capitalists, institutions, and biotech and pharmaceutical companies looking to find the best investment opportunities within the biotechnology market. See *www.biocapitaleurope.com* (accessed November 25, 2005).

38. See *http://www.biospace.com/news_story.aspx?StoryID=20035520&full=1* (accessed November 25, 2005). There are now many such Hotbed Maps, which can be found at *http://www.biospace.com/biotechhotbeds.aspx* (accessed November 26, 2005). The original 1985 Biotech Bay(TM) Map for the San Francisco Bay Area hangs on permanent display in the Smithsonian Institution's National Museum of American History. BioCapital was first launched in 1996.

39. See *http://www.imshealth.com/ims/portal/front/indexC/0,2605,6025_1825,00.htm* (accessed on November 21, 2005).

40. Data from IMS Health, at *http://open.imshealth.com/download/may2004.pdf*

41. Sarah Franklin and Margaret Lock, attributed the term "promissory capitalism" to Charis Thompson's then unpublished work on what she termed "the biotech mode of (re)production" (Franklin and Lock 2003a). See Thompson 2005, especially chapter 6. Thompson suggests the emergence, within contemporary capitalism, of what she terms a "biomedical mode of reproduction." She describes this as a mode that has a new emphasis on reproduction, and one in which human beings are not so much alienated from the products of their labor, as Marx suggested, but at risk of alienation from their bodies or body parts: here, she suggests, capital is not so much accumulated as "constitutively promissory" (p. 258). While Thomson's analysis is suggestive, and demands more attention than I can give here, the elements she picks seem specific to her own focus on reproductive technologies, and do not characterize most of contemporary bioeconomics. The idea that speculative risk and venture capital depend on issuing promissory notes against the hope of future returns has long had a central place in Marxist and other studies of the rise of capitalist economies. I draw here on arguments made in my "Clifford Barclay Memorial Lecture," at the London School of Economics and Political Science in February 2005.

42. See, for example, the way the OECD put it in 1996: "The OECD economies are increasingly based on knowledge and information. Knowledge is now recognized as the driver of productivity and economic growth, leading to a new focus on the role of information, technology and learning in economic performance. The term "knowledge-based economy" stems from this fuller recognition of the place of knowledge and technology in modern OECD economies" (Organization for Economic Co-Operation and Development 1996: 3).

43. Available from 10 Downing Street under Prime Minister's Speeches, at *http://www.number-10.gov.uk/output/Page1548.asp* (accessed August 11, 2005).

44. See also *http://sunsite.berkeley.edu/biotech/iceland/new.html* (accessed Jan. 12, 2004). In a press release from Reykjavik, Iceland dated August 2, 2005, DeCode Genetics put a brave face on their progress but nonetheless reported, for their second quarter 2005 financial results: "Net loss for the second quarter 2005 was US$13.3 million, unchanged from the second quarter 2004. For the six months of 2005, net loss was $30.3 million, compared to $25.3 million for the same period last year. This is the result principally of increased spending on the company's drug development programs. Basic and diluted net loss per share was $0.25 for the second quarter this year, unchanged from the same quarter last year. Basic and diluted net loss per share for the first six months of 2005 was $0.56, compared to $0.48 for the same period in 2004." See *http://www.decode.com/* (accessed on August 11, 2005).

45. For the Estonian Genome Project, see *http://www.geenivaramu.ee/index.php?show=main&lang=eng* (accessed on August 11, 2005).

46. See the U.K. House of Commons Trade and Industry Committee Report on Biotechnology, 2003.

47. And, of course, especially in the United States, one should not neglect the "opportunities and challenges in biodefense" following the terrorist attacks of September 11, 2001: the Department of Health and Human Services spending on biodefense increased almost fourteen-fold from 2001 to 2005, and the Bioshield Act of 2004 earmarked £5.6 billion for U.S. countermeasures against pathogens.

48. This outsourcing is driven by the great increase in numbers of clinical trials for product development, by competitive pressures for speed and economy in drug development, by the ethical and other restrictions that hamper the recruitment of trial subjects (for instance the ban on the use of prisoners for much testing of pharmaceutical products in the United States), and by the growth of commercial organizations contracting to carry out drug trials for profit. Many of these issues are insightfully analyzed in Petryna 2005. Some indication of the growth in this area is given in a report on the Globalization of Clinical Trials by the Office of the Inspector General of the U.S. Department of Health and Human Services, in 2001, which stated that just 41 foreign clinical investigators conducted drug research in 1980 under the FDA procedure known as "Investigational New Drug Applications." By 1990 that number had grown to 271, and by 1999, to 4,458 (Department of Health and Human Services Office of Inspector General 2001). Bonnie Brescia, writing in *Pharmaceutical Executive* in 2002 on the benefits to companies flowing from such exporting of trials, estimated that:

> [a]s of 2000, there were approximately 7,500 clinical projects in the R&D pipeline worldwide, according to IMS Health. By 2003, analysts estimate the total number will grow to more than 10,000 clinical projects worldwide. CenterWatch reports that for each new drug application to FDA, pharma companies must conduct an average of 68 studies (Phases I–III) that need a total of 4,300 volunteers. With the expected increase in global R&D projects over the next year, the number of patients needed for enrollment

is expected to grow by 15 percent annually. . . . In 2000 an estimated 86 percent of all clinical studies failed to enroll the required number of patients on time. And on average, US clinical studies are delayed by 366 days. With the daily out-of-pocket cost for a clinical study at roughly $37,000, those delays are significant. Combine those expenses with the cost of losing out on daily product sales averaging $1.3 million—up to $11 million for blockbuster drugs—and timely recruitment becomes a huge consideration. (Brescia 2002)

Many have been critical of such practices. In December 2000, researchers for the *Washington Post* published a series of six articles, under the title "The Body Hunters," documenting drug trials being carried out by U.S.-based pharmaceutical companies in less developed countries. Factors driving this outsourcing of trials included the difficulty and expense of recruiting patients in the United States, the desire for speed in drug development, the apparent laxity of regulations over questions of adverse effects and placebo arm trials, and the willingness of doctors and authorities in many countries to participate. They suggested that such trials were poorly regulated, often involved poorly educated patients (recruited by poorly paid doctors, who were promised handsome fees for each subject enrolled), who did not understand that they were being used as trial subjects, and were frequently not given proper medical care. They cases they describe—all dating from the 1990s—include Pfizer's testing of its potential blockbuster antibiotic Trovan on children in Nigeria during a meningitis epidemic in 1996, and its schizophrenia drug Zeldox in Bulgaria and Hungary to answer FDA concerns about adverse effects on heart rhythm; Maxim Pharmaceuticals trial of a drug for liver disease in Russia; Aventis's use of subjects in Argentina to test its experimental drug cariporide for the treatment of heart damage after surgery; the use by the Swiss research firm Van Tx of clinical trial subjects imported from Estonia with the promise of fees and hospitality to meet urgent demands from Roche and Novartis; and VaxGen's trials of its AIDS vaccine on drug addicts in Bangkok. See *http://www.washingtonpost.com/wp-dyn/world/issues/bodyhunters/*, accessed August 9, 2005. See also Kelleher 2004. Some of the complexities, including those of evaluating the Trovan case, are helpfully examined in the paper by Petryna cited at the start of this note: references in Petryna's paper guided me to the report of Office of Inspector General's and the Brescia article cited above.

49. For MND research at King's College, see *http://www.mndcentre.org.uk/Research/GeneticFront.html* (accessed Aug. 11, 2005).

50. In the 1990s, the research in China involving the French company GENSET and the U.S.-based Millennium Pharmaceuticals became particularly controversial (see Sleeboom 2005). However, commercial organizations have developed a business model based around such collections (see, for example, Asterand, based in Detroit, "the leading supplier of human tissue in the world, whose "worldwide network of donor sites has enabled collection of over 200,000 human tissue samples reflecting every area of pharmacologic interest" *http://www.asterand.com/Services/* [accessed Aug. 11, 2005]. In 2000, the Indigenous People's Council claimed that Autogen, an Australian biotechnology research and

development company, "committed to creating shareholder wealth through developing innovative products that improve the quality of life," and specializing in using gene discovery approaches to identify novel therapeutic targets, had "bought Tonga's gene pool" in its hunt for drugs to treat diabetes, cardiovascular disease, hypertension, cancers, and ulcers. The blood would be used to extract DNA from which to form genetic pedigrees of family members in the hunt for disease-causing genes. In an interview,

> Professor Collier denied the company was practicing "bio-piracy" and said that it had followed ethical guidelines set down by the World Health Organization. "The Tongan Government will get royalties if anything comes of it, there will be more jobs and the population will get any drugs that come of the research for free," he said. Patients would be asked for their full, informed consent before samples were taken. Autogen will begin collecting DNA samples from Tonga late this year or early 2001. The DNA of Tonga and other Polynesian nations are valuable to biotech companies because they are more genetically isolated than other populations, where families are made up of people of different ethnic backgrounds. "Tonga has a lot of history in their family groupings; they know who is related to whom," Professor Collier said. But like most Polynesians, as they became more exposed to the Western culture and diet, Tongans began to die of Western diseases.
> (*http://www.ipcb.org/issues/human_genetics/human_populations/tonga/autogen_buys.html* accessed on August 23, 2005).

The ethical issues involved in such sample collections, mainly those of informed consent and benefit sharing, became particularly fraught for the Human Genome Diversity Project and for the International HapMap project, which I discuss in chapter 6.

51. I discuss changing ideas of "life" in chapter 2.

52. All these issues run together in the famous case of John Moore and the cell lines isolated from his cancerous spleen (see John Moore, Plaintiff and Appellant, versus The Regents of The University of California et al., Defendants and Respondents. Supreme Court of California, No. S006987. July 9, 1990). Andrews and Nelkin cite John Moore's comment in a newspaper interview with approval: "My doctors are claiming that my humanity, my genetic essence, is their invention and their property. They view me as a mine from which to extract biological material. I was harvested" (in John Vidal and John Carvel, "Lambs to the gene market," *Guardian*, November 12, 1994: 25, quoted in Andrews and Nelkin 2001: 1). Yet it is not clear how much the issues here hinged on Moore's consent, the distribution of financial reward, or the "harvesting" of the cells. The same blurring occurs in many critical evaluations of the organ trade by sociologists and anthropologists.

53. Sarah Franklin has pointed acutely to the way in which biotechnology corporations themselves now seek to internalize these ethical considerations in their business models and their artifacts, in what she terms "ethical biocapital" (Franklin 2003). I return to biocapital in the afterword to this volume.

Chapter 2 Politics and Life

1. I am indebted to Sarah Franklin for discussion of this issue and for many useful insights and references. Much of my argument follows and develops hers in Franklin 2000.

2. Proposed in the same year by Jean Baptise Lamarck in France and by Gottfreid Reinhold Treviranus and Lorenz Oken in Germany.

3. Canguilhem's paper was based on two lectures given at the Ecole des sciences philosophiques et religieuses of the Faculté universitaire Saint-Louis in Brussels in February 1966; excerpts are contained in Canguilhem 1994.

4. See the excerpts in the section "Knowledge and the Living," in Canguilhem 1994.

5. Evelyn Fox Keller argues, I think correctly, that the key to this distinction between "living" and "nonliving," central to the establishment of biology as a distinctive scientific discipline, was "the establishment of an intellectual space from which the category of 'life' could be taken as a given, as a domain of natural phenomena declared to be itself 'natural,' and hence one to be investigated without calling the limits of that category into question" (Keller 2002: 15–16). However, she suggests, the question "What is life?" was posed with increasing urgency in the early decades of the twentieth century, provoked by such issues as "how did life begin," which led to a proliferation of definitions—life as metabolism, movement, responsiveness to environment, complexity, self-organization, or negative entropy—and to a particular concern with understanding and even replicating, the emergence of life from that which was nonliving.

6. For a powerful explication of this idea of life, but as the capacity for reproduction combined with a genetic program, along with the possibility of variation and the existence of competition, see Jacob 1974.

Of course, there were many disputes over these issues within the biological sciences from the 1960s to the 1990s; over the definition, borders, characteristics, and indeed unity or diversity of "life." This is not the place to document these disputes, a task insightfully commenced by Fox Keller (2002). To undertake this task one would also need to distinguish between the explicit reflections on this issue of "what is life" by philosophers of biology and biologists reflecting on their discipline, and the implicit presuppositions that guide working biomedical and biological researchers, that may not be much concerned with such questions (in the same way as practicing social scientist seldom worry too much about grand questions such as "what is society?").

7. So called "junk DNA"—that is to say, sequences of bases that do not comprise the triplets coding for amino acids—comprises about 97 percent of the human genome. Some have accorded it a protective function—the high proportion of such noncoding DNA provides a statistical reassurance that most random mutations in replication will not occur in coding sequences—but others have long suspected that some of this noncoding DNA plays a part in morphogenesis and regulation. Developments in sequencing techniques that have demonstrated that much of this noncoding DNA is evolutionarily preserved have supported the view

that it must have some function, for large sections of redundant DNA would have been selected against. Many now believe that some of these noncoding DNA sequences code for RNA that has a crucial regulatory function (Gibbs 2003).

In fact, the existence of such regulatory genes not coding for proteins but acting as switches, activating or inactivating gene expression, was demonstrated in bacteria by François Jacob and Jacques Monod who, in 1965, together with André Lwoff, were awarded the Nobel Prize in Physiology or Medicine for their discoveries concerning genetic control of enzyme and virus synthesis (Jacob and Monod 1961).

Forty years later, John Mattick's group, pointing out that most of these nonprotein-coding sequences that occur within genes (introns) and between genes (intergenic sequences) a significant proportion of which (around 45 percent) are actually expressed as nonprotein-coding RNAs, were

> working on the . . . hypothesis that the majority of the genomes of complex organisms are devoted to the regulation of development and that most of this information is transacted by noncoding RNAs. Both logic and the available evidence suggest that these RNAs form a highly parallel digital network that integrates complex suites of gene expression and controls the programmed responses required for the autopoeitic development of multicellular organisms. If this is correct, our current conceptions of the genomic information content and programming of complex organisms will have to be radically reassessed, with implications well beyond biology.
> (*http://jsm-research.imb.uq.edu.au/jsmgroup* [accessed August 11, 2005]); see also Mattick 2004 (Mattick's work is reviewed in Denis 2002).

In the popular media, this shift in thinking about junk DNA is sometimes dramatized in relation to the work of Malcolm J. Simons, founder of Simons Haplomics, largely because he sought to patent stretches of this junk DNA. His website summarizes the story thus:

> Dr. Malcolm J. Simons, the founder of Simons Haplomics, is the inventor of far reaching patents for noncoding DNA diagnostic and Gene discovery. Dr. Simons pioneered the concept that "Junk" noncoding DNA could not be "Junk" because the DNA sequence differences were ordered, and were conserved between humans of the same coding gene type. The ordered sequence patterns marked lengths of chromosomes, including adjacent and remote genes. Dr. Simons' vision sixteen years ago has led to patents which underpin DNA Diagnostics, and upon which genome-wide gene discovery in single patients is based. The gene mapping method, involving sequence variants in noncoding DNA, is being utilized by the National Institutes of Health-sponsored international consortium for haplotype mapping (HapMap). In 2001, Dr. Simons realized that the completion of the Human Genome sequence enabled an improved gene discovery strategy, particularly suited to those common diseases thought to involve many genes studded around the 23 pairs of chromosomes comprising the genome. These 23 pairs comprise 46 single chromosome haplomes. The Haplomics solution for Gene discovery involves the ultimate elements of genetics; single cells, single chromosomes, and their contained single Genes. Patent filings in 2003 concerning recovery of single foetal cells and describing the Haplomics strategy for multi-gene discovery, mark the commencement of the Haplomic

era in Genetic research and Gene Discovery. (*http://www.haplomics.com* [accessed on August 11, 2005]).

8. Margaret Lock provides an intriguing discussion of these issues (Lock 2005).

9. These senses of epigenesis relate to, but are distinct from, the classical argument in which epigenesis is opposed to doctrines of preformation. As Margaret Lock points out, the idea of epigenesis is interpreted in many ways in contemporary debates (Lock 2005). For a good introduction, see Speybroeck et al. 2002.

10. However, it is relevant to point out that "synthetic biologists" such as those in the team assembled by Craig Venter, were still, in 2005, aiming to create living organisms from minimum sets of artificially produced DNA sequences coding for protein; while synthetic cells have not yet been produced, they have claimed success in creating a fully synthetic chromosome for a bacteriophage (Smith et al. 2003).

In fact, they were not the first to achieve this: the first synthetic virus was created in 2002 by a team from the State University of New York at Stony Brook, led by Dr. Eckhard Wimmer, who built an active polio virus that seemed identical to the "natural" virus by downloading the gene sequence from the Internet and assembling it with oligonucleotides obtained by mail order. This fed into the anxiety, predominantly in the United States, about the dangers of bioterrorist attacks from groups using similar methods (Cello et al. 2002). See for discussion *http://news.bbc.co.uk/2/hi/science/nature/2122619.stm* (accessed Dec. 1, 2004).

For Venter's ambitions, in which artificially produced organisms will do everything from enabling the United States to protect itself against bioterrorism to reducing its dependence on foreign oil, see the website of his company Synthetic Genomics: (*http://www.syntheticgenomics.com/about.htm* [accessed Aug. 11, 2005]). However, even if Venter achieves synthetic assembly of one-celled organisms, his program may stall when it seeks to fabricate more complex forms of life. As Mattick makes clear, contrary to Jacques Monod's belief in the universality of the central dogma, and that what is true for bacteria will also be true for elephants, what is true for viruses and prokaryotes (simple one-celled organisms lacking a nucleus) is *not* true for living organisms classified as eukaryotes (multicellular organisms that have cells containing nuclei)—it is in eukaryotes that researchers now emphasize the crucial role of the noncoding world of RNA and epigenesis (Mattick 2004).

11. Stefan Helmreich has critically analyzed the assumptions embodied in one version of this informational metaphor, as it was played out in attempts to create artificial or "A-Life": (Helmreich 2000) To avoid misunderstandings, my remarks about life "resisting" are not a gesture to the ineffable, overflowing nature of the vital, but to the ways in which biology, like any other experimental science, can fail to dream up in reality what it has dreamed up in thought. My unfashionable reference to the capacity of the real to say "no" to thought is derived from my understanding of Gaston Bachelard's notion of the phenomenotechnical moment in science (see Bachelard 1984). See also Ian Hacking's emphasis on the impor-

tance of recognizing the experimental moment in those practices to which we give the name of science (Hacking 1983).

12. Embryoid bodies were discussed by Sarah Franklin in a paper entitled "Life Itself: Global Nature and the Genetic Imaginary," given at a symposium on "Life itself: Critical reflections on biopower," at Birkbeck College, University of London, October 22, 2004.

13. As reported in *Le Monde* (July 20, 2005, p. 24), the test claims to be able to identify a sequence on Chromosome 16 affecting the synthesis of the protein kinase C beta 1 in cells in specific brain regions linked to autism.

14. I have discussed Foucault's conception of biopower and biopolitics in more detail in a paper jointly written with Paul Rabinow (Rabinow and Rose forthcoming 2006).

15. In a conference paper, see Koch 2000. For examples of this rhetoric at work, see the papers collected together in a special issue of *Science in Context* 1998, 11: part (3/4).

See also Koch's later analysis of this relation, where she examines the Scandinavian case, showing that eugenic legislation was adopted by democratically elected parliaments, supported by large sections of the public, underpinned genetic research by leading scientists, was not primarily racist in inspiration but largely seen as important for reducing the costs of welfare and so integral to building Scandinavian welfare states: on this basis she argues that ill-informed accusations of eugenics in the present are unhelpful and that one can identify both similarities and differences between the use of genetic knowledge in biopolitical strategies of the first half of the twentieth century and those taking shape today (Koch 2004).

16. I am drawing directly here on my discussion of Galton's arguments and the style of thought that they exemplify in Rose 1985: chapter 3.

17. I discuss the English debate on degeneracy in detail in Rose, 1985, op. cit. For a useful overview of the European debate on degeneration from the 1840s to the early decades of the twentieth century, see Pick 1989.

18. For the United States, see Dowbiggin 1997, Kevles 1985, Kühl 1994, Larson 1995. For France, Carol 1995 and Drouard 1999. For Germany, see Burleigh 1994, Proctor 1988.

19. I am drawing here on an unpublished paper by Hugh Raffles, "J is for Jews," which is a section from his forthcoming *Illustrated Insectopedia*. See also Weindling 1999.

20. The quote is from a speech to SS officers, April 24, 1943, Kharkov, Ukraine. Reprinted in United States Office of Chief of Counsel for the Prosecution of Axis Criminality, *Nazi Conspiracy and Aggression, vol. 4*. Washington, D.C.: United States Government Printing Office, 1946, 572–578.

21. The Human Betterment Foundation reported that 50,707 Americans had been sterilized by Jan. 1, 1950, and that the pace of sterilization had increased since 1944 after a lull during the early war years (see Dickinson and Gamble 1950). Reilly reports cases of eugenic sterilization in the United States up into the 1970s, though on a relatively small scale, from the 1960s onwards, combined with a growth in the numbers of surgeons prepared to perform elective sterilization and the use of such sterilization for indigent women, those coming before the courts accused of child neglect and others (Reilly 1991). As late as 1975, West Virginia

enacted a sterilization law although this, like most of the other such laws in force into the late 1980s, applied to the mentally retarded or "developmentally disabled" rather than those deemed mentally ill.

For details of the postwar eugenic programs in the Nordic countries, see Broberg and Roll-Hansen 1996.

22. The research on this program by Maija Runcis, was publicized by articles in the leading Swedish newspaper *Dagens Nyheter* in August 1997 and widely reported in the English language newspapers; see *Guardian*, June 3, 1999. See also Broberg and Roll-Hansen 1996.

23. Sheldon Reed is usually credited with the invention of the term "genetic counseling" in 1947, and discusses the changing relations with eugenic thought in Reed 1974. There is no recent single, general history of genetic counseling, but see Fine 1993 and Kenen 1984.

24. Ludmerer points out that every member of the first editorial board of *Genetics* in 1916 participated in or supported the eugenics movement at some point in its early history (Ludmerer 1972: 34).

25. There was particular concern about chemical sterilization using the quinacrine method developed by Dr. Jaime Zipper in 1984 and distributed to 19 countries , including Bangladesh, Chile, China, Colombia, Costa Rica, Croatia, Egypt, India, Indonesia, Iran, Morocco, Pakistan, Philippines, Venezuela, Vietnam, the United States, Malaysia, and Romania, although subject to later banning in some. See, for example, *http://www.hsph.harvard.edu/Organizations/healthnet/contra/topic05.html#2* (accessed Jan. 12, 2004).

26. See, for example, Lee 1990, and Wang and Hull 1991.

27. Stone has argued that the Western critical focus on the eugenic basis of Chinese bioethics misunderstands the complexity of the Chinese situation (Stone 1996).

28. These issues were widely and critically reported in the Western science press, e.g. Coghlan 1998 and Dickson 1998.

29. Notably Qiu Renzong, Professor at the Philosophy Institute of the Chinese Academy of Social Sciences, who, from the 1980s onwards, "argued for the rights of individuals and their families to refuse sterilisation or the termination of pregnancy; explicitly extended these rights to those defined as "mentally retarded"; questioned the criteria whereby entire categories of people are dismissed as "unfit" or "deficient"; and denounced the stereotypical language used by medical experts when writing of people with mental or physical disabilities" (Dikötter 1998: 172–73). Qiu was a significant figure in the explicit rejection of eugenics in China at the end of the twentieth century, and in the embracing of a Euro-American bioethics based on principles of individual autonomy and rights.

30. Stone questions this interpretation, pointing to the fact that the Chinese emphasis on the need to maximize the health of the population, including the limiting of births of those with disabilities has gone alongside work to improve pre- and postnatal care, to explore and rectify the environmental and dietary causes of disability, and to work to integrate disabled people into society, including legislating for the equalization of opportunities and the protection of disabled people's rights and interests (Stone 1996).

31. Discussion of these developments in the English language literature has been very limited, and largely confined to newspaper reports. For example, those in the *People's Daily* (see *http://english.peopledaily.com.cn*). A particular concern in China has been the problem of sex selection in view of the imbalance of male to female births, which according to the fifth national census, conducted in 2000, averaged at around 117:100, and was as high as 130:100 in Guangdong and Hainan provinces. This is considered problematic less because of the differential valuation of male and female children it embodies, than because of the projected impact on the "marriage market" as these children grow up and seek partners.

32. Reported in the *People's Daily* online on December 6, 2002, available at http://english.people.daily.com.cn.

33. For a recent collection that elaborates on this way of approaching the issue of risk, see Baker and Simon 2002. I develop my own approach in more detail in relation to psychiatry in Rose 1998 and in relation to crime control in Rose 2000b.

34. However, as Ewald has argued, there are significant differences between the principles of risk spreading and indifference to fault that characterized prevention and compensation in twentieth-century social insurance and health policies and the kinds of contemporary risk thinking that underpin the "precautionary principle"—where a demand for the total elimination of risk goes along with attempts to make someone bear responsibility for any untoward event, and where the almost incalculable uncertainty of the consequences of every new development is nonetheless linked to demands for assurances of total safety (Ewald 2001).

35. Others have argued that eugenic considerations also underpin the quasi-voluntary practice of sterilization of women on welfare in some cities in the United States (Horsburgh 1996).

36. Over the closing decades of the twentieth century, such forms of group-based risk management spread to other fields, notably those concerning pathologies of conduct: families suspected of committing child abuse (Castel 1991, Parton 1991), the risk profiling of those convicted of certain types of sexual or violent offences (Pratt 2000), and of psychiatric patients (Rose 1998). In each case a combination of demographic, biographical, lifestyle, and other factors are used to identify those thought to be "high risk" who are then placed on risk registers, subject to surveillance and reporting by a whole range of authorities and may be subject to other restrictive measures, such as having their children taken into care or subject to preventive detention.

37. A recent debate, actually about predictive genetic testing, illustrates this well. In 1987 Margery W. Shaw, a professor of health law at the University of Texas, suggested that a predictive genetic test for Huntington's meant that it was "now possible to begin to eradicate the Huntington gene from our species" within the next two generations and suggested that "it is necessary that those who possibly or certainly carry the gene take positive steps to prevent its transmission" (Shaw 1987: 243).

Her argument brought immediate and heated rejections from those involved in developing preclinical testing programs: "Whilst all would welcome a reduction in the gene frequency for HD this eugenic argument is both impractical and certainly not a primary goal for the preclinical testing programs in Canada and the United Kingdom . . . [the] major goal of preclinical detection of HD is the improvement of the quality of life for persons at risk" (Hayden et al. 1987: 752). I owe these references to Carlos Novas's research on the history of genetic testing and counseling for Huntington's Disease (Novas 2003).

38. I am leaving to one side the areas of crime control and psychiatry where these famous ethical principles have only limited purchase.

Chapter 3 An Emergent Form of Life?

1. This chapter is developed from a paper presented at the first Blankensee Conference, Emergent Forms of Life: Towards an Anthropology of Life Sciences (December 11–13, 2003), Schloß Blankensee, Berlin, organized by Stefan Beck and Michi Knecht. This version has benefited from comments made to me then, and at presentations to the Department of Social Science and Medicine at McGill University and the Department of Sociology and Anthropology at Carleton University.

2. There are dozens, perhaps hundreds, of websites devoted to this proposition, which has become an article of faith and intense desire among "transhumanists." See, for example *http://transhumanism.org/index.php/WTA/* (accessed June 26, 2004).

3. See note 1 above. Stefan Beck tells me that they took the phrase from Michael Fischer's 2003 collection of essays that used it in the title (Fischer 2003) In the essay that addresses these issues, entitled "Emergent Forms of Life: Anthropologies of Late or Post-Modernities," Fischer suggests that this phrase acknowledges "an ethnographic datum, a social theoretic heuristic, and a philosophical stance regarding ethics" (p. 37). He argues that practitioners in many areas of life feel that traditional ways of thinking and doing no longer work, that "life is outrunning the pedagogies in which we have been trained," that we live among societies that are a complex of "emergent, dominant and fading historical horizons," and that our form of life entails a "sociality of action, that always contains within it ethical dilemmas" (p. 37). While I am not convinced by the global form of this diagnosis—which appears to embrace phenomena ranging from the changing organization of science, through the rise of computer-mediated perception, to civil and ethnic wars—or the belief that we live in times so radically different socially or ethically from those of our predecessors, I agree with Fischer's identification of changes in the life sciences and biotechnology as significant, and am sympathetic to his belief that ethnography provides one important investigative resource to "explore connections between changing subjectivities, social organization, modes of production and symbolic forms" (p. 57) that are emerging today. A more modest, ethnographically informed but historically grounded empiricism might enable us to chart, and to locate, some of the specificities of these connections and mutations, and the mixtures of continuity and novelty that they embody.

4. The paper upon which this chapter is based was called "The politics of life in the twenty first century," to emphasis this contrast.

5. See chapter 1 of this book.

6. See chapter 4 of this book.

7. In a personal communication in 2003: thanks to Charles Rosenberg for his very helpful comments. In discussion following a recent seminar presentation on this theme of emergent forms of life, another excellent historian of science, John Pickstone, made very similar points to me about my urge to diagnose "novelty."

8. The first part of the quotation apparently comes from Leriche's contribution to Volume 6 of the *Encyclopédie Française* published in 1936, the second is referred to his *Physiologie et pathologie du tissue osseux* (Paris: Masson), 1939.

9. Lindsey Prior and his colleagues have examined the role of visualization technologies that make risk "visible" to clinicians and patients (Prior et al. 2002).

10. Data from *http://www.rxlist.com/top200_sales_2003.htm* (accessed June 26, 2004).

11. All quotes from *http://www.systemsbiology.org/* (accessed June 26, 2004).

12. At *http://www.celeradiagnostics.com/cdx/pr_1086215171* (accessed June 26, 2004).

13. Chitwood & Harley is one of the leading class action law firms in the United States, and they provide this helpful account at *http://classlaw.com/CM/Articles/articles6.asp* (accessed June 26, 2004).

14. Quoted in Chitwood & Harley, op. cit.

15. See, for example, David Ewing Duncan's article in *Wired* magazine, on his genetic testing for a series of genetic disease markers at Sequenom in San Diego in late 2002, under the title "DNA as Destiny," with the subhead "DNA is the book of life. It's also the book of death. In the future we'll all be read cover to cover. Here's what it's like to take the world's first top-to-bottom gene scan." At *http://www.wired.com/wired/archive/10.11/dna_pr.html* (accessed May 15, 2004).

16. Note that the report itself retailed at US$4,450! The players identified were: Acadia Pharmaceuticals; Affymetrix, Inc.; Amgen; AP Biotech; Applera Corp.; Arius Research; ArQule; AstraZeneca; Aventis (USA); Base4, Inc.; Bayer (USA); BioMérieux; BioTechnology General Corp.; Boehringer Ingelheim (USA); Bristol-Myers Squibb Co.; Clontech Laboratories, Inc.; CuraGen Corp.; Élan (USA); Eli Lilly and Co.; EM Industries (USA); ExonHit Therapeutics; Ferring Pharmaceuticals, Inc. (USA); Genaissance Pharmaceuticals, Inc.; Genentech, Inc.; Genta; GlaxoSmithKline; Hoffmann-La Roche; Hybrigenics; Incyte Pharmaceuticals; InterMune Pharmaceuticals; Invitrogen Corp.; Lorus Therapeutics, Inc.; Matrix Pharmaceuticals, Inc.; MDS Ocata, Inc.; Medicis; Merck; Microarray Centre; Molecular Templates; Nabi; Nova Biomedical; Novartis (USA); OmniViz, Inc.; Ontario Cancer Institute; Orchid BioSciences, Inc.; Orphan Australia Pty Ltd.; Orphan Europe SARL; Orphan Medical, Inc.; Orphan Pharmaceuticals; Orphan Pharma International; Packard Instrument Co.; Pierre Fabre SA; Pfizer Pharmaceuticals Group; PPGx, Inc.; Qiagen Genomics; Sanofi-Syntélabo (USA); Sequenom, Inc.; Shire Pharmaceuticals Group (USA); Sigma-Tau Pharmaceuticals, Inc. (USA); SignalGene, Inc.; Swedish Orphan; Teva Pharmaceuticals (USA); Therapeutic Goods Administration; Third Wave Technologies; TM BioScience Corp.; Visible Genetics, Inc.; YM Biosciences, Inc. (Frost & Sullivan 2001).

17. For details of the program, see *http://www.kcl-phs.org.uk/haemscreening/default.htm* (accessed December 1, 2005).

18. Alpha and Beta thalassemia arise from mutations in the genes that code for the alpha protein chain of hemoglobin (HBA on Chromosome 16) or the beta protein chain (HBB on Chromosome 11). Alpha thalassemia major—arising from two copies of the mutation—is fatal at birth. Beta thalassemia major is a condition requiring repeated blood transfusions. Those affected with a single copy of the mutation have less severe, but still disabling, symptoms. Sickle-cell anemia is an autosomal recessive genetic disorder caused by a mutation in the sequence that produces an abnormal version of the hemoglobin beta chain called hemoglobin S.

19. Much publicity has been given to the test patented by Myriad Genetics for variants of the BRCA1 and BRCA2 sequences that have been associated with breast and ovarian cancer in high risk families. However, most cases of breast and ovarian cancers are not associated with variants in the BRCA1 or BRCA2 genes, and the genetic test has not been recommended for routine screening. *Genomics and Population Health* 2005, the most recent report of the U.S. Centers for Disease Control and Prevention on this issue, is typical when it states: "Results of *BRCA1* and *BRCA2* testing can only be interpreted in the context of family history. An affected person usually must first be tested to determine whether a *BRCA1* or *BRCA2* mutation can be identified within the person's family. Only if a mutation is identified will testing unaffected family members be informative for predicting their cancer risk. Both *BRCA1* and *BRCA2* mutations are not found in all women with family histories of breast or ovarian cancer, and not all women who have a *BRCA1* or *BRCA2* mutation will develop either breast or ovarian cancer. . . . Most breast and ovarian cancers occur in women who have no family histories of either of these cancers or only a single affected relative. *BRCA1* and *BRCA2* mutations occur in approximately 1 in 400 women and account for, at most, 5%–10% of all cases of breast and ovarian cancers; there are probably other, as yet unidentified, inherited breast cancer susceptibility genes" (Centers for Disease Control and Prevention Office of Genomics and Disease Prevention 2005, chapter 4). The report is responding critically to Myriad's use of direct to consumer marketing campaigns to promote this test.

20. The Huntington's Disease gene was mapped to Chromosome 4p 16.3 in 1983 but not sequenced until 1993. When finally identified, the gene (IT15) was found to contain a CAG repeat within its 5'-end coding sequence. This CAG repeat is expanded in individuals with HD who may or may not be symptomatic. The presence of a CAG repeat expansion is found in virtually all symptomatic HD individuals. In the normal HD gene, CAG repeats range from 10–29. Less than 1 percent of normal individuals have intermediate HD gene CAG repeats of between 30 and 35 repeats. Individuals affected with HD typically have at least one HD gene CAG repeat of 36 repeats or greater. Almost no affected individuals have less than 36 CAG repeats. However, in a few rare instances individuals having repeats of 36–39 are found to have remained asymptomatic by standard clinical criteria at advanced age. In one exceptional case, a 95-year-old patient had 39 repeats, see Rubinsztein et al. 1996.

21. Available empirical evidence as to the consequences for individuals of being designated "at risk" on the basis of a genetic test is ambiguous and somewhat contradictory. Most suggest that this information is interpreted in the context of individuals' knowledge and belief about genetics, heredity, and conditions that

"run in the family" and that its significance depends on culturally specific "risk portfolios" (Cox and McKellin 1999, Santos and Bizzo 2005) (Robertson 2000, Sanders et al. 2003, Van Dijk et al. 2004).

In a recent study, Scott, Prior and their colleagues found that users of a cancer genetics service told that they were at risk often see themselves in a liminal position caught between the healthy and the sick, and seek recourse to systems of medical surveillance that can continuously monitor their state of health; they also found that such surveillance may actually be sought, in that many of those deemed by professionals to be at low risk of inheriting cancer-related mutations subsequently strove to be recategorized as being at moderate or high risk of an adverse outcome (Scott et al. 2005).

22. Now see *http://www.ubimon.org/* (accessed November 25, 2005).

23. The first International Workshop on Wearable and Implantable Body Sensor Networks was held at Imperial College London in April 2004.

24. I have derived the following paragraph from Rose 1994.

25. I have discussed these issues in a little more detail in chapter 1.

26. In this respect, these critics echo the concerns of an earlier generation, critical of the "narcissistic" uses of the psychotherapies in the "Me Generation" of the 1960s. Those critics, for example Phillip Rieff and Christopher Lasch, were similarly concerned about the ethical implications of the turn to happiness that they thought was indexed by the rise of the psychotherapies. Like Kass and his colleagues, they misunderstood the ethical implications of these psychotechnologies and their links with the forms of life that were coming into existence in North America, Europe, and Australasia at that time: see my discussion in Rose 1989.

27. In fact, there is good evidence to suggest that the SSRIs are no more efficacious in managing mood than an older class of psychiatric drugs, the tricyclics (Faravelli et al. 2003).

28. I have adapted this argument from an editorial cowritten with Ilina Singh, who has specifically studied the evidence on Ritalin (Rose and Singh 2006).

29. Singh nicely illuminates the dilemmas about selfhood here, showing that many parents also feel that, on weekends, the drug should not be administered, as the child should have the chance to "just be himself." Even in a regime of the self stressing authenticity—perhaps especially here—we may be inhabited by many different and even contradictory notions of the reality of ourselves and of the selves of others: see my own discussion of these issues in Rose 1996b.

30. These now abound—following on from Elizabeth Wurtzel's bestseller (Wurtzel 1995). See, for example, Solomon 2001, Styron 1990, and Wolpert 1999.

31. For other evidence of the wishes, and experiences, of those who take SSRI antidepressants, see the numerous accounts on email lists on the Internet and the accounts collected in Karp 1996.

32. See the editorial cited above, note 26.

33. This, of course, was the central argument of Howard Becker's classic essay, "On Becoming a Marijuana User," in Becker 1963.

34. See *http://www.paxilcr.com/index.jsp* (accessed Aug. 20, 2004).

35. These have been beautifully analyzed by Emily Martin in a number of papers that will be published together (Martin forthcoming).

36. It is a source of amazement to me that so many neuroscientists still operate in terms of this distinction, so much so that Antonio Damasio becomes something of a celebrity by suggesting that emotion and reason, feeling and thinking intertwine. See, e.g. Damasio 1994, 1999, and 2003).

37. The best known of those who has discussed this is Laurence Diller, author of *Running on Ritalin* (Diller 1998); see also his website at *http://www.docdiller.com/*.

38. The subject of a CBS news story entitled "A Dream Come True? New Drug Tricks Brain to Be Awake: at *http://www.cbsnews.com/stories/2002/01/14/eveningnews/main324299.shtml* (accessed on August 20, 2004). Sales of modafinil were estimated at AU$290m in 2003 and the *Economist* estimated their likely growth to be 30 percent in 2004.

39. "Mild Cognitive Impairment" is a rapidly growing diagnosis in the United States argued by some to be a presymptomatic indicator for the development of Alzheimer's Disease.

40. E.g. in "Supercharging the brain," in *The Economist*, September 16, 2004.

41. For Helicon, see *http://www.helicontherapeutics.com/Scientific/Helicon_CREB.htm* (accessed June 26, 2004).

42. See *http://www.forbes.com/global/2002/0204/060_print.html* (accessed June 26, 2004).

43. See an article on their website by Will Block, entitled "Academic Doping: The Wave of the Future: Drugs to Improve the Learning Process," at *http://www.life-enhancement.com/LE/article_template.asp?ID=872* (accessed June 26, 2004).

44. Although here I have focused on issues of enhancement of mental capacities, analogous issues arise in relation to surgical modification of different parts of the body: see Frank 2004.

Chapter 4 At Genetic Risk

1. This is a revised and edited version of a paper jointly written with Carlos Novas, and published as C. Novas and N. Rose (2000) Genetic risk and the birth of the somatic individual, *Economy and Society*, 2000, 29 (4): 484–513. I thank Carlos Novas for allowing me to publish it here, in this revised form. I have reframed and properly referenced the empirical material he contributed, in particular the account of the history of genetic counseling and the analysis of the ethical and life strategy issues in the Huntington's Disease Webforum. However I should stress that the paper was jointly written and the concepts and arguments developed collaboratively.

2. See the discussion in chapter 3 of this volume. Biobanks and DNA databases have been the subject of extensive social scientific research, most focusing on such issues at the political disputes over giving private companies the licenses to develop such banks, and ethical issues such as informed consent and ethical regulation of such biobanks. See, for example Høyer 2003, Palsson and Rabinow 1999, and Rose 2003.

3. Of course, less widely noted is the possibility that as genetic diagnostic tests become available for susceptibility to such conditions as lung cancer and heart disease, those not carrying the gene sequences in question may be free to eat, drink, and smoke their way to a happy old age!

4. Nancy Wexler's work on genetic linkages in a community of Venezuelan families with a very high incidence of Huntington's disease, which led to the location of its genetic basis to the short arm of Chromosome 4, was perhaps the first such gene hunting exercise (see Wexler 1996). Since that time, there have been many publicly funded and / or commercially driven projects to identify, sequence, and test for genetic susceptibilities; I discuss some of these in chapter 3.

5. Novelty seeking was linked to variations in the D4DR site on the short arm of Chromosome 11 (Ebstein et al. 1996), although this has not been reliably replicated in many later studies; bipolar affective disorder was linked to DNA markers on Chromosome 11 in the Old Order Amish (Kidd et al. 1987), but the correlation was not replicated in later studies and proved to be an artifact. Attention has now switched to Chromosome 17, home of the serotonin transporter gene e.g. (see Caspi et al. 2002).

6. Huntington's Disease is a late onset genetic disorder characterized by progressive neurological deterioration that results in choreic movements, mood swings, and depression for which treatment is solely palliative. Long known to run in families, and roughly mapped in 1983, the gene itself and the mutation involved that identified in 1993, and direct mutation DNA tests have been developed that can predict with virtual certainly whether an individual will develop the disease, and also give some indication of the likely age and severity of onset.

7. Others have also been critical of the way the term "geneticization" forecloses debate; see Hedgecoe 1999.

8. See *http://www.dnadirect.com/* (accessed August 20, 2005).

9. For further examples, see the website of the Genetic Alliance, which "fosters a dynamic coalition of consumers and professionals to promote the interests of children, adults and families living with genetic conditions. For twelve years the Alliance has brought together support groups, consumers and health care professionals, creating partnership solutions to common concerns about access and availability of quality genetics services. Currently numbering 287 support groups and 214 consumers and professional members, the Alliance was founded in 1986—propelled by the energy of the self-help and support group movements" *http://www.geneticalliance.org/allianceinfo.html* (accessed July 22, 2004).

10. See my discussion in chapter 8 of this book.

11. In the original paper this section was written by Carlos Novas, and this part of the present chapter is based on his work, reported in detail in Novas 2003.

12. This history focuses on the United States and United Kingdom; there are similarities and differences in other national histories within Europe, and between European nations and those, for example, in Asia. For one European comparison, Mianna Meskus has traced a rather similar history in Finland, finding that the mutations occur about a decade later (Meskus 2003).

13. In fact such concerns are far from universal, and there is evidence of considerable national variation in the extent and focus of anxieties about genetic tests. See, for example, the research undertaken by Darryl Macer, Director of the Eubios

Ethics Institute, much of which is available at *http://www.csu.edu.au/learning/eubios/Papers.html* (accessed August 23, 2004).

14. Quoted from a Report on Reuter's Health Information, (July 22, 2004).

15. In November 2005, *New Scientist* reported the results of a survey conducted by a campaigning organization in Australia, the Genetic Discrimination Project, which claimed that 7.3 percent of their survey of over 1,000 people who had taken a predictive genetic test for a serous disease felt that they had suffered specific instances of negative treatment. In no case had a formal complaint been made. However the researchers claim to have found evidence to support such claims, for example a case where a woman claimed to have been denied life insurance coverage after testing positive for the variant of BRCA1 that puts her at higher risk of developing breast cancer. (*New Scientist*, November 5, 2005, at *http://www.newscientist.com*). Long-standing critics of genetic discrimination have suggested that such practices are widespread among insurers and employers despite the regulations and agreements in place in many countries.

16. Reported in an opinion column in *AMNews* on March 21, 2005, at *http://www.ama-assn.org/amednews* (accessed March 17, 2005). For the Coalition for Genetic Fairness, see *http://www.geneticfairness.org/*. One of the founders of this coalition is Nancy Buelow, who became an advocate for the rights of people with genetic conditions since she was diagnosed in 1993 with alpha1-antitrypsin deficiency (see her own discussion of this issue at *http://www.ramazziniusa.org/geneticdiscrim.htm* [accessed August 23, 2005]). Sharon Terry, president and CEO of the Genetic Alliance and the founding executive director of PXE International, a lay advocacy group for the genetic condition pseudoxanthoma elasticum (PXE), of whom we will hear more about in chapter 5, is chair of their executive committee.

17. See *http://www.genome.gov/11510230* (accessed August 23, 2005).

18. For official evidence of the increasing diagnosis of mental disorders and the increasing use of psychiatric drugs in children, see World Health Organization 2004. And for the view of the industry about the potential psychiatric drug marked offered by children, see Business Communications Company 2005. I discuss this in more detail in chapter 7.

In late 2003, there was a flurry of concern about such prescribing, as claims were made of adverse effects such as suicidal thoughts and behavior, together with accusations that the pharmaceutical companies withheld crucial information. In both the United States and the United Kingdom physicians were warned not to prescribe a number of these drugs to children. This issue is further discussed in chapter 7.

19. For a discussion of the rise and fall of the XYY paradigm, and the role of recent biological developments in strategies of crime control, see chapter 8.

20. The moratorium is overseen by the Genetics and Insurance Committee (GAIC), a body formed in 1999, that is also responsible for evaluating genetic tests and their applicability for insurance underwriting purposes and approving their use in this manner. By 2005, GAIC had only approved one predictive genetic test, that of Huntington's Disease for use in determining premiums for life insurance policies over £500,000.

21. Health Secretary John Reid said, "choosing to have a predictive genetic test can be life saving, and nobody should be put off having such a test because of fears it will be used against them by insurers." See *http://www.dh.gov.uk/PublicationsAndStatistics/PressReleases/PressReleasesNotices/fs/en?CONTENT_ID=4106051&chk=2CNwmM* (accessed on August 23, 2005).

22. Actually, this debate may have paradoxical effects. Some suggest that, far from leading to wholesale genetic discrimination, it will reveal the superiority of a national, compulsory scheme of health care funded through taxation over all individual and group based schemes: as it becomes possible to predict those who may have an increased likelihood to develop Alzheimer's or Parkinson's disease, as there are increased pressures to exclude those at high risk from cover by fiat or by the costs of purchase, as individuals at no or low risk begin to exit the private insurance system rather than subsidize those at high risk, the strategy of pooling of risk across a national population will begin to seem attractive to those who wish to be insured without a knowledge of what the future might hold for them, to governments seeking to maximize cover because of fears about the implications of any public safety-net system being overwhelmed by demands from those deemed uninsurable in the private sector, and to the insurers themselves, on the basis that there can be no individual risk classification without access to full predictive information.

23. Erving Goffman provided the classic analysis of the management of "spoiled identities" in an earlier age of personhood. A comparison with the present examples is instructive (Goffman 1968).

24. The data and analysis in this section of the chapter is derived from a study that Novas conducted of a Huntington's Disease (HD) webforum that was downloaded from Massachusetts General Hospital webforum. The study covered a period of two years ranging from May 6, 1995 to January 27, 1997.

25. My thoughts on this "flattening" were helped by hearing Scott Lash give a lecture on "Technological Forms of Life," his inaugural lecture for a Chair in the Department of Sociology at Goldsmiths College, University of London, delivered on February 22, 2000.

Chapter 5 Biological Citizens

1. The original version of this chapter was written with Carlos Novas and published as "Biological Citizenship" in Ong and Collier 2005: 439–463. I would like to thank Carlos Novas for allowing me to publish this revised version. Although I have tried to give proper credit in what follows to the aspects of the argument based on his empirical research, and edited out much of his empirical material presented in the original paper, I must point out that the concepts and argument of the paper were developed jointly.

2. I had first started to think in terms of biological citizenship when writing my paper, "The Politics of Life Itself," in 1999 (Rose 2001). As Carlos Novas and I were developing the concept in 2001, writing a paper with this title for a conference held in Prague in April 2002, we discovered that others used this term, although with different resonances. In particular, we learned of Adriana Petryna's

work on post-Chernobyl Ukraine (now published as Petryna 2002), and I would like to thank her for generously letting me see this in manuscript. In our use of the term, we embraced her rather specific usage in a more general context. A Web search in 2001 picked up a paper by Chris Latiolais called "The Body Politic: Naturalizing Biological Citizenship and Philosophical Reservations," delivered at a University of Chicago Midwest Seminar, in March 1998, and we wrote to the author, who told us in an email that "I don't take the term seriously at all. The term 'biological citizenship' is patently oxymoronic, the conflation of theoretical, natural-scientific categories with the quite different practical, moral-political categories of elective allegiance. The 'naturalizing' qualifier ambiguously hints at the requisite process that might span the gap between organism and political identity . . . such classifications are wholly irrelevant to legal standing."

The idea of genetic citizenship is more widely used (see, for example, Peterson and Bunton 2002), and has been particularly developed in the work of Deborah Heath, Rayna Rapp, and Karen-Sue Taussig (see Heath et al. 2004), and Novas and I benefited from seeing some of their work in draft. Others have sought to apply the distinction between passive and active citizenship to health provision (see, for example Abraham and Lewis 2002), but the plausibility of this mapping of the distinction of passive and active onto different stages of late capitalism depends on a prior acceptance of this periodization.

In addition, Chetan Bhatt and Engin Isin of York University, Toronto, provided initial guidance on the history of citizenship, and the paper benefited from comments from participants at the Prague conference where it was first given, especially Aihwa Ong and Stephen Collier, and from participants in a CRICT Seminar at Brunel University in November 2002. Amaya Carmen Novas-Peña arrived while Carlos and I were writing the initial version of this paper, and increased our understanding of many contemporary aspects of biological citizenship.

3. For conceptions of citizenship, and projects of citizen building, in earlier periods, see Isin 2002.

4. The situation is different in China and perhaps other countries in South East Asia, see Dikötter 1998.

5. As I have argued in chapter 2 of this book.

6. As I have said, I take this term from Catherine Waldby, who introduces it in her work on the Visible Human Project (Waldby 2000); however I use it in a looser and wider manner.

7. For a thought provoking set of reflections on the ethic of hope, see Zournazi 2002.

8. This seems to have been the first use of the phrase "political economy of hope," although many others have used it in a range of contexts; see, for example, Andrews 1999. Mary-Joe Delvecchio Good develops her own analysis in several later texts, although not giving priority to this notion (for example, Good 2001, 2003).

9. They term this "eugenic," but as I have argued elsewhere I think it more helpful to restrict the term "eugenics" to calculated attempts to improve the quality of the nation or the race through acting on reproduction.

10. I am grateful to Carlos Novas for directing me to this text.

11. C. Joppke, "Mobilization of Culture and the Reform of Citizenship Law: Germany and the United States," at *http://www.europanet.org/conference2000/papers/a-2_joppke1.doc* (accessed Sept. 23, 2001).

12. While references to Chinese people as the "yellow race" in the West have usually been pejorative and negative, in China, the phase carries the opposite connotation. Yellow was the color of the "Yellow Emperor" in ancient China, who according to legend was the common ancestor of the tribes occupying the Central Plain and, in some versions, of all Chinese people. He was called the "Yellow Emperor" because he possessed all the virtues of the yellow soil. Only the king, queen, prince, princess, and the relatives of the royal family were permitted to use or wear yellow. Yellow is the name of the Yellow River, which some Chinese credit for creating a nation identified by its yellow skin. Yellow also means powerful and prosperous; although in some case it has the same meaning as blue (that is to say, erotic) in English. Some of these issues are discussed in Su 1991; the author was also involved in producing a controversial Chinese television series in 1988 under the title *Yellow River Elegy.* In chapter 6 I say a little more about the complicated and shifting ways in which the fifty-six official ethnicities that are recognized in China have been unified or differentiated in terms of culture, lineage, and genetics. Thanks to my Chinese colleagues for their advice on the complexity of the meaning of color terms in Chinese, and for making it clear that they do not have the same resonance in China as they do in the United States.

13. See Eeva Bergelund, "Biotechnology as a Finnish National Imperative" (paper presented at BIOS Research Group, Goldsmiths College, February 28, 2002).

14. For a popular overview, see Wheelwright available online at *http://www.discover.com/issues/apr-05/features/finlands-fascinating-genes/?page=1* (accessed July 23, 2005).

15. Notably, of course, Lancelot Hogben's *Science for the Citizen* (Hogben 1938).

16. I adopt Ian Hacking's nice phrase, "making up" here; see, for example, Hacking 2002.

17. See, for example, Claeson et al. 1996.

18. This discussion is based on my analysis of the website, *www.prozac.com* in 2001.

19. See *http://www.prozac.com/HowProzacCanHelp.jsp* (accessed Feb. 17, 2001).

20. See *http://www.prozac.com/DiseaseInformation/Recovery.jsp* (accessed Feb. 17, 2001).

21. See *http://www.prozac.com/generic_info.jsp* (accessed Feb. 17, 2001).

22. See *http://www.mdf.org.uk/about/* (accessed Feb. 17, 2001).

23. Ibid.

24. See *http://www.pendulum.org/* (accessed Oct. 17, 2001).

25. See *http://www.searchingwithin.com/bipolar/* (accessed Oct. 17, 2001).

26. I discuss these disputes in chapter 7.

27. On Africa, see V.-K. Nguyen, "Antiretrovirals, Globalism, Biopolitics and Therapeutic Citizenship," in Ong and Collier 2005.

28. Hunt-Dis is an electronic mailing list where persons affected by HD, those at risk, and carers can discuss any topic relating to Huntington's Disease.

29. Carmen Leal also maintains a website called writerspeaker.com, which aims to help aspiring writers and speakers to learn how to use the Internet for research and bring their products to market (see *http://www.writerspeaker.com/*).

30. This coalition was formed on April 1, 2000 and is designed to provide support for HD families by HD families, in addition to providing a range of information and answers to those affected by this disease. Source: *http://www.hdac.org/about.html* (accessed Oct. 17, 2001).

31. Carmen Leal, "The Last Generation," Huntington's Disease Advocacy Center, April 8, 2001, *http://www.hdac.org/features/article.php?p_articleNumber=13* (accessed Oct. 17, 2001).

32. M. L. Miller, "Reasons for Hope," ibid., *http://www.hdac.org/features/article.php?p_articleNumber=23*.

33. See the reports, available at *http://www.swissinfo.org/sen/Swissinfo.html*.

34. On biovalue, once more, see Waldby 2000, 2002; for analogous concepts, see Franklin 2005 and Nguyen 2005.

35. For an account of developments in Iceland, see Palsson and Rabinow 1999.

36. Source: *http://www.decode.com/*. The company website further states that "Iceland makes an ideal home for the company, as the Icelandic population is, genetically speaking, relatively homogeneous. The country has a sophisticated, high-quality healthcare system and extensive genealogical records. Through these, resources can be generated to identify genes associated with a multitude of diseases. Research based on this unique population provides distinctive insights into the pathogenesis of these diseases, and the depth and comprehensiveness of deCODE's genealogical database are unrivalled worldwide" (*http://www.decode.com/company/profile/*) (accessed Oct. 17, 2001).

37. Source: *http://www.umangenomics.com/index2.asp* (accessed Oct. 17, 2001). The recognition of the potential of the Medical Biobank of Umeå for the production of biovalue was made by the technology transfer unit at the University of Umeå, called the Technology Bridge Foundation (see Abbott 1999).

38. Source: *http://www.umangenomics.com/index2.asp*.

39. PXE International has also been extensively studied by Karen Sue Taussig, see her paper "Genetics and its publics: Crafting genetic literacy and identity in the Early 21st Century," delivered at a conference on Medicine as Culture—Cultural Studies of Medicine, organized by the Swiss Academies of Medical Sciences and Humanities & Social Science in Zurich in November 2004. I am grateful to her for letting me see a manuscript of this paper.

40. Quoted from the paper by Taussig cited above.

41. See Taussig, ibid. and M. Fleischer, "Patent Thyself (on-line version)," *The American Lawyer*, June 21, 2001, *http://www.americanlawyer.com/newcontents06.html* (accessed Oct. 17, 2001).

42. Taussig, n. 39, has described some of the controversies that arise, however, as for example in the engagement of Native Americans in such genomic research.

43. Ibid.

44. Ibid.

45. In this context it is interesting to note that the 2005 Annual Conference of the Genetic Alliance was sponsored by a number of pharma and biotech companies, including Novartis, Millennium Pharmaceuticals, GlaxoSmithKline, Affymetrix, Genzyme, and DNA Direct. See *http://www.geneticalliance.org/* (accessed August 23, 2005).

46. Source: *http://www.umangenomics.com/index2.asp*. Apart from gaining the informed consent of research participants, novel uses of the tissue and information stored in UmanGenomics's database require approval from a regional ethics committee in addition to the Swedish Medical Research Council. Sune Rosell, temporary chairman of UmanGenomics, in an article in *Science*, states that the company created a unique model for the handling of ethical issues: "There is control at the individual level through informed consent, at the social level through the regional ethics committee which screens all research proposals, and at the population level, since local politicians sit as non-voting members on the boards of both the company and the Medical Bank" (quoted in Abbott, "Sweden Sets Ethical Standards"). Klaus Høyer's ethnographic fieldwork indicates how many participants in UmanGenomics's database do not actually read the informed consent sheet that is provided to them, tacitly consenting to participate in this study, and only engaging in the public arena of informed consent when confronted in the context of the anthropological interview (see Høyer 2003).

47. See Taussig, n. 39.

48. N. Rose, "Do Psychiatric Drugs Have Ethics" (paper presented at Brunel University, workshop on "Do Bio Artifacts Have Ethics," February 15, 2002).

Chapter 6 Race in the Age of Genomic Medicine

1. Versions of this chapter were given as papers to the London Medical Sociology Group, to a research seminar at the University of Cambridge, and in the seminar series on Race, Health and Medicine in the Department of African American Studies at Yale University: thanks to all those who gave me critical comments. I have also benefited from discussion at the symposium on Race in the Age of Genomic Medicine organized by the BIOS Centre at the LSE and partly funded by a grant from the Wellcome Trust's Biomedical Ethics Program, and by LSE. It brought together leading international social scientists, geneticists, and those involved in drug development and DNA banking, to explore the extent to which we are witnessing the emergence of a new biological understanding of "race" in the era of genomics and in particular, genomic medicine. Participants included scientists from China, Japan, the United States, and the United Kingdom, as well as a bioethicist from South Africa, and social scientists and ethicists from North America and Europe. The aim was to evaluate the current state of debate within genomics on the biological status of racial and ethnic categories and to lay the groundwork for international collaboration in a program of future research. Of course, the usual disclaimers apply.

2. Quoted from *http://www.bio-itworld.com/archive/files/043003.html* (accessed August 23, 2005).

3. See *http://www.aaanet.org/stmts/racepp.htm* (accessed August 23, 2005).

4. See the contributions by Pilar Ossorio, Troy Duster, Richard Lewontin, Evelynn Hammonds, Steven Jay Gould, Joseph Graves Jnr, Alan Goodman, Jonathan Marks, and others that are collected together on the website entitled "Race: The Power of an Illusion": *http://www.pbs.org/race/000_General/000_00-Home.htm* (accessed August 23, 2005).

5. This comparison is explicitly made by Larry Adelman, Executive Producer of the 2003 U.S. television series *Race—The Power of an Illusion*. See *http://www.newsreel.org/guides/race/whatdiff.htm* (accessed August 23, 2005).

6. Drift (or random genetic drift) refers to the fluctuation of gene frequencies from one generation to another as a result of random sampling of sperm and eggs. Admixture refers to genetic exchange between groups from two or more ancestral populations.

7. See also the editorial "Genes, drugs and race" in *Nature Genetics* in 2001 (Anon 2001).

8. Recently the subject of a workshop in London, see *http://www.londonideas.org/internet/professionals/ethics_resources/EthnicityWorkshop/EthnicityWorkshop.htm* (accessed August 23, 2005). For a critical assessment of the view that "sickling is a disease specific to the black body—a 'racial disease,' " see Tapper 1999: 2.

9. In China, the issue of "Chinese DNA" is a controversial one, and there is ongoing debate as to whether the fifty-six officially recognized Chinese ethnicities are or are not genetically unified, whether *Zhonghua Minzu* (the Chinese people) are distinct from other populations, and whether the Han majority are themselves a genetically unified population. This debate was stimulated in the mid-1990s, when reports in the U.S. media led to alarm in China that Harvard University was obtaining access to Chinese DNA through the work of the Chinese born director of their program of population genetics, Xu Xiping—see the *China Daily* of July 10, 2003, accessible via *http://www2.chinadaily.com.cn/english/home/index.html* (accessed Aug. 23, 2005) and (Stephens et al. 2000), available at *http://www.washingtonpost.com/wp-dyn/world/issues/bodyhunters/*. In response, many Chinese scientists argued for the need of legislation to "protect China's genomic resources" as they feared that the genetic heritage of the Chinese population was being appropriated by foreign pharmaceutical companies; regulations were passed in 1997 to "protect the Chinese genetic heritage" (see Li and Wang 1997). There was further debate about the distinctiveness of Chinese DNA and its protection from international or American exploitation when some scientists argued that China should participate in the Human Genome Project. These issues are currently being studied by Wen-Ching Sung at Harvard University (see Sung 2005).

10. Donna Haraway has proposed something similar, making a division into three periods: race as blood, body, kinship, nation, language, and culture from the mid nineteenth century through the eugenics of the first half of the twentieth; race as population group with certain gene frequencies, shaped in part by mutation and natural selection in particular geographical areas, but relatively permeable, interacting and intermingling with other population groups through migration and interbreeding—the liberal humanism of the family of man in the aftermath of World War II; race as genome at the turn of the millennium—molecular, informational, technological, manipulable, exploitable in the politics of diversity and multiculturalism. While, for her, the politics of the contemporary operate

in terms of an account of human-ness very different from the organicist ideas underpinning eugenics, she is struck, among much else, by the way in which the racial categories of the nineteenth century reappear as entries in contemporary electronic datatabases with paradoxical consequences (Haraway 1995).

11. This is now available on-line at *http://www.chaf.lib.latrobe.edu.au/dcd/census.htm* (accessed August 23, 2005). All spelling as in the original text.

12. I have drawn here on Snipp 2003.

13. I discuss the history of the U.S. census in Rose 1991, and in chapter 6 of Rose 1999.

14. Ian Hacking has also argued that, if we are careful in our language and our understanding of statistics, it is sometimes "useful to speak in terms of the category of race, on the grounds that the races in some contexts are not only statistically significant but also statistically useful classes" (Hacking 2005: 108). He distinguishes cases where differences are *statistically significant*—where the distribution of a characteristic in one population is significantly different from that in another population—from those where they are *statistically meaningful*—when the causes of the difference are known—but he suggests that we can also consider them *statistically useful*, even if causes are not known, when the statistics of population differences can be used as indicators to inform an immediate practical concern such as risk of illness or death in a population.

15. In this context it is relevant to note that Ian Hacking, who brought into circulation the distinction between "natural kinds" and "human kinds," along with the looping processes by which human kinds shape and reshape the humans who are caught up by them, had abandoned this terminology by 2005, "partly, and only partly, because it was modelled on the unsatisfactory idea of a natural kind" (Hacking 2005: 114).

16. United Nations Declaration on the Elimination of All Forms of Racial Discrimination, 1963, Preamble.

17. The completion of the first draft sequence of the chimpanzee genome was announced in August 2005 and confirms this similarity, showing that the average number of protein-changing mutations per gene is just two, and 29 percent of human genes are absolutely identical. What is more, only a handful of genes present in humans are absent or partially deleted in chimps. However, of course, "there are about 35 million nucleotide differences, 5 million indels and many chromosomal rearrangements to take into account," not to mention epigenesis and much more (Li and Saunders 2005).

18. Quoted in A. M'charek, 2000, *Technologies of Similarity and Difference*, University of Amsterdam, pp. 5–6. This thesis is now published as M'charek 2005.

19. The ELSI arm of the International HapMap Consortium focused on such issues, together with the staples of U.S. bioethics, such as informed consent, confidentiality, venturing further from the norm in the area of "community consultation"—adopted in order to ensure that there was no local opposition to the collection of samples and the ways in which they were described—and "benefit sharing" not in the sense of direct participation in the putative benefits of the HapMap, but in the form of funding provided by the project to improve health facilities in the areas where samples were collected. For details of the project see, *http://*

www.wellcome.ac.uk/en/genome/thegenome/hg04f002.html (accessed August 23, 2005).

20. See *http://www.wellcome.ac.uk/en/genome/thegenome/hg04f001.html* (accessed August 23, 2005).

21. See *http://www.wellcome.ac.uk/en/genome/thegenome/hg04f002.html* (accessed August 23, 2005).

22. Latest estimates revise this figure downwards to from 20,000 to 25,000 coding sequences (see the discussion in chapter 2 of this book).

23. See *http://www.netlink.de/gen/Zeitung/1114.htm* (accessed August 23, 2005). The development was reported in *New Scientist* in November 1996. See also *http://fpeng.peopledaily.com.cn/200112/05/eng20011205_86010.shtml* (accessed August 23, 2005). See n. 9 above for some of the controversies generated in China in the 1990s by developments of this sort, and the subsequent restrictions on foreign access to "Chinese DNA."

24. For example, in South Africa, from the 1960s, medical geneticists argued that genetic information was important to understand the different frequencies of a number of diseases in the different population groups in southern Africa (see the discussion in Soodyall 2003).

25. Ethnographic studies of genomic research on race and health appears to bear out the arguments of the critics about the instability of the categories that are used. For example, Duana Fullwiley's current research on practices in the clinics and labs shows that there is considerable uncertainly and divergence between the researchers themselves, at all levels, from selection of samples to formulation of hypotheses, and from the conduct of experiments to the interpretation of results.

26. See *http://grants.nih.gov/grants/guide/notice-files/NOT-OD-01–053.html* (accessed August 23, 2005).

27. See *http://www.census.gov/population/www/socdemo/race/racefactcb.html* (accessed August 23, 2005).

28. My reference here, obviously, is to the classification scheme, attributed to "a passage in Borges," and thence to "a certain Chinese encyclopedia" by Michel Foucault at the start of the preface to *The Order of Things* (Foucault 1989).

29. I discuss these in detail in chapters 4 and 5 of this book.

30. See *http://www.founders.howard.edu/presidentReports/Mission.htm* (accessed August 23, 2005).

31. Quoted from the website of Research Centers in Minority Institutions at *http://rcmi.rcm.upr.edu/* (accessed August 23, 2005). The press release announcing the linkup is at *http://www.howard.edu/newsevents/Capstone/2003/June/news2.htm* (accessed August 23, 2005).

32. See *http://www.canavanfoundation.org/familyfacts.php* (accessed August 23, 2005).

33. I discuss the history and repudiation of eugenics in China in chapter 2; See also Dikötter 1998.

34. Research in progress by Dr. Stefan Beck of Institut für Europäische Ethnologie, Berlin. See chapter 3 for a discussion of the national screening program for sickle cell and the thalassemias in England.

35. See the discussion in chapters 4 and 5 of this book.

36. Broadcast on BBC2 on February 14, 2003, and supported by web pages at *http://www.takeawaymedia.com/motherlandhomepage.htm* (accessed August 23, 2005).

37. See *http://www.oxfordancestors.com/background.html* (accessed August 23, 2005).

38. See *http://www.oxfordancestors.com/index.html* (accessed August 23, 2005).

39. Work by Katherine Tyler, "The Genealogical Imagination: the inheritance of inter-racial identities," presented at a BIOS seminar at the LSE in February 2004, makes some similar arguments.

40. This is illustrated, I think, by the reflections by Charles Rotimi of Howard University, on the attractions and anxieties generated by such ancestry tracing in relation to himself, his wife, and their children (Rotimi 2003).

41. Reported at the international symposium on Race in the Age of Genomic Medicine, referenced in n.1 above.

42. For more recent results from the same team, see Persson et al. 1999, Roh et al. 1996, and Yasar et al. 1999.

43. Such evidence of variations in the activity levels of enzymes involved in drug metabolism among population groups raises questions for pharmacogenomic drug development as to the capacity of drugs developed for one group, for example Caucasians, effectively and safely to treat diseases in other population groups. See, e.g., Nuffield Council on Bioethics 2003.

44. See *http://www.travatan.com/professional/africanAmericans.asp* (accessed December 29, 2004).

45. See *http://www.travatan.com/media/TravatanPR121603.pdf* (accessed December 29, 2004).

46. See *http://www.travatan.com/media/TravatanPR121603.pdf* (accessed December 29, 2004).

47. As, for example, reported by Anjana Ahuja in *The Times* of London on October 29, 2004, under the headline "We can treat your heart disease . . . if you're black." The article reported the findings of David Goldstein's group at University College London, in the special issue of *Nature Genetics*, as showing that "29 medicines have safety or efficacy profiles that vary between ethnic or racial groups. The list is peppered with familiar names: the anticoagulant Warfarin, for which ideal dosage differs across ethnicity; interferon, which shows a lesser response in African Americans; insulin, to which Hispanics and African Americans show lower sensitivity than Europeans" (available on-line at *http://www.timesonline.co.uk/article/0,8123–1334916,00.html* [accessed August 23, 2005]).

48. See *http://www.ornl.gov/sci/techresources/Human_Genome/elsi/minorities.shtml* (accessed August 23, 2005).

Chapter 7 Neurochemical Selves

1. This chapter is an extended and revised version of "From the Psychological Self to the Somatic Individual," in P. Gardner, J. Metzl and J. Dumit, eds, *Psychiat-*

ric Culture: Disordered Mood, Remedies, and Everyday Life, under consideration by Duke University Press. Versions of the papers on which it is based were presented as "Normality and Pathology in a Biological Age," public lecture in the Faculty of Humanities, University of Copenhagen, March 9, 2001, and at the Conference on Risk and Morality held at the University of British Columbia in May 2001, and as a Keynote Address to the Australian Critical Psychiatry Conference, Sydney, November 2002. It also contains material taken from Rose 2004.

2. "Discovering Our Selves: The Science of Emotion" brought together some of America's leading neuroscientists to describe cutting-edge research that demonstrates the connections between the brain and the body, between nature and nurture, and ultimately, between illness and health. See *http://lcweb.loc.gov/loc/brain/activity.html* (accessed April 27, 2002).

3. See, for instance, the chronology given on E. H. Chudler's web page, "Milestones in Neuroscience Research," at *http://faculty.washington.edu/chudler/hist.html* (accessed April 27, 2002). My discussion of Trimble comes from "Biological Psychiatry as a Style of Thought," prepared for a symposium on models and cases in scientific thought, Princeton University, December 2000.

4. The stethoscope was invented in 1816 by Rene Theophile Hyacinthe Laennec in order to listen to the heart of an overweight female patient; he published his findings *in De l'Auscultation Médiate* in 1819.

5. These new ways of visualizing the body were beautifully illustrated in the exhibition Spectacular Bodies at the Hayward Gallery in London in 2000–2001 (see Kemp and Wallace 2000).

6. I analyze these transformations in an unpublished paper, "The Psychiatric Gaze," first given to the London History of the Present Research Network in 1998.

7. Brainviews: *http://www.brainviews.com/abFiles/ImgPet.htm* (accessed April 27, 2002).

8. Brainplace.com, presented by the Amen Clinic for Behavioral Medicine. See *http://www.brainplace.com/bp/atlas/ch7.asp* (accessed April 27, 2002).

9. Kiran van Rijn pointed out to me that this use of visualization has a long history: within a few months of Roentgen's announcement of his discovery of X-Rays, the following report appeared: "The Roentgen Ray as a Moral Agent.—*The Union Médicale*, March 28, reports the case of a young woman who applied for an operation on account of pains in her arm, as she was convinced that there was some abnormal condition of the bone. The surgeon diagnosed the case as the effect of some slight traumatism on a hysterical subject, and by taking a photograph of the arm proved to her that it was perfect. Once convinced of this, the patient left entirely cured." ("Miscellany," *JAMA*, April 25, 1896, 843–844). Thanks to Kiran van Rijn for this reference.

10. See *http://www.alcoholmd.com/* (accessed April 27, 2002).

11. See, for example, the Clinical Trial of Fluoxetine in Anxiety and Depression in Children, and Associated Brain Changes reported at *http://www.clinicaltrials.gov/ct/show/NCT00018057* (accessed December 1, 2005).

12. Brain scans have also been used in the courtroom in attempts to demonstrate that mental disorders should be regarded as physical illnesses. Thus Jane Fitts, denied the same insurance benefits that would be due to a physically disabled

employee when she was forced to retire from her bipolar disorder, introduced scans showing brain abnormalities into her suit in the federal court of the District of Columbia, arguing that the physical changes to the brain shown by the scans meant her biopolar was not a "mental disorder" as defined in her benefit plan. The court heard evidence from many other expert witnesses—including Frederick Goodwin, author of the standard text on manic depressive illness (who we will meet again in chapter 8) arguing that bipolar was a physical illness, like any other disease, and ruled in her favor, apparently as a result of the very vague and loose phrasing of the definition of mental illness in the policy (as "mental, nervous or emotional disease or disorder of any type"). The judgment can be found at *http://www.dcd.uscourts.gov/Opinions/2002/Kennedy/98-617.pdf* (accessed December 1, 2005). The National Alliance for the Mentally Ill reported the case, which was decided in February 2002, under the heading, "Federal court strikes down boundary between physical and mental illness," *http://www.namiscc.org/newsletters/February02/CourtDecision.htm* (accessed December 1, 2005).

13. Compare them, for example, to the much more complex three-dimensional visualizations of the chloride ion channel generated by X-ray crystallography, in the award-winning work of a team of scientists (led by Howard Hughes Medical Institute investigator Roderick MacKinnon) at The Rockefeller University reported in the January 17, 2002 issue of the journal *Nature*, which claim to reveal an entirely new type of protein architecture designed to be an efficient conductor of chloride anions across the membrane of cells.

14. See *http://www.sciam.com/askexpert/medicine/medicine43/medicine43.html* (accessed on November 1, 2002).

15. See *http://www.prozac.com/HowProzacCanHelp/HowItWorks.jsp* (accessed November 1, 2002).

16. See *http://www.rxlist.com/cgi/generic/parox_cp.htm* (accessed August 12, 2002).

17. Elizabeth Wilson's account of the mode of action of serotonin on the gut highlights the way in which, in this style of thought, action of drugs outside the brain is ignored or minimized (Wilson 2004a).

18. See the discussion in chapters 2 and 3.

19. See chapter 4.

20. See chapter 3.

21. See *http://www.sciam.com/article.cfm?articleID=0004F784-F698–1C5A-B88 2809EC588ED9F* (accessed August 18, 2004).

22. On September 15, 2004, ChemGenex Pharmaceuticals announced that it had filed for patent protection for five novel depression-associated gene targets discovered by its depression and anxiety research group. According to the press release:

> The discovery follows the recent signing of a major partnership with the leading UK biopharmaceutical company Vernalis plc, a deal worth A$2 million in the first year pending milestones. . . . ChemGenex Pharmaceuticals researchers based at Deakin University in Geelong have identified five genes that are differentially expressed over a period of 8 days as the company's unique animal model undergoes behavioural changes after separation from littermates. The discoveries take to ten the number of genes that ChemGenex Phar-

> maceuticals has protected in the field of CNS diseases. . . . Greg Collier PhD, ChemGenex Pharmaceuticals Chief Executive Officer says the new discoveries reinforce the strength of the company's depression and anxiety gene discovery program. "This is a timely discovery, coming so soon after we have signed a major partnership deal with a strong international player in the field of CNS disorders. The discoveries include some genes that have never before been reported and a known receptor that has previously not been associated with depression. With the global anti-depressant market worth more than A$24 billion, there is a strong push for novel leads and new therapeutic approaches. We are confident that our novel animal model, linked with outstanding validation technologies and robust human genetics analysis will provide a suite of very attractive leads to Vernalis."

See *http://www.chemgenex.com/wt/page/pr_1109122078* (accessed August 23, 2005). Greg Collier is also named as Chief Operating Officer of Autogen Limited, whose "purchase" of the gene pool of the entire population of Tonga is discussed in chapter 1.

23. Such screening was proposed in the United States by President George W. Bush's New Freedom Commission on Mental Health although not using genetic technologies (see chapter 8 of this book, and Lenzer 2004.

24. From their website at *http://www.aspectmedical.com/professionals/neuroscience/default.mspx* (accessed August 20, 2003).

25. The figures that follow are from a commissioned study carried out by IMS Health for "The Age of Serotonin" research project at Goldsmiths College, University of London, funded by the Wellcome Trust Programme in Biomedical Ethics. More detailed analysis is given in Rose 2004.

26. Mental Health legislation proposed by the British government in 2004—almost universally condemned by the professional bodies of psychiatry, mental health organizations, and groups of users, consumers, and survivors of psychiatry—contained clauses enabling those psychiatric patients being treated "in the community" who were not "medicine compliant" to be taken compulsorily to a medical facility where the drugs could be administered.

27. See also the quotes from British school children presented by Cooper (Cooper 2004). However, these are interpreted differently by Steven Rose, who suggests that whatever the drugs do for the children, they make life more comfortable for harassed parents and teachers. He suggests that much of the pressure to use the diagnosis of ADHD and the drugs that treat it, has come from them, rather than from the medical profession. See the discussion in chapter 10 of Rose, S. 2005.

28. See *http://www.sarafem.com/* (accessed November 1, 2002).

29. See the Doctor's Guide website, *http:// www.pslgroup.com/dg/1f8182.htm*, (accessed Dec. 8, 2002).

30. I would like to thank Joseph Dumit for providing me (and many other researchers on this issue) with copies of many of these television advertisements.

31. See *http://www.policyalternatives.ca/publications/articles/article315.html* (accessed August 18, 2004).

32. See *http://www.pmlive.com/pharm_market/prac_guides.cfm* (accessed August 18, 2004).

33. See *http://www.wyeth.com/news/Pressed_and_Released/pr08_16_2005_07_43_19.asp* (accessed August 23, 2005).

34. See *http://www.pfizer.com/pfizer/are/news_releases/2005pr/mn_2005_0811.jsp* (accessed August 23, 2005).

35. As discussed in chapters 4 and 5.

36. SANE is supported by many celebrities and has a high media presence (see *http://www.sane.org.uk*). NAMI is most explicit, declaring mental illnesses to be "biologically based brain disorders," carrying synopses of studies on its website that claim that there is strong evidence for genetic predispositions involving multiple genes for schizophrenia and mood disorders, and supporting genetic research conducted by the National Institute for Mental Health (*http://www.nami.org/*).

37. See *http://www.help4adhd.org/en/about/causes* (accessed August 26, 2005).

38. Thanks to Chloe Silverman for information in this area, and also for suggestions concerning CHADD and other parents groups for childhood mental disorders that follow the now familiar model of "genetic citizenship."

39. See *http://www.geneticalliance.org/* (accessed August 18, 2004).

40. For MIND see *http://mind.org.uk*, and for the Mental Health Foundations see *http://www.mentalhealth.org.uk*. See also Rethink—formerly the National Schizophrenia Fellowship—at *http://www.rethink.org.uk* for similar formulations.

41. For example, *Pendulum*, the magazine of The Manic Depression Fellowship, in summer 2005, carried an account of research that might lead to a blood test to detect gene expression that could distinguish those diagnosed with schizophrenia from those diagnosed with bipolar disorder, suggesting that such a test would allow early diagnosis and more timely treatment (they were referring to Tsuang et al. 2005). For the Manic Depression Fellowship, see *http://www.mdf.org.uk/*.

42. The Depression Alliance is typical of these see *http://www.depressionalliance.org/*).

43. This is not the place to trace the history of these critical movements, which I have discussed in Miller and Rose 1986, and Rose 1986. As is well known, the critical psychiatry movement in the United Kingdom and Europe drew on the work of Laing, Szasz, a certain reading of the work of Michel Foucault, and the influence of the Italian psychiatrist Franco Basaglia. Key organizations in the United Kingdom history were the Mental Patient's Union in the 1970s, CAPO (the Campaign Against Psychiatric Oppression) in the early 1980s, the British Network for Alternatives to Psychiatry, and Survivors Speak Out. At the European level, some of this campaigning work is still undertaken by the European Network of (ex-)Users and Survivors of Psychiatry (ENUSP) (see *http://www.enusp.org/*).

44. For example, MindFreedom brings together an international coalition of such organizations committed to challenging not only the genetic basis of mental disorders, but the "biomedical model" itself and the "mental health industry," notably the pharmaceutical companies (see *http://www.mindfreedom.org/*).

45. It can be found on his website at *http://www.moshersoteria.com/*. From 1968 to 1980 Mosher was the first chief of NIMH's Center for Studies of Schizophrenia, where he founded and served as first editor-in-chief of the *Schizophrenia Bulletin*. From 1988 to 1996 he was chief medical director of Montgomery County Maryland's Department of Addiction, Victim and Mental Health Services, and a clinical professor of psychiatry. From 1996 to 1998 he was clinical director of Mental Health Services for San Diego, California and a clinical professor of psychiatry at the School of Medicine, University of California at San Diego. In his role in Montgomery County, he helped establish a number of innovative programs, including a consumer owned and operated computer company and a new residential alternative to psychiatric hospitalization for persons in crisis. From 1970 to 1992 he was a collaborating investigator, then research director, of the Soteria Project—Community Alternatives for the Treatment of Schizophrenia. Lauren Mosher died in 2004.

46. The exemplar of toxicity in these arguments is the long term effects of the "traditional" antipsychotics or neuroleptic drugs, such as Largactil / Thorazine, which, when used as they were in the mental hospitals in the 1950s, 1960s, and 1970s, led to an irreversible condition of tardive dyskinesia (see Gelman 1999, Healy 2001, and Rose 2004).

47. Many have also looked to the law to support these arguments. The first lawsuit against Prozac to reach the courtroom in Louisville in 1994, concerned Joseph Wesbecker who some five years earlier, shortly after being prescribed Prozac, had shot twenty-eight people at the printing plant where he worked, killing eight before shooting himself. This case brought into the public domain longstanding concerns about adverse effects of these drugs—about increases in agitation (akathesia) and suicidal ideation in a small but significant number of those administered Prozac. This had led the German licensing authorities to insist upon product warning in 1984 before they would issue a license. As the first generation of the drugs goes out of patent, the manufacturers are also fighting against many analogous cases. In June 2001, a court in Cheyenne, ordered GlaxoSmithKline to pay US$6.4m (£4.7m) to the family of Donald Schell who shot his wife, daughter, and granddaughter and then killed himself—two days after his GP prescribed Paxil (paroxetine, known as Seroxat in Europe) for depression. The jury decided that the drug was 80 percent responsible for the deaths. And two weeks earlier, in May 2001, an Australian judge ruled that a larger than recommended dose of sertraline, Zoloft, which is Australia's most widely used antidepressant, caused David Hawkins to murder his wife and attempt to kill himself: "I am satisfied that but for the Zoloft he had taken he would not have strangled his wife" (Justice Barry O'Keefe). Quoted at *http://www.antidepressantsfacts.com/David-JohnHawkins.htm* (accessed August 24, 2005).

48. Marcia Angell, former Editor of the prestigious New *England Journal of Medicine*, has produced some of the most telling assaults on the pharmaceutical industry, (see Angell 2004, and Relman and Angell 2002.)

49. Emily Martin has demonstrated this clearly in her study of the marketing of antidepressants, showing how Frank Ayd, central to the clinical trials for amitryptiline with Merck, and author of a book sometimes thought to have played a major role in increasing the diagnosis of depression and its treatment with the

drug, was committed evangelically to the struggle to get this disorder recognized and treated (Ayd 1961) I have drawn on her paper, "Pharmaceutical virtue: an ethnographic approach to the meaning of psychotropic drugs for marketers and patients," given at Vital Matters: Biotechnology and its Social and Ethical Implications, Sept. 30–Oct. 2, 2005, Bergen, Norway.

50. Taken from one of a large number of medical information company websites that promote such reports, in this case, that of Lead Discovery: *http://www.lead discovery.co.uk/reports/Anxiety%20Disorders%20%20More%20Than%20Just%20a%20Comorbidity.html* (accessed December 1, 2005).

51. I have argued this at length in Rose 1996.

Chapter 8 The Biology of Control

1. This is an extensively revised version of "The biology of culpability: pathological identities in a biological culture," *Theoretical Criminology*, 2000, 4 (1): 5–34. The paper originated from a keynote address to the 18th Annual Congress of the Australian and New Zealand Association of Psychiatry, Psychology and Law, in Melbourne, Victoria, Australia, April 16–19, 1998. Other versions were given to the Department of Sociology at the University of Durham in March 1999, and to the Centre of Criminology at the University of Toronto and the Law and Society Association Annual Meeting in Chicago in May 1999. Thanks to Deborah Denno, Dorothy Nelkin, and David Wasserman for help in finding material for this original paper and for very helpful comments on an early draft, to anonymous reviewers for *Theoretical Criminology* for useful suggestions, and to Pat O'Malley and Mariana Valverde, as ever, for sound advice. Responsibility for the argument is mine alone.

2. For a more detailed account of control practices in "advanced liberal" societies, see Rose 2000.

3. For an excellent account of these developments, focusing on the United States, see Rafter 1997. While attention was first on the external shapes, proportions, and surfaces of the body, later techniques sought to render visible the marks of uniqueness in the invisible interior of the body; the issue of nonsymptomatic "carriers" was of particular concern (Paul 1998g). Note that, while it is possible to trace a line from Lombroso through Bertillon to DNA fingerprinting, each worked within a distinct problem space concerning the individuation of the criminal.

4. I discuss eugenics in more detail in chapter 2. As far as crime is concerned, sterilization of male and female convicted criminals in the United States began as early as 1898, and laws permitting sterilization of habitual or confirmed criminals or sex offenders were passed in sixteen states by 1917: a *Fortune Magazine* poll in 1937 showed that 63 percent of Americans were in favor of compulsory sterilization of habitual criminals (Kevles 1985: 114). The genetics of crime was a specialty in 1920s Germany, flourished after the Nazis came to power, and by 1939 examination of the genetics and genealogy of criminal suspects had become a routine part of criminal investigations (Proctor 1988). In England, the criminal was not a privileged object of eugenic discourse, which focused on the issue of the feebleminded and hesitated about the uses of compulsory interventions into

reproductive freedoms (Garland 1994, Rose 1985). In the Nordic countries, which adopted eugenic sterilization for the mentally ill and the feebleminded with various degrees of enthusiasm in the 1930s, eugenics took a kind of "pastoral" or "welfare" form: while sterilization did take place on the basis of "antisocial" indicators, as far as criminal conduct was concerned it tended to focus on the sex offender, no doubt because of confused reasons concerning castration and sexuality (Broberg and Roll-Hansen 1996).

5. In previous chapters I have criticized Lippman's account of "geneticization," as set out, for example, in Lippman 1992. Spallone develops a more complex argument (Spallone 1998); she uses the term "geneticism," which she draws from the earlier criticisms of misapplied genetics made by Medewar (1984).

6. The claim was withdrawn after a number of individuals without the marker developed the disorder, undermining the statistical significance of the association.

7. The dispute is described in Wasserman 1996. Many thanks to David Wasserman for providing me with papers and information on this conference and the surrounding events. Many of the papers from the conference are now published as Wasserman and Wachbroit 2001. Note that similar, less controversial events, were held in Europe at around the same time; see for example, Bock and Goode 1996, and Crusio 1996.

8. The controversy is discussed by one of the organizers, in Wasserman and Wachbroit 2001.

9. There are many national differences, and, in the United States, differences among the various state and federal jurisdictions. While the legal outcomes differ from jurisdiction to jurisdiction, there are many similarities in the problems raised—it is the problems, rather than the specific outcomes, that interest me here. I make no claims to comprehensive coverage, but select some exemplary cases whether they arise in the United States, England, or Australia. Useful accounts of U.S. cases are given in Coffey 1993, Denno 1988, 1996, and Dreyfuss and Nelkin 1992.

10. For juveniles, see the excellent summary of the position up until then in Freeman 1983. For the complex position with regard to the agency and responsibility of criminal women, see Allen 1988.

11. In many cases, the mere fact that an XYY male was an inmate in a mental hospital or penal institution was taken as evidence of a history of violence, and the researchers did not distinguish property crimes from violent crimes, or confinement for criminal offenses from confinement for reasons of mental ill health. When better studies were undertaken, the only correlations between XYY status and criminality concerned petty property offenses. The rise and fall of the defenses is discussed in Saulitis 1979.

12. As far as I am aware, this defense has not been used in the U.S. courts (see Downs 2002, Lewis 1990.)

13. Studies of electroencephalographs of delinquents and criminals from the 1940s onwards attempted to discover specific abnormalities in offenders as compared to the general population, and to develop such techniques for diagnostic and prescriptive purposes, with conflicting and disputable success. See the reviews in Fishbein and Thatcher 1986, and Pollock et al. 1983.

14. The Hinckley case and subsequent cases using brain scans are discussed in Denno 1988.

15. In fact, the implications of the Hinckley case are complicated because, under U.S. federal law, the prosecution had to prove beyond reasonable doubt that the defendant was sane at the time of the offense; in most individual states and other jurisdictions the defense must prove by the preponderance of evidence that the defendant is insane—a standard that would probably have led to a conviction for Hinckley.

16. *People vs. Weinstein*, 591 NYS.2d 715 (Sup. Ct. 1992). The court concluded that expert evidence and consideration of the results of a PET scan and other physiological tests—to indicate a cyst and metabolic imbalances in the defendant's brain—was not unreasonable in making a diagnosis of insanity, but agreed to negotiate a reduced charge from murder to manslaughter, rather than going to trial. In the early 1998 trial of Michael Person, in New Haven, Connecticut, prosecutors contested the attempts of defense lawyers to present the jury with PET scans showing brain abnormalities, and the findings of Adrian Raine on the increased prevalence of abnormal brain scans in convicted murders in seeking to reduce the charge from murder to manslaughter. Raine's work is discussed below. Another early case involving a request to introduce brain scans was that of Jack Dempsey Ferrell, who was convicted in 1995 of the first degree murder of his girlfriend in 1992. At his appeal hearing before the Supreme Court of Florida in 2005, Ferrell contended that his counsel at his first trial should have supported arguments that he had suffered neurological impairment with a brain scan. At hearings between the original trial and the final appeal, the state had objected to the request for SPECT scans to be conducted, not only because there was already a diagnosis of frontal lobe brain damage that had been taken into account as a mitigating factor, but because no scan could show how the physical brain affected Ferrell's capacity to function. The Supreme Court of Florida, in its ruling of June 16, 2005 (No. SC03–218) was of the view that "a particularized showing of necessity is the polestar for whether any diagnostic test should be authorized," and that this had not been demonstrated in Ferrell's case.

17. Note that there is a distinction between the use of scans to demonstrate brain *damage* and the use of scans to demonstrate mental disorder. It has certainly been claimed that scan evidence of brain damage confirms a clinical diagnosis of schizophrenia, as in the failed appeal to the Supreme Court of Florida in 2001 against the sentence of death imposed on Jonathan Huey Lawrence (as of 2005 further appeals are still pending in this case). It is rare for scans to be used in an attempt to demonstrate what used to be termed "functional" disorder, that is mental disorder without concomitant brain lesions.

18. Details of the trial and excepts from transcripts of the evidence presented are at *http://www.pbs.org/wgbh/pages/frontline/shows/kinkel/trial/* (accessed August 29, 2005).

19. For NAMI's position paper, see *http://www.nami.org/Content/ContentGroups/Policy/WhereWeStand/The_Criminalization_of_People_with_Mental_Illness___WHERE_WE_STAND.htm* (accessed August 29, 2005).

20. More recently some have suggested that Kinkel was suffering from adverse effects of Prozac and Ritalin that he had allegedly been taking at the time of the murders and extended this claim to the young people responsible for a number of other school shootings in the United States. See, for example, Dan Edwards, at

http://www.geocities.com/StNektarios/BIOPSYCH.html (accessed on August 29, 2005).

21. The fullest account of this case that I have been able to find is that provided in 2005 by Court TV at *http://www.crimelibrary.com/serial_killers/predators/stayner/* (accessed August 30, 2005). I have drawn on that report here.

22. See *http://www.crimelibrary.com/serial_killers/predators/stayner/* under Updates (accessed August 29, 2005). Dr. McInnes describes her current research area as neurobehavioral genetics.

23. Brunner discusses the implications of his research in rather different terms, as we shall see later.

24. Other cases sometimes cited in support of the argument for genetic essentialism concern the effects of alcoholism on the free will of attorneys subject to disciplinary proceedings or disbarment. Nelkin supports her claim that genetic essentialism is on the rise by reference to two such cases in the late 1980s, in which Ewaniszyk, whose alcoholism led to embezzlement, was disbarred while Baker, whose conduct was similar but claimed a genetic predisposition, was not (Dreyfuss and Nelkin 1992). But the paradoxical status of "alcoholism"—as both willed and unwilled, biological and voluntary—raises peculiar problems for the law: see the debate in the similar case of Kersey: 520 A. 2d 321 (D.C. App. 1987) and, more generally, Valverde 1998.

25. Despite the worries of the neuroethicists, the same will almost certainly be the case if "brain fingerprinting"—a technique hoping to supplement or displace the lie detector or polygraph by using brain scans that claims to be able to demonstrate if an individual is lying, or if he or she has previously seen a face, place, or crime scene—is used in the courts. This technique is based on EEG monitoring of a specific, electrical brain wave response, known as a P300, that is emitted by the brain within a fraction of a second when an individual recognizes and processes an incoming stimulus that is significant or noteworthy. Brain Fingerprinting was developed by Dr. Larry Farwell, whose organization, Brain Fingerprinting[R] Laboratories, has patented it (see *http://www.brainwavescience.com/* [accessed August 23, 2005]). This technique was deployed in a number of court cases in 2002 and 2003, including the successful appeal of Terry Harrington convicted of murder in Iowa in 1978. It has received much publicity in the U.S. media. By 2005 it was still not clear to what extent, and in what circumstances, evidence from this technique would be admissible in the U.S. courts. Some of the jurisprudential issues are reviewed in Keckler 2005.

26. Of course, the limits that circumscribe those thought to be capable of bearing the obligations of free will have long been contested. There is a long history of debates in the criminal justice system and among legal philosophers about the capacity of women and children to be fully responsible for their actions. From the mid-nineteenth century, claims to scientific knowledge about their distinction from fully rational adult males have often been deployed—for example in relation to issues of mothers' responsibility in cases of infanticide, and in relation to the menstrual period as we saw earlier in this chapter. For children, much of the debate has turned on the age of criminal responsibility, but once more, the courts have been resistant to adjusting their own forms of reasoning according the dictates of positive knowledge. The recent U.S. campaign against the death penalty

for young people has drawn on evidence from brain studies to show that the adolescent brain does not reach maturity until 18, 20, or even 25 years of age (see, for example, Beckman 2004). However there is debate about this within the scientific community itself. In an article in *Science News Online* in May 2004, some researchers, notably David Fassler, a psychiatrist from the University of Vermont and Rueben Gur, a neuropsychologist from the University of Pennsylvania, argued that the law must take account of the fact that the brains of adolescents function differently from those of adults, while others argued that there was "no evidence linking the specific character of the brains of adolescents to legally relevant conditions such as impaired moral judgment or an inability to control murderous impulses. 'Juvenile death sentences bother me, but this is an ethical issue,' remarks Harvard University psychologist Jerome Kagan. 'The brain data don't show that adolescents typically have reduced legal culpability for crimes.'" (quoted from Bruce Bower, Teen Brains on Trial in *Science News Online* for the week of May 8, 2004, Vol. 156, No. 19, from *http://www.sciencenews.org* [accessed August 28, 2005]).

27. See, for example, the cases of Sean Sellars and Brent Ullery, reported by the Oklahoma Coalition to Abolish the Death penalty, available at http://www.ocadp.org/.

28. For cases debating this issue, see *Lucas v. State*, 887 S. W.2d 315 (Tex. Crim. App. 1994), drawing on *Johnson v. Texas*, 125 L. Ed. 2d 290, 113 S. Ct. 2658 (1993) and *Penry v. Linaugh*, 492 U.S, 915 (1989). Many U.S. states have introduced similar "guided discretion" clauses in death penalty cases, in which, after a verdict, a trial goes into a separate phase, in front of a jury, for deciding on the penalty: juries are required to consider a range of aggravating and mitigating factors in determining sentencing, which in some states includes the likelihood of reform.

29. This is an international phenomenon. In the United Kingdom, at the time of writing (July 2005) proposals are being debated for the detention of untreatable individuals thought to be predisposed to violence and to present a high risk to public safety. In Victoria, Australia, in April 1990, the Community Protection Act was passed to legitimate detention of one individual, Garry David, considered to be dangerous but who did not fall under the ambit of either criminal or mental health law. In related quasi-psychiatric areas, notably pedophilia, preventive detention is being discussed in many national contexts: it appears that the conventions of "rule of law" must be waived for the protection of the community against a growing number of "predators," who do not conform to either legalistic or psychiatric models of subjectivity (Pratt 1998; Simon 2000; Scheingold et al. 1994).

30. That was also the view of most of the contributors to the volume of essays containing Morse's chapter. However, Joshua Green and Jonathan Cohen, in a paper in a special themed issue of *Philosophical Transactions on Law and the Brain*, take a different view, suggesting that evidence from the neuroscience may not trouble the law directly, but it will do so indirectly by transforming people's moral intuitions about freedom and responsibility, and supporting arguments against retribution as the logic of punishment, and in favor of consequentialist approaches aimed at promoting future welfare rather than meting out just deserts

(Greene and Cohen 2004). It has to be said that all the evidence runs in the other direction, as we shall see, or rather, that the future welfare that the criminal justice system seeks to promote is increasingly that of "society," as in social protection and social defense, rather than the reformation of the individual (see, for example, Garland 2001, and Rose 2000).

31. Linus Pauling was an enthusiast for the utopian possibilities of rational control of human capacities through molecular genetics as early as the 1950s (see Duster 1990 46). James Watson and Robert Sinsheimer of Caltech both provide quotable quotes illustrating the view that "our fate is in our genes," (see Nelkin and Lindee 1995). For the racists, quotes are usually found from William Shockley, the Pioneer Fund that has been associated with racist eugenics since its establishment in 1937, or from J. Phillipe Rushton, professor of psychology at the University of Western Ontario in Canada, who has argued that behavioral differences among blacks, Asians, and whites result from evolutionary variations in their reproductive strategies.

32. Finland has something of a specialism in this area. Ian Hacking has discussed the sequence of papers on the neurochemistry of aggression produced by Markku Linnoilla and Matti Virkkunnen and their colleagues (Hacking 2001). He points out that Frederick K. Goodwin, later central to the Violence Initiative controversy, was a coauthor on some of these papers.

33. On the specific idea of "therapeutic jurisprudence," which is one element within this more general strategy, see Carson 1995. In therapeutic jurisprudence, the law, and the criminal justice system more generally, are to be used in order to produce a therapeutic effect upon the actual or potential offender, where that therapeutic effect is largely understood in terms of a reintegration of the individual into the moral and behavioral norms of their community.

34. Piers Beirne points out that this was argued by Charles Goring in 1913 (Beirne 1988).

35. Summarized from an article he wrote for the BBC website linked to this program, at *http://news.bbc.co.uk/1/hi/programmes/if/4102371.stm* (accessed August 31, 2005).

36. On "world making by kind making," see Hacking 1992.

37. This strategy is not identical to the actuarial form of risk management suggested by Malcolm Feeley and Jonathon Simon (in Feeley and Simon 1992, 1994). The calculation of riskiness may be factorial, and may try to identify particular subgroups of the population where factors associated with high risk are concentrated, but the aim is still to identify and neutralize specific risky individuals.

38. I discuss some of these studies in chapter 7.

39. See *http://www.sciencedaily.com/releases/2003/01/030123072840.html* (accessed December 29, 2004).

40. For instance, Cornelius Gross and his colleagues at the European Molecular Biology Laboratory in Rome, Italy are using sophisticated designs to try and explore gene environment interaction in mice in relation to anxiety and aggression (see for example Gross and Hen 2004).

41. They helpfully provide an analysis of candidate genes for antisocial behaviors and a glossary at *http://www.aic.gov.au/publications/tandi2/tandi263.html* (accessed Dec. 29, 2004).

42. Goodwin was previously head of the Alcohol, Drug Abuse and Mental Health Administration and was the author (with Kay Redfield Jamison) of the standard textbook on manic depression (Goodwin and Jamison 1990). In the spring of 1992, while discussing the initiative before the U.S. National Mental Health Advisory Council, Goodwin drew analogies between inner city violence and events in some monkey populations, where males kill other males, and then, with competition limited, proceed to copulate prolifically with females. According to Goodwin, these "hyperaggressive" males also seem to be "hypersexual." He seemed to be connecting this monkey behavior to that of young violent men involved in the criminal justice system, who were, of course, overwhelmingly black. As the *Los Angeles Times* put it, Goodwin "made comparisons between inner-city youths and violent, oversexed monkeys who live in the wild." He angered many with his references to genetic factors inclining human beings toward violence and the need to look for biological markers of violent disposition. Within weeks, the Violence Initiative was abandoned amid charges of racism, and Goodwin moved to become Director of the National Institute of Mental Health, leaving in 1994 for a position at George Washington University Medical Center. He continued to win awards and plaudits, and from 1998 through 2004, was host of *The Infinite Mind* a weekly U.S. public radio program on the art and science of the human mind.

43. See, for example, the overview prepared by the NIH in 2000 on their research on child and adolescence violence, at *http://www.nimh.nih.gov/publicat/violenceresfact.cfm* (accessed Dec. 29, 2004) and the 2004 NIH State of the Science Consensus statement, at *http://consensus.nih.gov/ta/023/023youthviolencepostconfintro.htm* (accessed Dec. 29, 2004).

44. To that extent, these initiatives are entirely compatible with the new forms of communitarianism that are on the rise in the United States and United Kingdom (see my discussion in Rose 1999). I am grateful to Daniel Wasserman for comments that clarified this point for me.

45. This program followed up on the earlier research and the recommendations of the 1994 panel on NIH Research on Anti-social, Aggressive, and Violence Related Behaviors and their Consequences. In view of the accusations of racism and sexism that were leveled at the earlier research program, it is significant that it recommends that advisory committees for research projects contain community representatives, and that "special attention should be directed towards the unique needs and special concerns of racial and ethnic minority group members, so that services and opportunities are appropriate and acceptable to these individuals."

46. There is a growing literature on these issues. See in particular Pratt 1995 and O'Malley 1998. Of course one should avoid suggesting a misleading coherence: as Garland and O'Malley have both pointed out, crime control tactics are heterogeneous, mutually contradictory, and fluctuate rapidly according to local political agendas (Garland 1996, O'Malley 1999.) In particular, the risk prevention strategies that I have mentioned, which entail adjusting the punishment to the criminal, are accompanied by an apparently contradictory move toward determinate sentencing, where punishment is calculated solely in response to the nature of the offense.

47. To discuss the impact of the recent rise of evolutionary paradigms in psychology and sociology would require a separate paper.

48. On such scenarios in relation to genetic screening in employment and insurance, although with reference to disease rather than to crime or behavioral disorder, see Gostin 1991.

Afterword Somatic Ethics and the Spirit of Biocapital

1. I am obviously referring here to the understanding of ethics developed by Michel Foucault and Gilles Deleuze—a way of thinking about ethics and self-technologies that I have discussed extensively in my earlier work (I draw in particular on the ways these are framed in Deleuze 1988, Foucault 1985, Rose 1996b).

2. As Weber himself argues in the final paragraph of *The Protestant Ethic and the Spirit of Capitalism* (Weber 1930: 183).

Bibliography

Abbott, Alison (1999) Sweden Sets Ethical Standards for the Use of Genetic "Biobanks." *Nature*, 400 July 1, 1999: 3.

Abraham, J., & Lewis, G. (2002) Citizenship, Medical Expertise and the Capitalist Regulatory State in Europe. *Sociology*, 36 (1): 67–88.

Adams, M. D., et al. (2000) The Genome Sequence of Drosophila Melanogaster. *Science*, 287 (5461): 2185–2195.

Agamben, Giorgio (1998) *Homo Sacer: Sovereign Power and Bare Life*. Stanford, Calif.: Stanford University Press.

Alldridge, P. (1992) Recognising Novel Scientific Techniques: DNA as a Test Case. *Criminal Law Review*, 687–698.

——— (1994) Forensic-Science and Expert Evidence. *Journal of Law and Society*, 21 (1): 136–150.

Allen, Hilary (1984) At the Mercy of Her Hormones: Premenstrual Tension and the Law. *m/f*, 9 19–44.

——— (1988) *Justice Unbalanced*. Milton Keynes: Open University Press.

American College of Obstetricians and Gynecologists, Cystic Fibrosis Steering Committee (2001) Preconception and Prenatal Carrier Screening for Cystic Fibrosis: Clinical and Laboratory Guidelines. Washington: American College of Obstetricians and Gynecologists.

American Psychiatric Association (1952) *Diagnostic and Statistical Manual of Mental Disorders*. Washington, D.C.: American Psychiatric Association.

——— (1968) *Diagnostic and Statistical Manual of Mental Disorders: DSM II*. Washington, D.C.: American Psychiatric Association.

——— (1980) *Diagnostic and Statistical Manual of Mental Disorders: DSM III*. Washington, D.C.: American Psychiatric Association.

——— (1987) *Diagnostic and Statistical Manual of Mental Disorders: DSM III-R*. Washington, D.C.: American Psychiatric Association.

——— (1994) *Diagnostic and Statistical Manual of Mental Disorders: DSM IV*. Washington, D.C.: American Psychiatric Association.

Anderson, C. (1992) Brain Scans Deemed Admissible at Trial. *New York Law Journal*, 210 (1).

Andrews, Lori B., & Nelkin, Dorothy (2001) *Body Bazaar: The Market for Human Tissue in the Biotechnology Age*. New York: Crown Publishers.

Andrews, Marcellus (1999) *Political Economy of Hope and Fear: Capitalism and the Black Condition in America*. New York: New York University Press.

Angell, Marcia (2004) *The Truth About the Drug Companies: How They Deceive Us and What to Do About It*. New York: Random House.

Anon (1919) Training Course for Field Workers in Eugenics. *Eugenical News*, 4 (5): 40.

Anon (1998) Genome Sequence of the Nematode C-Elegans: A Platform for Investigating Biology. *Science*, 282 (5396): 2012–2018.

Anon (2001) Genes, Drugs and Race. *Nature Genetics*, 29 (3): 239–240.

Armstrong, David (1983) *Political Anatomy of the Body: Medical Knowledge in Britain in the Twentieth Century*. Cambridge: Cambridge University Press.

——— (1984) The Patient's View. *Social Science and Medicine*, 18, 737–744.

——— (1995) The Rise of Surveillance Medicine. *Sociology of Health & illness*, 17 (3): 393–404.

Armstrong, David, Michie, Susan & Marteau, Theresa (1998) Revealed Identity: A Study of the Process of Genetic Counselling. *Social Science and Medicine*, 47 (11): 1653–1658.

Arney, William Ray, & Bergen, Bernard J. (1984) *Medicine and the Management of Living: Taming the Last Great Beast*. Chicago: University of Chicago Press.

Arnold, David (2002) Madness, Cannabis and Colonialism: The 'Native-Only' Lunatic Asylums of British India, 1857–1900. *Social History of Medicine*, 15 (3): 519–520.

——— (2005) Legible Bodies: Race, Criminality and Colonialism in South Asia. *English Historical Review*, 120 (486): 555–556.

Association of British Insurers (1997) *Genetic Testing: ABI Code of Practice*. London.

Aviles-Santa, L., Maclaren, N., & Raskin, P. (2004) The Relationship between Immune-Mediated Type 1 Diabetes Mellitus and Ethnicity. *Journal of Diabetes and its Complications*, 18 (1): 1–9.

Ayd, Frank J. (1961) *Recognizing the Depressed Patient; with Essentials of Management and Treatment*. New York: Grune & Stratton.

Bachelard, Gaston (1984) *The New Scientific Spirit*. Boston, Mass.: Beacon Press.

Baker, Robert B., et al. (Eds.) (1999) *The American Medical Ethics Revolution. How the AMA's Code of Ethics Has Transformed Physicians' Relationships to Patients, Professionals, and Society*. Baltimore, MD: Johns Hopkins University Press.

Baker, Tom, & Simon, Jonathan (2002) *Embracing Risk: The Changing Culture of Insurance and Responsibility*. Chicago; London: University of Chicago Press.

Balaban, E. (1996) Reflections on Wye Woods: Crime, Biology, and Self-Interest. *Politics and the Life Sciences*, 15 (1): 86–88.

Balaban, E., Alper, J. S., & Kasamon, Y. L. (1996) Mean Genes and the Biology of Aggression: A Critical Review of Recent Animal and Human Research. *Journal of Neurogenetics*, 11 (1–2): 1–43.

Balick, Michael J., & Cox, Paul Allan (1996) *Plants, People and Culture: Science of Ethnobotany.* New York: Scientific American Library.

Baron, M. (1989) The Genetics of Manic-Depressive Illness. *Journal of Nervous and Mental Disease*, 177 (10): 645–645.

Baron, M., et al. (1987) Genetic-Linkage between X-Chromosome Markers and Bipolar Affective-Illness. *Nature*, 326 (6110): 289–292.

Barry, Andrew, Osborne, Thomas & Rose, Nikolas (1996) *Foucault and Political Reason: Liberalism, Neo-Liberalism and Rationalities of Government.* Chicago: University of Chicago Press.

Bates, B. R., et al. (2004) Evaluating Direct-to-Consumer Marketing of Race-Based Pharmacogenomics: A Focus Group Study of Public Understandings of Applied Genomic Medication. *Journal of Health Communication*, 9 (6): 541–559.

Bateson, Patrick, & Martin, Paul R. (1999) *Design for a Life: How Behaviour Develops*. London: Jonathan Cape.

Bauman, Zygmunt (1989) *Modernity and the Holocaust*. Cambridge: Polity.

Beaulieu, Anne (2000) The Space inside the Skull: Digital Representations, Brain Mapping and Cognitive Neuroscience in the Decade of the Brain. Ph.D. Dissertation, University of Amsterdam.

Beauregard, M., et al. (1998) The Functional Neuroanatomy of Major Depression: An FMRI Study Using an Emotional Activation Paradigm. *Neuroreport*, 9 (14): 3253–3258.

Beck, Ulrich, Giddens, Anthony & Lash, Scott (1994) *Reflexive Modernization: Politics, Tradition and Aesthetics in the Modern Social Order.* Cambridge: Polity Press.

Becker, Howard Saul (1963) *Outsiders. Studies in the Sociology of Deviance.* New York: Free Press of Glencoe; London: Collier-Macmillan.

Beckman, M. (2004) Neuroscience—Crime, Culpability, and the Adolescent Brain. *Science*, 305 (5684): 596–599.

Beirne, P. (1988) Heredity Versus Environment—a Reconsideration of Goring, Charles—the English Convict (1913). *British Journal of Criminology*, 28 (3): 315–339.

Bergen, A. A. B., et al. (2000) Mutations in ABCC6 Cause Pseudoxanthoma Elasticum. *Nature Genetics*, 25 (2): 228–231.

Berlinguer, G. (2004) Bioethics, Health, and Inequality. *Lancet*, 364 (9439): 1086–1091.

Bevan, J. L. et al. (2003) Informed Lay Preferences for Delivery of Racially Varied Pharmacogenomics. *Genetics in Medicine*, 5 (5): 393–399.

Bhopal, R. (2002) Revisiting Race/Ethnicity as a Variable in Health Research. *American Journal of Public Health*, 92 (2): 156–157.

Bhopal, R., & Donaldson, L. (1998) White, European, Western, Caucasian, or What? Inappropriate Labeling in Research on Race, Ethnicity, and Health. *American Journal of Public Health*, 88 (9): 1303–1307.

Bhopal, R., & Rankin, J. (1999) Concepts and Terminology in Ethnicity, Race and Health: Be Aware of the Ongoing Debate. *British Dental Journal*, 186 (10): 483–484.

Bibby, Cyril (1939) *Heredity, Eugenics and Social Progress*. London: Gollancz.

Biesecker, Barbara Bowles & Marteau, Theresa M. (1999) The Future of Genetic Counselling: An International Perspective. *Nature Genetics*, 22 (2): 133–137.

Billings, Paul R., et al. (1992) Discrimination as a Consequence of Genetic Testing. *American Journal of Human Genetics*, 50 (3): 476–482.

Bishop, W. H. (1909) *Education & Heredity; or, Eugenics: A Mental, Moral and Social Force*. London.

Bloch, M. et al. (1993) Diagnosis of Huntington Disease: A Model for the Stages of Psychological Response Based on the Experience of a Predictive Testing Program. *American Journal of Medical Genetics*, 47, 368–374.

Blum, K., et al. (1990) Allelic Association of Human Dopamine-D2 Receptor Gene in Alcoholism. *Journal of the American Medical Association*, 263 (15): 2055–2060.

Bock, Gregory, & Goode, Jamie (1996) *Genetics of Criminal and Antisocial Behavior.* Chichester/New York: Wiley.

Boring, Edwin Garrigues (1929) *A History of Experimental Psychology.* New York/London: Century Co.

Boston Women's Health Book Collective., Phillips, Angela & Rakusen, Jill (1978) *Our Bodies Ourselves: A Health Book by and for Women*. Harmondsworth: Penguin.

Bouchard, T. J., et al. (1990) Sources of Human Psychological Differences—the Minnesota Study of Twins Reared Apart. *Science*, 250 (4978): 223–228.

Bovet, P., & Paccaud, F. (2001) Race and Responsiveness to Drugs for Heart Failure. *New England Journal of Medicine*, 345 (10): 766–766.

Bowcock, A. M., et al. (1991) Drift, Admixture, and Selection in Human Evolution: A Study with DNA Polymorphisms. *Proceedings of the National Academy of Sciences of the United States of America*, 88 (3): 839–843.

Bowcock, A. M., et al. (1994) High Resolution of Human Evolutionary Trees with Polymorphic Microsatellites. *Nature*, 368 (6470): 455–457.

Bowker, Geoffrey C., & Star, Susan Leigh (1999) *Sorting Things Out: Classification and its Consequences*. Cambridge, Mass./London: MIT Press.

Bradby, Hannah (1995) Ethnicity—Not a Black-and-White Issue—a Research Note. *Sociology of Health & illness*, 17 (3): 405–417.

——— (1996) Genetics and Racism. In Marteau, T. & Richards, M. (Eds.) *The Troubled Helix: Social and Psychological Aspects of the New Human Genetics*. Cambridge: Cambridge University Press.

——— (2003) Describing Ethnicity in Health Research. *Ethnicity & Health*, 8 (1): 5–13.

Braidotti, Rosi (1994) *Nomadic Subjects: Embodiment and Sexual Difference in Contemporary Feminist Theory.* New York: Columbia University Press.

——— (2002) *Metamorphoses: Towards a Materialist Theory of Becoming*. Cambridge, UK/Malden, Mass.: Published by Polity Press in association with Blackwell Publishers.

Braun, L. (2002) Race, Ethnicity, and Health: Can Genetics Explain Disparities? *Perspectives in Biological Medicine*, 45 (2): 159–174.

Breggin, Peter (1995–96) Campaigns against Racist Federal Programs by the Centre for The Study of Psychiatry and Psychology. *Journal of African American Men*, 1 (3): 3–22.

Breggin, Peter, & Breggin, Ginger Ross (1994) *The War against Children*. New York: St. Martin's Press.

Brenner, Sydney (2000) Genomics—the End of the Beginning. *Science*, 287 (5461): 2173–2174.

Brescia, Bonnie (2002) Better Budgeting for Patient Recruitment. *Pharmaceutical Executive*, 1 May 2002.

Broberg, Gunnar & Roll-Hansen, Nils (1996) *Eugenics and the Welfare State: Sterilization Policy in Denmark, Sweden, Norway, and Finland*. East Lansing: Michigan State University Press.

Brody, A. L., et al. (2002) Brain Metabolic Changes During Cigarette Craving. *Archives of General Psychiatry*, 59 (12): 1162–1172.

Brown, Nik (1998) Ordering Hope: Representations of Xenotransplantation: An Actor-Network Account. Ph.D. thesis. Lancaster: University of Lancaster.

——— (2003) Hope against Hype—Accountability in Biopasts, Presents and Futures. *Science Studies*, 16 (2): 3–21.

Brown, Nik, Rappert, Brian, & Webster, Andrew (2000) *Contested Futures: A Sociology of Prospective Techno-Science*. Aldershot: Ashgate.

Brown, Nik, & Webster, Andrew (2004) *New Medical Technologies and Society: Reordering Life*. Cambridge: Polity.

Brown, Wendy (1995) *States of Injury: Power and Freedom in Late Modernity*. Princeton, N.J.: Princeton University Press.

Brunner, H. G. (1996) MAOA Deficiency and Abnormal Behavior: Perspectives on an Association. *Genetics of Criminal and Antisocial Behavior*, CIBA Foundation Symposium 194: 155–164.

Brunner, H. G., et al. (1993) Abnormal Behavior Associated with a Point Mutation in the Structural Gene for Monoamine Oxidase-A. *Science*, 262 (5133): 578–580.

Bucknill, John Charles, & Tuke, Daniel Hack (1874) *A Manual of Psychological Medicine, by John Charles Bucknill and Daniel*. London: X. J. & A. Churchill.

Burchard, E. G., et al. (2003) The Importance of Race and Ethnic Background in Biomedical Research and Clinical Practice. [See Comment]. *New England Journal of Medicine*. 348 (12): 1170–1175.

Burleigh, Michael (1994) *Death and Deliverance: "Euthanasia" in Germany C.1900–1945*. Cambridge: Cambridge University Press.

Burroughs, V. J., Maxey, R. W., & Levy, R. A. (2002) Racial and Ethnic Differences in Response to Medicines: Towards Individualized Pharmaceutical Treatment. *Journal of The National Medical Association*, 94 (10 Suppl): 1–26.

Burwick, Frederick, & Douglass, Paul (1992) *The Crisis in Modernism: Bergson and the Vitalist Controversy*. Cambridge/New York: Cambridge University Press.

Business Communications Company (2005) *Emotional and Behavioral Disorders in Children and Adolescents*. Business Communication Company Inc.

Butcher, James (2003) Cognitive Enhancement Raises Ethical Concerns: Academics Urge Pre-Emptive Debate on Neurotechnologies. *The Lancet*, 362 (9378): 132–133.

Calafell, F., et al. (1998) Short Tandem Repeat Polymorphism Evolution in Humans. *European Journal of Human Genetics*, 6, 38–49.

Callahan, L. A., et al. (1995) The Hidden Effects of Montana's "Abolition" of the Insanity Defense. *Psychiatric Quarterly*, 66 (2): 103–117.

Callon, Michel, & Rabeharisoa, Vololona (1999) Gino's Lesson on Humanity. *Third WTMC-CSI Workshop: Producing taste, configuring use, performing citizenship*. Maastricht.

——— (2004) Gino's Lesson on Humanity: Genetics, Mutual Entanglement and the Sociologist's Role. *Economy and Society*, 33 (1): 1–27.

Canguilhem, Georges (1966) Le Concept et la Vie. *Revue philosophique de Louvain*, 64, 193–233.

——— (1968) La Concept et la Vie. *Etudes d'histoire de Philosophie des Sciences*. Paris: Vrin.

——— (1978) *On the Normal and the Pathological*, Dordrecht: Reidel.

——— (1980) What Is Psychology? Lecture Delivered in the College Philosophique on 18 December 1956. *I&C*, 7, 37–50.

——— (1994) *A Vital Rationalist: Selected Writings from George Canguilhem, Edited by François Delaporte with an Introduction by Paul Rabinow*. New York: Zone Books.

Caplan, Arthur L., and Farah, Martha J. (2003) Emerging Ethical Issues in Neurology, Psychiatry and the Neurosciences, in R. N. Rosenberg, S. Prusiner, S. DiMauro, R. L. Barchi and E. J. Nesler (Eds). *Molecular and Genetic Basis of Neurology and Psychiatric Disease*. 3rd. ed., Philadelphia, PA: Butterworth-Heinemann.

Carey, G., & Gottesman, II (1996) Genetics and Antisocial Behavior: Substance Versus Sound Bytes. *Politics and the Life Sciences*, 15 (1): 88–90.

Carmalt, William H., & Connecticut State Medical Society (1909) *Heredity and Crime: A Study in Eugenics*. Hartford: Connecticut State Medical Society.

Carol, Anne (1995) *Histoire De L'eugenisme En France: Les Medecins Et La Procreation, XIXe–XXe Siecle*. Paris: Seuil.

Carr-Saunders, A. M. (1926) *Eugenics*. London: Butterworth.

Carson, D. (1995) Therapeutic Jurisprudence for the United Kingdom? *Journal of Forensic Psychiatry*, 6 (3): 463–466.

Carter, C. O., et al. (1971) Genetic Clinic: A Follow-Up. *The Lancet*, 281–285.

Carter, Rita (1998) *Mapping the Mind*. London: Weidenfeld & Nicholson.

Cartwright, Lisa (1995a) An Etiology of the Neurological Gaze. *Screening the Body: Tracing Medicine's Visual Culture*. Minneapolis: University of Minnesota Press.

——— (1995b) *Screening the Body: Tracing Medicine's Visual Culture*. Minneapolis: University of Minnesota Press.

Caspi, A., et al. (2002) Role of Genotype in the Cycle of Violence in Maltreated Children. *Science*, 297 (5582): 851–854.

Castel, Robert (1991) From Dangerousness to Risk. In Burchell, G., Gordon, C., & Miller, P. (Eds.) *The Foucault Effect: Studies in Governmentality*. London: Harvester Wheatsheaf.

Castells, Manuel (2000) *The Rise of the Network Society*. Oxford: Blackwell Publishers.

Castle, William Ernest (1916) *Genetics and Eugenics. A Text-Book for Students of Biology and a Reference Book for Animal and Plant Breeders. (Third Impres-*

sion.) [with a Bibliography and Plates.]. Cambridge, Mass.: Harvard University Press.

Cavalli-Sforza, L. L., & Feldman, M. W. (2003) The Application of Molecular Genetic Approaches to the Study of Human Evolution. *Nature Genetics*, 33 (Suppl) 266–275.

Cavalli-Sforza, L. L., Menozzi, Paolo, & Piazza, Alberto (1994) *The History and Geography of Human Genes*. Princeton, N.J.: Princeton University Press.

Cello, J., Paul, A. V., & Wimmer, E. (2002) Chemical Synthesis of Poliovirus cDNA: Generation of Infectious Virus in the Absence of Natural Template. *Science*, 297 (5583): 1016–1018.

Centers for Disease Control and Prevention Office of Genomics and Disease Prevention (2005) Genomics and Population Health 2005. Atlanta: Centers for Disease Control and Prevention.

Chadarevian, Soraya, de, & Kamminga, Harmke (1998) *Molecularizing Biology and Medicine: New Practices and Alliances*. Australia/United Kingdom: Harwood Academic.

Chamberlin, J. Edward, & Gilman, Sander, L. (1985) *Degeneration: The Dark Side of Progress*. New York: Columbia University Press.

Charney, Dennis S., et al. (2002) Neuroscience Research Agenda to Guide Development of a Pathophysiologically Based Classification System. In Kupfer, D. J., First, M. B. & Regier, D. A. (Eds.) *A Research Agenda for DSM V*. Washington, D.C.: American Psychiatric Association.

Cheng, T. O. (2003) Ethnic Differences in Incidence of Diseases and Response to Medicines. *Journal of The National Medical Association*, 95 (5): 404–405.

China Concept Consulting & The Information Centre of the State Drug Agency (2000) *China Pharmaceuticals Guide: New Policy and Regulation*. Beijing.

Citizens Commission on Human Rights (1996) *The Violence Initiative*. Los Angeles: Citizens Commission on Human Rights.

Claeson, Bjorn, et al. (1996) Scientific Literacy, What It Is, Why It's Important, and Why Scientists Think We Don't Have It? The Case of Immunology and the Immune System. In Nader, L. (Ed.) *Naked Science: Anthropological Inquiry into Boundaries, Power, and Knowledge*. New York: Routledge.

Clarke, A. E., et al. (2003) Biomedicalization: Technoscientific Transformations of Health, Illness, and US Biomedicine. *American Sociological Review*, 68 (2): 161–194.

Cloninger, C. Robert (2002) The Discovery of Susceptibility Genes for Mental Disorders. *Proceedings of the National Academy of Science*, 99 (21): 13365–13367.

Coccaro, E. F. (1992) Impulsive Aggression and Central Serotonergic System Function in Humans—an Example of a Dimensional Brain-Behavior Relationship. *International Clinical Psychopharmacology*, 7 (1): 3–12.

Coffey, Maureen P. (1993) The Genetic Defense: Excuse or Explanation? *William and Mary Law Review*, 35: 353–399.

Coghlan, Andy (1998) Perfect People's Republic. *New Scientist*, October 24, 1998.

——— (2001) Patient Power. *New Scientist*, February 21, 2001.

Cohen, L. E. (1987) Review Essay of J. Q. Wilson and R. J. Herrnstein's *Crime and Human Nature*. *Contemporary Sociology*, 16 (2): 202–205.

Collins, F. S. (2004) What We Do and Don't Know About "Race," "Ethnicity," Genetics and Health at the Dawn of the Genome Era. *Nature Genetics*, 36 (11): S13–S15.

Collins, F. S., Lander, E. S., Rogers, J., & Waterston, R. H. (2004) Finishing the Euchromatic Sequence of the Human Genome. *Nature*, 431 (7011): 931–945.

Comings, D. E., et al. (2000) Multivariate Analysis of Associations of 42 Genes in ADHD, ODD and Conduct Disorder. *Clinical Genetics*, 58 (1): 31–40.

Condit, C., et al. (2003) Attitudinal Barriers to Delivery of Race-Targeted Pharmacogenomics among Informed Lay Persons. *Genetics in Medicine*, 5 (5): 385–392.

Conrad, Peter (1976) *Identifying Hyperactive Children: The Medicalization of Deviant Behavior*. Lexington, Mass./London: Heath.

Conrad, Peter, & Potter, D. (2000) From Hyperactive Children to ADHD Adults: Observations on the Expansion of Medical Categories. *Social Problems*, 47 (4): 559–582.

Cooper, Paul (2004) Education in the Age of Ritalin. In Rees, D., & Rose, S. P. R. (Eds.) *The New Brain Sciences: Prospects and Perils*. Cambridge: Cambridge University Press.

Cooper, R. S., & Freeman, V. L. (1999) Limitations in the Use of Race in the Study of Disease Causation. *Journal of The National Medical Association*, 91 (7): 379–383.

Cooper, R. S., Kaufman, J. S., & Ward, R. (2003) Race and Genomics. *New England Journal of Medicine*, 348 (12): 1166–1170.

Cooter, Roger (2004) Historical Keywords: Bioethics. *The Lancet*, 364 (9447): 1749.

Corrigan, Oonagh, & Tutton, Richard (Eds.) (2004) *Donating, Collecting and Exploiting Human Tissue*. London: Routledge.

Cox, Susan M., & McKellin, William (1999) 'There's This Thing in Our Family': Predictive Testing and the Construction of Risk for Huntington Disease. *Sociology of Health & illness*, 21 (5): 622–646.

Criqui, M. H., et al. (2005) Ethnicity and Peripheral Arterial Disease—the San Diego Population Study. *Circulation*, 112 (17): 2703–2707.

Crookshank, F. G. (1924) *The Mongol in Our Midst: A Study of Man and His Three Faces*. London: K. Paul Trench Trubner & Co.; New York: E. P. Dutton & Co.

Crusio, W. E. (1996) The Neurobehavioral Genetics of Aggression. *Behavior Genetics*, 26 (5): 459–461.

Daily Telegraph (2000) Insurers to DNA Test for Genetic Illnesses. *Daily Telegraph* (March 20, 2000).

Damasio, Antonio R. (1994) *Descartes' Error: Emotion, Reason, and the Human Brain*. New York: Putnam.

——— (1999) *The Feeling of What Happens: Body and Emotion in the Making of Consciousness*. New York: Harcourt Brace.

——— (2003) *Looking for Spinoza: Joy, Sorrow, and the Feeling Brain*. Orlando, Fla.: Harcourt.

Dampier, William Cecil Dampier, & Whetham, Catherine Durning (1909) *The Family and the Nation: A Study in Natural Inheritance and Social Responsibility*. London: Longmans Green and Co.

Darwin, Charles Robert (1871) *The Descent of Man, and Selection in Relation to Sex. With Illustrations*: 2 vols. John Murray: London.

Darwin, Leonard (1928) *What Is Eugenics?* London: Watts.

Davis, Dena S. (2003) Genetic Research and Communal Narratives. *Hastings Center Report*, 34, 40–49.

Decruyenaere, Marleen, et al. (1999) Psychological Functioning before Predictive Testing for Huntington's Disease: The Role of the Parental Disease, Risk Perception, and Subjective Proximity of the Disease. *Journal of Medical Genetics*, 38 (12): 897–905.

Decruyenaere, Marleen, et al. (1996) Prediction of Psychological Functioning One Year after the Predictive Test for Huntington's Disease and the Impact of the Test Result on Reproductive Decision Making. *Journal of Medical Genetics*, 33 (9): 737–743.

Deleuze, Gilles (1988a) *Bergsonism*. New York: Zone Books.

——— (1988b) *Foucault*. Minneapolis: University of Minnesota Press.

——— (1995) Postscript on Control Societies. *Negotiations*. New York: Columbia University Press.

Denis, Carina (2002) Gene Regulation: The Brave New World of RNA. *Nature*, 418, 122–124.

Denno, D. W. (1988) Human-Biology and Criminal Responsibility—Free Will or Free Ride. *University of Pennsylvania Law Review*, 137 (2): 615–671.

——— (1996) Legal Implications of Genetics and Crime Research. *Genetics of Criminal and Antisocial Behavior*, CIBA Foundation Symposium 194, 258–296.

Department of Health and Human Services Office of Inspector General (2001) The Globalization of Clinical Trials: A Growing Challenge in Protecting Human Subjects. Boston: Department of Health and Human Services Office of Inspector General.

Dice, Lee R. (1952) Heredity Clinics: Their Value for Public Service and for Research. *American Journal of Human Genetics*, 4 (1): 1–13.

Dickinson, Robert, & Gamble, Clarence (1950) *Human Sterilization*. np.

Dickson, D. (1998) Congress Grabs Eugenics Common Ground. *Nature*, 394 (6695): 711.

Didi-Huberman, Georges (1982) *Invention De L'hystérie: Charcot et L'iconographie Photographique de la Salpêtrière*. Paris: Macula.

Didi-Huberman, Georges, & Charcot, J. M. (2003) *The Invention of Hysteria: Charcot and the Photographic Iconography Of the Salpêtrière*. Cambridge, Mass./London: MIT.

Dikötter, Frank (1998) *Imperfect Conceptions: Medical Knowledge, Birth Defects, and Eugenics in China*. New York: Columbia University Press.

Diller, Lawrence H. (1998) *Running on Ritalin: A Physician Reflects on Children, Society, and Performance in a Pill*. New York: Bantam Books.

Dinwiddie, S. H. (1996) Genetics, Antisocial Personality, and Criminal Responsibility. *Bulletin of The American Academy of Psychiatry and The Law*, 24 (1): 95–108.

Diprose, Rosalyn (1998) Sexuality and the Clinical Encounter. In Shildrick, M., & Price, J. (Eds.) *Vital Signs: Feminist Reconfigurations of the Bio/Logical Body*. Edinburgh: Edinburgh University Press.

Donzelot, Jacques (1979) *The Policing of Families*. New York: Pantheon Books.

Dowbiggin, Ian Robert (1997) *Keeping America Sane: Psychiatry and Eugenics in the United States and Canada, 1880–1940*. Ithaca, N.Y./London: Cornell University Press.

Downs, L. L. (2002) PMS, Psychosis and Culpability: Sound or Misguided Defense? *Journal of Forensic Sciences*, 47 (5): 1083–1089.

Doyle, Richard (1997) *On Beyond Living: Rhetorical Transformations of the Life Sciences*. Stanford, Calif.: Stanford University Press.

Dreyfuss, Rochelle Cooper, & Nelkin, Dorothy (1992) The Jurisprudence of Genetics. *Vanderbilt Law Review*, 45 (2): 313–348.

Drouard, Alain (1999) Eugenics and Bioethics in Today's France. *Tartu University History Museum Annual Report 1998*. Tartu: History Museum of Tartu University.

Dumit, Joseph (1997) A Digital Image of the Category of the Person: Pet Scanning and Objective Self-Fashioning. In Downey, G. L., & Dumit, J. (Eds.) *Cyborgs and Citadels: Anthropological Interventions in Emerging Sciences*. Sante Fe, N.M.: School of American Research Press.

——— (2003) *Picturing Personhood: Brain Scans and Biomedical Identity*. Princeton NJ: Princeton University Press.

Duster, Troy (1990) *Backdoor to Eugenics*. New York: Routledge.

——— (2003) Buried Alive: The Concept of Race in Science. In Goodman, A. H., Heath, D., & Lindee, M. S. (Eds.) *Genetic Nature/Culture*. Berkeley: University of California Press.

Dyson, S. M. (1998) "Race," Ethnicity and Haemoglobin Disorders. *Social Science and Medicine*, 47 (1): 121–131.

Ebstein, Richard P., et al. (1996) Dopamine D4 Receptor (D4dr) Exon Iii Polymorphism Associated with the Human Personality Trait of Novelty Seeking. *Nature Genetics*, 12 (1): 78.

Eddy, Sean R. (2001) Non-Coding RNA Genes and the Modern RNA World. *Nature Reviews Genetics*, 2, 919–929.

Editorial (1995) News from the Field. *Psychiatric Genetics*, 5, 3–4.

Elliot, Carl (1999) *A Philosophical Disease: Bioethics, Culture and Identity*. New York: Routledge.

——— (2001) Pharma Buys a Conscience. *The American Prospect*, 12 (17): 16–20.

——— (2003) *Better Than Well: American Medicine Meets the American Dream*. New York: W. W. Norton.

——— (2004) When Pharma Goes to the Laundry: Public Relations and the Business of Medical Education. *Hastings Center Report*, 34 (5): 18–23.

Elliott, R. et al. (1997) Effects of Methylphenidate on Spatial Working Memory and Planning in Healthy Young Adults. *Psychopharmacology*, 131 (2): 196–202.

Ellison, George T. H., & Rees Jones, Ian (2002) Social Identities and the "New Genetics": Scientific and Social Consequences. *Critical Public Health*, 12 (3): 265–282.

Elwyn, Glyn, Gray, Jonathon, & Clarke, Angus (2000) Shared Decision Making and Non-Directiveness in Genetic Counselling. *Journal of Medical Genetics*, 37 (2): 135–138.

Engelhardt, H. Tristram, & Towers, Bernard (Eds.) (1979) *Clinical Judgment: A Critical Appraisal: Proceedings of the Fifth Trans-Disciplinary Symposium on Philosophy and Medicine Held at Los Angeles, California, April 14–16, 1977*, Dordrecht/London: Reidel.

Epstein, Steven (1996) *Impure Science: Aids, Activism, and the Politics of Knowledge*. Berkeley: University of California Press.

——— (1997) Activism, Drug Regulation, and the Politics of Therapeutic Evaluation in the Aids Era: A Case Study of ddC and the "Surrogate Markers" Debate. *Social Studies of Science*, 27 691–726.

Ericson, Richard, Barry, Dean, & Doyle, Aaron (2000) The Moral Hazards of Neoliberalism: Lessons from the Private Insurance Industry. *Economy and Society*, 29 (4): 532–558.

Ernst & Young (2003a) *Beyond Borders: The Global Biotechnology Report*. Ernst & Young.

——— (2003b) *Resilience: America's Biotechnology Report*. Ernst & Young.

——— (2005) *Beyond Borders: Global Biotechnology Report 2005*. Ernst & Young.

Esquirol, Jean Etienne Dominique (1838) *Des Maladies Mentales, consideres sous les Rapports Medical, Hygienique et Medico-Legal. Accompagnees de 27 Planches Gravès*. Paris: Baillière.

European Commission Health and Consumer Protection Directorate-General (2005) *Improving the Mental Health of the Population: Towards a Strategy on Mental Health for the European Union (Green Paper)*. Brussels: European Commission.

Ewald, François (1986) *L'ètat Providence*. Paris: Grasset.

——— (1991) Insurance and Risk. In Burchell, G., Gordon, C., & Miller, P. (Eds.) *The Foucault Effect: Studies in Governmentality*. London: Harvester Wheatsheaf.

——— (2001) The Return of Descartes' Malicious Demon: An Outline of a Philosophy of Precaution. In Baker, T., & Simon, J. (Eds.) *Embracing Risk*. Chicago: Chicago University Press.

Exner, D. V., Domanski, M. J., & Cohn, J. N. (2001a) Race and Responsiveness to Drugs for Heart Failure. Reply. *New England Journal of Medicine*, 345 (10): 767–768.

Exner, D. V., et al. (2001b) Lesser Response to Angiotensin-Converting-Enzyme Inhibitor Therapy in Black as Compared with White Patients with Left Ventricular Dysfunction. *New England Journal of Medicine*, 344 (18): 1351–1357.

Falek, Arthur, & Glanville, Edward V. (1962) Investigation of Genetic Carriers. In Kallmann, F. J. (Ed.) *Expanding Goals of Genetics in Psychiatry.* New York: Grune & Stratton.

Farah, Martha J. (2002) Emerging Ethical Issues in Neuroscience. *Nature Neuroscience*, 5, 1123–1129.

Farah, Martha J., et al. (2004) Neurocognitive Enhancement: What Can We Do and What Should We Do? *Nature Reviews Neuroscience*, 5 (5): 421–425.

Faravelli, C., et al. (2003) A Self-Controlled, Naturalistic Study of Selective Serotonin Reuptake Inhibitors Versus Tricyclic Antidepressants. *Psychotherapy and Psychosomatics*, 72 (2): 95–101.

Farrer, L. A., et al. (1997a) Effects of Age, Gender and Ethnicity on the Association between APOe Genotype and Alzheimer Disease. *American Journal of Human Genetics*, 61 (4): A45–A45.

——— (1997b) Effects of Age, Sex, and Ethnicity on the Association between Apolipoprotein E Genotype and Alzheimer Disease. A Meta-Analysis. APOe and Alzheimer Disease Meta Analysis Consortium. *Journal of the American Medical Association*, 278 (16): 1349–1356.

Feeley, Malcolm, & Simon, Jonathan (1992) The New Penology: Notes on the Emerging Strategy of Correction and Its Implications. *Criminology*, 30 (4): 449–474.

——— (1994) Actuarial Justice: Power/Knowledge in Contemporary Criminal Justice. In Nelken, D. (Ed.) *The Futures of Criminology.* London: Sage.

Feinstein, Alvan R. (1967) *Clinical Judgment.* Baltimore: Williams & Wilkins.

Fine, Beth A. (1993) The Evolution of Nondirectiveness in Genetic Counseling and Implications of the Human Genome Project. In Bartels, D. M., Leroy, B. S. & Caplan, A. L. (Eds.) *Prescribing Our Future: Ethical Challenges in Genetic Counseling*. New York: Aldine de Gruyter.

Fischer, Michael M. J. (2003) *Emergent Forms of Life and the Anthropological Voice.* Durham, N.C.: Duke University Press.

Fishbein, Diane H. (1992) The Psychobiology of Female Aggression. *Criminal Justice and Behavior*, 19 (2): 99–126.

——— (1996) Prospects for the Application of Genetic Findings to Crime and Violence Prevention. *Politics and the Life Sciences*, 15 (1): 91–94.

——— (2000) *The Science, Treatment, and Prevention of Antisocial Behaviors: Application to the Criminal Justice System*, Kingston, N.J.: Civic Research Institute.

Fishbein, Diane H., & Thatcher, R. W. (1986) New Diagnostic Methods in Criminology—Assessing Organic Sources of Behavioral-Disorders. *Journal of Research in Crime and Delinquency*, 23 (3): 240–267.

Fleck, Ludwik (1979) *Genesis and Development of a Scientific Fact.* Chicago: Chicago University Press.

Fleischer, Matt (2001) Patent Thyself (Online Version). *The American Lawyer* (http://www.americanlawyer.com/newcontents06.html), (June 21).

Flower, Michael J., & Heath, Deborah (1993) Micro-Anatomo Politics: Mapping the Human Genome Project. *Culture, Medicine and Psychiatry*, 17 (1): 27–41.

Fogelman, Y., et al. (2003) Prevalence of and Change in the Prescription of Methylphenidate in Israel over a 2-Year Period. *CNS Drugs*, 17 (12): 915–919.

Forrester, John (1996) If *P*, Then What? Thinking in Cases. *History of the Human Sciences*, 9 (3): 1–25.
Foucault, Michel (1970) *The Order of Things: An Archaeology of the Human Sciences*. London: Tavistock Publications.
——— (1973) *The Birth of the Clinic: An Archaeology of Medical Perception*. London: Tavistock Publications.
——— (1978) *The History of Sexuality, Vol. 1. The Will to Knowledge*. London: Penguin.
——— (1985) *The History of Sexuality, Vol. 2. The Use of Pleasure*. London: Penguin.
——— (1986) *The History of Sexuality, Vol. 3. The Care of the Self*. London: Penguin.
——— (1999) The Politics of Health in the Eighteenth Century. In Faubion, J. D. (Ed.) *Michel Foucault: The Essential Works, Volume 3: Power*. New York: New Press.
——— (2001) "Omnes Et Singulatim": Towards a Critique of Political Reason. In Rabinow, P. (Ed.) *Power: The Essential Works*. London: Allen Lane.
——— (2002) *Society Must Be Defended: Lectures at the Collège De France, 1975–76*. New York: Picador.
Frank, Arthur (2004) Emily's Scars: Surgical Shapings, Technoluxe, and Bioethics. *Hastings Center Report*, 34 (2): 18–29.
Franklin, Sarah (1995) Life. In Reich, W. T. (Ed.) *The Encyclopedia of Bioethics, Revised Edition*. New York: Simon and Schuster.
——— (1997) *Embodied Progress: A Cultural Account of Assisted Conception*. London: Routledge.
——— (2000) Life Itself: Global Nature and the Genetic Imaginary. In Franklin, S., Lury, C. & Stacey, J. (Eds.) *Global Nature, Global Culture*. London: Sage.
——— (2003) Ethical Biocapital. In Franklin, S. & Lock, M. (Eds.) *Remaking Life and Death: Toward and Anthropology of the BioSciences*. Santa Fe, N.M.: School of American Research Press.
——— (2005) Stem Cells R Us: Emergent Life Forms and the Global Biological. In Ong, A. & Collier, S. J. (Eds.) *Global Assemblages: Technology, Politics, and Ethics as Anthropological Problems*. Malden, Mass.: Blackwell Publishing.
——— (2006) *Dolly Mixtures*. Durham, N.C.: Duke University Press.
Franklin, Sarah, & Lock, Margaret (2003a) Animation and Cessation: The Remaking of Life and Death. In Franklin, S. & Lock, M. (Eds.) *Remaking Life and Death: Toward an Anthropology of the BioSciences*. Santa Fe: Society of American Research Press.
Franklin, Sarah, & Lock, Margaret (Eds.) (2003b) *Remaking Life and Death: Toward an Anthropology of the BioSciences*. Santa Fe, N.M.: School of American Research Press.
Fraser, Mariam (2001) The Nature of Prozac. *History of the Human Sciences*, 14 (3): 56–84.
——— (1999) Prozac: A Story of Chemical Femininity. Keele University.
Freeman, Michael, D. A. (1983) *The Rights and Wrongs of Children*. London: Pinter.

Freeman, Walter Jackson, & Watts, James Winston (1942) *Psychosurgery*, Springfield, Ill.: C. C. Thomas.

——— (1950) *Psychosurgery. On the Treatment of Mental Disorders and Intractable Pain.* Oxford/Fort Worth printed: Blackwell Scientific Publications.

Frost & Sullivan (2001) *US Pharmacogenomic Markets.* New York: Frost & Sullivan.

Fukuyama, Francis (2002) *Our Posthuman Future: Consequences of the Biotechnology Revolution.* London: Profile.

Fullilove, M. T. (1998) Comment: Abandoning "Race" as a Variable in Public Health Research—an Idea Whose Time Has Come. *American Journal of Public Health*, 88 (9): 1297–1298.

Galton, Francis (1883) *Inquiries into Human Faculty and Its Development.* London: Macmillan.

——— (1889) *Natural Inheritance.* London: Macmillan.

Galton, Francis, & Eugenics Education Society (Great Britain) (1909) *Essays in Eugenics.* London: Eugenics Education Society.

Garland, David (1994) of Crime and Criminals: The Development of Criminology in Britain. In Maguire, M., Morgan, R., & Reiner, R. (Eds.) *The Oxford Handbook of Criminology.* Oxford: Oxford University Press.

——— (1996) The Limits of the Sovereign State—Strategies of Crime Control in Contemporary Society. *British Journal of Criminology*, 36 (4): 445–471.

——— (2001) *The Culture of Control: Crime and Social Order in Contemporary Society.* Oxford: Oxford University Press.

Gazzaniga, Michael S. (2005) *The Ethical Brain.* New York: Dana Press.

Gelman, Sheldon (1999) *Medicating Schizophrenia: A History.* New Brunswick, N.J.: Rutgers University Press.

Gibbs, J. P. (1985) Review Essay of J. Q. Wilson and R. J. Herrnstein's Crime and Human Nature. *Criminology*, 23 (3): 381–388.

Gibbs, W. Wayt (2003) The Unseen Genome: Gems among the Junk. *Scientific American*, 289 (5): 46–53.

Gigerenzer, Gerd (1991) From Tools to Theories: A Heuristic of Discovery in Cognitive Psychology. *Psychological Review*, 98, 254–257.

Gilbert, Walter (1993) A Vision of the Grail. In Kevles, D. J., & Hood, L. (Eds.) *The Code of Codes: Scientific and Social Issues in the Human Genome Project.* Cambridge, Mass.: Harvard University Press.

Gilman, Sander (1982) *Seeing the Insane.* New York: Wiley.

Gilroy, Paul (2000) *Between Camps: Race, Identity and Nationalism at the End of the Colour Line.* London: Allen Lane.

Glass, Bentley (1971) Science: Endless Horizons or Golden Age. *Science*, 171, 23–29.

Glueck, Sheldon, & Glueck, Eleanor Touroff (1930) *500 Criminal Careers.* New York: Knopf.

——— (1934) *One Thousand Juvenile Delinquents; Their Treatment by Court and Clinic.* Cambridge, Mass.: Harvard University Press.

Glueck, Sheldon, Glueck, Eleanor Touroff, & Commonwealth Fund. (1943) *Criminal Careers in Retrospect.* New York: The Commonwealth Fund.

Goffman, Erving (1968) *Stigma: Notes on the Management of Spoiled Identity.* Harmondsworth: Penguin.

Goldberg, David Theo (2001) *The Racial State.* Malden, Mass.: Blackwell Publishers.

Golden, R. N., et al. (1991) Serotonin, Suicide, and Aggression—Clinical-Studies. *Journal of Clinical Psychiatry,* 52, 61–69.

Good, Mary-Jo Delvecchio (2001) The Biotechnical Embrace. *Culture, Medicine and Psychiatry,* 25, 395–410.

——— (2003) The Medical Imaginary and the Biotechnical Embrace: Subjective Experience of Clinical Scientists and Patients. *Russell Sage Working Papers.* Russell Sage Foundation.

Good, Mary-Jo Delvecchio, et al. (1990) American Oncology and the Discourse on Hope. *Culture, Medicine and Psychiatry,* 14 (1): 59–79.

Goodwin, Frederick K., & Jamison, Kay R. (1990) *Manic-Depressive Illness.* New York: Oxford University Press.

Gostin, Larry (1991) Genetic Discrimination: The Use of Genetically Based Diagnostic Tests by Employers and Insurers. *American Journal of Law and Medicine,* 17 (1–2): 109–144.

Gottweis, Herbert (2002a) Governance and Bioethics. *Conference on Biomedicalization, Social Conflicts and the New Politics of Bioethics.* Vienna.

——— (2002b) Stem Cell Policies in the United States and in Germany: Between Bioethics and Regulation. *Policy Studies Journal,* 30 (4): 444–469.

Gray, Chris Hables (2000) *Cyborg Citizen: Politics in the Posthuman Age.* New York: Routledge.

Gray, Chris Hables, Mentor, Steven & Figueroa-Sarriera, Heidi J. (1995) *The Cyborg Handbook.* New York; London: Routledge.

Greco, Monica (1993) Psychosomatic Subjects and "the Duty to Be Well"—Personal Agency within Medical Rationality. *Economy and Society,* 22 (3): 357–372.

Greene, J., & Cohen, J. (2004) For the Law, Neuroscience Changes Nothing and Everything. *Philosophical Transactions of the Royal Society of London Series B-Biological Sciences,* 359 (1451): 1775–1785.

Greenhalgh, Susan (1986) Shifts in China Population-Policy, 1984–86—Views from the Central, Provincial, and Local-Levels. *Population and Development Review,* 12 (3): 491–515.

——— (2003) Science, Modernity, and the Making of China's One-Child Policy. *Population and Development Review,* 29 (2): 163–196.

——— (2005) Globalization and Population Governance in China. In Ong, A., & Collier, S. (Eds.) *Global Assemblages: Technology, Governmentality, Ethics.* Malden, Mass.: Blackwell.

Gross, Cornelius, & Hen, Rene (2004) The Developmental Origins of Anxiety. *Nature Reviews Neuroscience,* 5, 545–552.

Gupta, A., Elheis, M. & Pansari, K. (2004) Imaging in Psychiatric Illnesses. *International Journal of Clinical Practice,* 58 (9): 850–858.

Gurling, H. (1990) Genetic-Linkage and Psychiatric Disease. *Nature,* 344 (6264): 298–298.

Habermas, Jürgen (2003) *The Future of Human Nature.* Cambridge: Polity.

Hacking, Ian (1983) *Representing and Intervening: Introductory Topics in the Philosophy of Science*. Cambridge: Cambridge University Press.

——— (1986) Making up People. In Heller, T. C., Sosna, M., & Wellbery, D. E. (Eds.) *Reconstructing Individualism: Autonomy, Individuality and the Self in Western Thought*. Stanford, Calif.: Stanford University Press.

——— (1990) *The Taming of Chance*. Cambridge: Cambridge University Press.

——— (1992a) "Style" for Historians and Philosophers. *Studies in the History and Philosophy of Science*, 23 (1): 1–20.

——— (1992b) World-Making by Kind-Making: Child Abuse for Example. In Douglas, M., & Hull, D. (Eds.) *How Classification Works: Nelson Goodman among the Social Sciences*. Edinburgh: Edinburgh University Press.

——— (1995) Self-Improvement. In Smart, B. (Ed.) *Michel Foucault: Critical Assessments*. London: Routledge.

——— (2001) Degeneracy, Criminal Behavior, and Looping. In Wasserman, D. A., & Wachbroit, R. (Eds.) *Genetics and Criminal Behavior.* Cambridge: Cambridge University Press.

——— (2002) *Historical Ontology.* Cambridge, Mass./London: Harvard University Press.

——— (2005) Why Race Still Matters. *Daedalus*, Winter 2005, 102–116.

——— (1998) Canguilhem Amid the Cyborgs. *Economy and Society*, 27 (2/3): 202–216.

Hahn, R. A. (1992) The State of Federal Health-Statistics on Racial and Ethnic-Groups. *Journal of the American Medical Association*, 267 (2): 268–271.

Hahn, R. A., Mulinare, J., & Teutsch, S. M. (1992) Inconsistencies in Coding of Race and Ethnicity between Birth and Death in United-States Infants—a New Look at Infant-Mortality, 1983 through 1985. *Journal of the American Medical Association*, 267 (2): 259–263.

Haigh, Elizabeth (1984) *Xavier Bichat and the Medical Theory of the Eighteenth Century.* London: Wellcome Institute for the History of Medicine.

Hall, M. A. (2000) The Impact of Genetic Discrimination on Law Restricting Health Insurer's Use of Genetic Information. *American Journal of Human Genetics*, 66, 293–307.

Hallowell, Nina, & Richards, Martin P.M. (1997) Understanding Life's Lottery: An Evaluation of Studies of Genetic Risk Awareness. *Journal of Health Psychology*, 2 (1): 31–43.

Hamer, Dean H., & Copeland, Peter (1994) *The Science of Desire: The Search for the Gay Gene and the Biology of Behavior.* New York: Simon & Schuster.

Haraway, Donna, J. (1991a) A Cyborg Manifesto: Science, Technology, and Socialist-Feminism in the Late Twentieth Century. *Simians, Cyborgs and Women: The Reinvention of Nature*. New York: Routledge.

——— (1991b) *Simians, Cyborgs, and Women: The Re-Invention of Nature*. London: Free Association.

——— (1995) Universal Donors in a Vampire Culture: It's All in the Family: Biological Kinship Categories in the Twentieth-Century United States. In Cronon, W. (Ed.) *Uncommon Ground: Toward Reinventing Nature*. New York: W. W. Norton and Company.

Hardt, Michael, & Negri, Antonio (2000) *Empire*. Cambridge, Mass.: Harvard University Press.

——— (2004) *Multitude: War and Democracy in the Age of Empire*. New York: Penguin.

Harris, M., et al. (2003) Mood-Stabilizers: The Archeology of the Concept. *Bipolar Disorder*, 5 (6): 446–452.

Hayden, Michael R., et al. (1987) Ethical Issues in Preclinical Testing in Huntington Disease: Response to Margery Shaw's Invited Editorial Comment (Letter). *American Journal of Medical Genetics*, 28 (3): 761–763.

Hayles, N. Katherine (1999) *How We Became Posthuman: Virtual Bodies in Cybernetics, Literature, and Informatics*. Chicago, Ill./London: University of Chicago Press.

Healy, David (1997) *The Antidepressant Era*. Cambridge, Mass.: Harvard University Press.

——— (2001) *The Creation of Psychopharmacology*. Cambridge, Mass.: Harvard University Press.

——— (2002) SSRIs and Deliberate Self-Harm. *British Journal of Psychiatry*, 180, 547–548.

——— (2004) *Let Them Eat Prozac: The Unhealthy Relationship between the Pharmaceutical Industry and Depression*. New York: New York University Press.

Healy, David, Langmaak, C., & Savage, M. (1999) Suicide in the Course of the Treatment of Depression. *Journal of Psychopharmacology*, 13 (1): 94–99.

Heath, Deborah, Rapp, Rayna, & Taussig, Karen-Sue (2004) Genetic Citizenship. In Nugent, D., & Vincent, J. (Eds.) *Companion to the Anthropology of Politics*. Oxford: Blackwell.

Hedgecoe, Adam M. (1999) Reconstructing Geneticization: A Research Manifesto. *Health Law Journal*, 7, 5–18.

——— (2005) *The Politics of Personalised Medicine*. Cambridge: Cambridge University Press.

Helgadottir, Anna, et al. (2005) A Variant of the Gene Encoding Leukotriene A4 Hydrolase Confers Ethnicity-Specific Risk of Myocardial Infarction. *Nature Genetics*, Nov 10: EPub ahead of print.

Helmreich, Stefan (2000) *Silicon Second Nature: Culturing Artificial Life in a Digital World*. Berkeley/London: University of California Press.

Hen, Rene (1996) Mean Genes. *Neuron*, 16 (1): 17–21.

Hendricks, T. J., et al. (2003) Pet-1 Ets Gene Plays a Critical Role in 5-Ht Neuron Development and Is Required for Normal Anxiety-Like and Aggressive Behavior. *Neuron*, 37 (2): 233–247.

Herndon, C. Nash (1955) Heredity Counseling. *Eugenics Quarterly*, 2 (2): 83–89.

Hickey, Sonja Sherry (1986) Enabling Hope. *Cancer Nursing*, 9 (3): 133–137.

Hinds, Pamela S., (1984) Inducing a Definition of "Hope" through the Use of Grounded Theory Methodology. *Journal of Advanced Nursing*, 9 357–362.

Hinds, Pamela S., & Martin, Janni (1988) Hopefulness and the Self-Sustaining Process in Adolescents with Cancer. *Nursing Research*, 37 (6): 336–340.

Ho, Mae Wan, Meyer, Harmut, & Cummins, Joe (2003) The Biotech Bubble. *The Ecologist*, 28 (3): 146–153.

Hogben, Lancelot Thomas (1938) *Science for the Citizen*. London: Allen & Unwin.

Hoggart, C. J., et al. (2004) Design and Analysis of Admixture Mapping Studies. *American Journal of Human Genetics*, 74 (5): 965–978.

Holder, Indrani, & Shriver, Mark D. (2003) Measuring and Using Admixture to Study the Genetics of Complex Diseases. *Human Genomics*, 1 (1): 52–62.

Horgan, John (1993) Eugenics Revisited. *Scientific American*, 268 (6): 122–131.

Horsburgh, Beverley (1996) Schrödinger's Cat: Eugenics and the Compulsory Sterilization of Welfare Mothers. *Cardozo Law Review*, 17, 531–582.

Horton, Richard (2004) *Health Wars: On the Global Front Lines of Modern Medicine*. New York: New York Review of Books.

Høyer, Klaus (2002) Conflicting Notions of Personhood in Genetic Research. *Anthropology Today*, 18 (5): 9–13.

——— (2003) 'Science Is Really Needed That's All I Know.' Informed Consent and the Non-Verbal Practices of Collecting Blood for Genetic Research in Northern Sweden. *New Genetics and Society*, 22 (3): 229–244.

Hubbard, Ruth, & Ward, Elijah (1999) *Exploding the Gene Myth*. New York: Beacon.

Hughes, J. R., & John, E. R. (1999) Conventional and Quantitative Electroencephalography in Psychiatry. *Journal of Neuropsychiatry and Clinical Neurosciences*, 11 (2): 190–208.

Human Genetics Advisory Committee (1997) *The Implications of Genetic Testing for Insurance*. London: Human Genetics Advisory Committee.

Illes, Judy (2005) *Neuroethics: Defining the Issues in Theory, Practice, and Policy*. New York: Oxford University Press.

Isin, Engin F. (2002) *Being Political: Genealogies of Citizenship*. Minneapolis: University of Minnesota Press.

Itakura, H. (2005) Racial Disparities in Risk Factors for Thrombosis. *Current Opinion in Hematology*, 12 (5): 364–369.

Jacob, François (1974) *The Logic of Living Systems. A History of Heredity. [La Logique Du Vivant.]* Trans. by Betty E. Spillmann. London: Allen Lane.

Jacob, François, & Monod, Jacques (1961) Genetic Regulatory Mechanisms in the Synthesis of Proteins. *Journal of Molecular Biology*, 3, 318–356.

Jennings, Bruce (2003) *Knowledge and Empowerment in Personal and Civic Health*. A Concept Paper Prepared for the March of Dimes/Health Resources and Services Administration/Genetic Services Branch Project on Genetic Literacy. New York: Hastings Center.

Jensen, Arthur Robert, & Miele, Frank (2002) *Intelligence, Race, and Genetics: Conversations with Arthur R. Jensen*. Boulder, Colo.: Westview Press.

Jensen, Uffe Juul (1987) *Practice & Progress: A Theory for the Modern Health-Care System*. Oxford: Blackwell Scientific.

Joad, C.E.M. (1928) *The Future of Life, a Theory of Vitalism*. London/New York: G. P. Putnam's Sons.

Johansson, I., et al. (1994) Genetic Analysis of the Chinese Cytochrome P4502D Locus: Characterization of Variant CYP2D6 Genes Present in Subjects with

Diminished Capacity for Debrisoquine Hydroxylation. *Molecular Pharmacology*, 46 (3): 452–459.

Jones, William Llewellyn Parry (1972) *The Trade in Lunacy: A Study of Private Madhouses in England in the Eighteenth and Nineteenth Centuries*. London: Routledge and Kegan Paul.

Jonsen, Albert R. (1998) *The Birth of Bioethics*. New York: Oxford University Press.

Kahn, Jonathan (2003) Getting the Numbers Right—Statistical Mischief and Racial Profiling in Heart Failure Research. *Perspectives in Biology and Medicine*, 46 (4): 473–483.

——— (2004) How a Drug Becomes "Ethnic": Law, Commerce, and the Production of Racial Categories in Medicine. *Yale Journal of Health Policy, Law, and Ethics*, 4, 1–46.

Kallmann, Franz J. (1956) Psychiatric Aspects of Genetic Counseling. *American Journal of Human Genetics*, 8 (2): 97–101.

——— (1961) New Goals and Perspectives in Human Genetics. *Acta Geneticae et Gemellogiae*, 10, 377–388.

——— (1962) Genetic Research and Counseling in the Mental Health Field, Present and Future. In Kallmann, F. J. (Ed.) *Expanding Goals of Genetics in Psychiatry*. New York: Grune & Stratton.

Kalow, W. (1982) Ethnic-Differences in Drug-Metabolism. *Clinical Pharmacokinetics*, 7 (5): 373–400.

——— (1989) Race and Therapeutic Drug Response. *New England Journal of Medicine*, 320 (9): 588–590.

——— (1991) Interethnic Variation of Drug-Metabolism. *Trends in Pharmacological Sciences*, 12 (3): 102–107.

Kamin, Leon J. (1974) *The Science and Politics of I.Q.*, Potomac, Md.: L. Erlbaum Associates; distributed by Halsted Press, New York.

Kaplan, J. B., & Bennett, T. (2003) Use of Race and Ethnicity in Biomedical Publication. *Journal of the American Medical Association*, 289 (20): 2709–2716.

Karp, David Allen (1996) *Speaking of Sadness: Depression, Disconnection, and the Meanings of Illness*. Oxford: Oxford University Press.

Kass, Leon (2002) *Life, Liberty, and the Defense of Dignity: The Challenge for Bioethics*. San Francisco: Encounter Books.

Kass, Leon R. (1997) The Wisdom of Repugnance. *New Republic*, 216 (22): 17–26.

Kaufman, J. S., & Cooper, R. S. (2001) Commentary: Considerations for Use of Racial/Ethnic Classification in Etiologic Research. *American Journal of Epidemiology*, 154 (4): 291–298.

Kay, Lily E. (1993) *The Molecular Vision of Life: Caltech, the Rockefeller Foundation, and the Rise of the New Biology*. New York/Oxford: Oxford University Press.

——— (2000) *Who Wrote the Book of Life? A History of the Genetic Code*. Stanford, Calif.: Stanford University Press.

Keckler, Charles N. (2005) Cross-Examining the Brian: A Legal Analysis of Neural Imaging for Credibility Impeachment. *George Mason University School of Law Working Paper Series*. George Mason University School of Law.

Kelleher, F. (2004) The Pharmaceutical Industry's Responsibility for Protecting Human Subjects of Clinical Trials in Developing Nations. *Columbia Journal of Law and Social Problems*, 38 (1): 67–106.

Keller, Evelyn Fox (2000) *The Century of the Gene*. Cambridge, Mass.: Harvard University Press.

——— (2002) *Making Sense of Life: Explaining Biological Development with Models, Metaphors, and Machines*. Cambridge, Mass.: Harvard University Press.

Kelsoe, et al. (1989) Re-Evaluation of the Linkage Relationship between Chromosome- 11p Loci and the Gene for Bipolar Affective-Disorder in the Old Order Amish. *Nature*, 342 (6247): 238–243.

Kemp, Martin, & Wallace, Marina (2000) *Spectacular Bodies: The Art and Science of the Human Body from Leonardo to Now*. London: Hayward Gallery.

Kenen, Regina H. (1984) Genetic Counseling: The Development of a New Interdisciplinary Occupational Field. *Social Science and Medicine*, 18 (7): 541–549.

——— (1994) The Human Genome Project: Creator of the Potentially Sick, Potentially Vulnerable and Potentially Stigmatized? In Robinson, I. (Ed.) Life *and Death under High Technology Medicine*. Manchester: Manchester University Press.

Kennedy, Ian (1981) *The Unmasking of Medicine*. London: George Allen & Unwin.

Kevles, Bettyann (1997) *Naked to the Bone: Medical Imaging in the Twentieth Century*. New Brunswick, N.J.: Rutgers University Press.

Kevles, Daniel J. (1985) in *the Name of Eugenics: Genetics and the Uses of Human Heredity*. New York: Knopf.

Kidd, J. R., et al. (1987) Searching for a Major Genetic-Locus for Affective-Disorder in the Old Order Amish. *Journal of Psychiatric Research*, 21 (4): 577–580.

Kim, K., Johnson, J. A., & Derendorf, H. (2004) Differences in Drug Pharmacokinetics between East Asians and Caucasians and the Role of Genetic Polymorphisms. *Journal of Clinical Pharmacology*, 44 (10): 1083–1105.

Kittles, Rick A., & Weiss, Kenneth M. (2003) Race, Ancestry, and Genes: Implications for Defining Disease Risk. *Annual Review of Genomics and Human Genetics*, 4, 33–67.

Koch, Lene (2000) Eugenics and Genetics. *Conference on Ethos of Welfare*. Helsinki.

——— (2004) The Meaning of Eugenics: Reflections on the Government of Genetic Knowledge in the Past and Present. *Science in Context*, 17 (3): 315–331.

Kohler, Robert E. (1994) *Lords of the Fly: Drosophila Genetics and the Experimental Life*. Chicago/London: University of Chicago Press.

Koivukoski, L., et al. (2004) Meta-Analysis of Genome-Wide Scans for Hypertension and Blood Pressure in Caucasians Shows Evidence of Susceptibility Regions on Chromosomes 2 and 3. *Human Molecular Genetics*, 13 (19): 2325–2332.

Konrad, Monica (2005) *Narrating the New Predictive Genetics*. Cambridge: Cambridge University Press.

Kraepelin, Emil (1903) *Psychiatrie: Ein Lehrbuch Für Studierende Und Arzte*. Leipzig: Verlag Von Johann Ambrosius Barth.

Kramer, Peter D. (1994) *Listening to Prozac*. London: Fourth Estate.

Krausz, Y., et al. (1996) Brain Spect Imaging of Neuropsychiatric Disorders. *European Journal of Radiology*, 21 (3): 183–187.

Kühl, Stefan (1994) *The Nazi Connection: Eugenics, American Racism, and German National Socialism*. New York/Oxford: Oxford University Press.

Kumar, Sanjay (2004) Victims of Gas Leak in Bhopal Seek Redress on Compensation. *British Medical Journal*, 329 (7462): 366.

Kupfer, David J., First, Michael B., & Regier, Darell A. (Eds.) (2002) *A Research Agenda for DSM V*. Washington, D.C.: American Psychiatric Association.

Lancet (Anon) (1996) Have You Had a Gene Test? *The Lancet*, 347, 133.

Lander, E. S., et al. (2001) Initial Sequencing and Analysis of the Human Genome. *Nature*, 409 (6822): 860–921.

Lapham, E. V., Kozma, C., & Weiss, J. O. (1996) Genetic Discrimination: Perspectives of Consumers. *Science*, 274 (5287): 621–624.

Larson, Edward J. (1995) *Sex, Race, and Science: Eugenics in the Deep South*. Baltimore/London: Johns Hopkins University Press.

Le Saux, O., et al. (2000) Mutations in a Gene Encoding an Abc Transporter Cause Pseudoxanthoma Elasticum. *Nature Genetics*, 25 (2): 223–227.

Leal-Pock, Carmen (Ed.) (1998) *Faces of Huntington's*, Belleville, Ontario: Essence Publishing.

Lee, L. (1990) The National Sampling of Disability in China. *American Journal of Epidemiology*, 134, 757.

Lemmens, Trudo, & Poupak, Bahamin (1998) Genetics in Life, Disability, and Additional Health Insurance in Canada: A Comparative Legal and Ethical Analysis. In Knoppers, B. M. (Ed.) *Socio-Legal Issues in Human Genetics*. Cowansville, Quebec: Les editions Yvon Blais.

Lenoir, N. (2000) Europe Confronts the Embryonic Stem Cell Research Challenge. *Science*, 287 (5457): 1425–1427.

Lenzer, Jeanne (2004) Bush Plans to Screen Whole US Population for Mental Illness. *British Medical Journal*, 328 (7454): 1458.

Levit, K., et al. (2004) Trends—Health Spending Rebound Continues in 2002. *Health Affairs*, 23 (1): 147–159.

Lewis, J. W. (1990) Premenstrual Syndrome as a Criminal Defense. *Archives of Sexual Behavior*, 19 (5): 425–441.

Lewontin, Richard C., et al. (1984) *Not in Our Genes: Biology, Ideology, and Human Nature*. New York: Pantheon Books.

Li, Hui, & Wang, Jue (1997) Backlash Disrupts China Exchanges. *Science*, 278 (5337): 376–377.

Li, Wen-Hsiung, & Saunders, Matthew A. (2005) News & Views: The Chimpanzee and Us. *Nature*, 437 (7055): 50.

Lidbetter, Ernest James (1933) *Heredity and the Social Problem Group*. London: E. Arnold.

Lippman, Abby (1991) Prenatal Genetic Testing and Screening: Constructing Needs and Reinforcing Inequities. *American Journal of Law and Medicine*, 17 (1–2): 15–50.

——— (1992) Led (Astray) by Genetic Maps: The Cartography of the Human Genome and Health Care. *Social Science and Medicine*, 35 (12): 1469–1476.

Lock, Margaret (2005) The Eclipse of the Gene and the Return of Divination. *Current Anthropology*, 46: 547–570.

Lock, Margaret M. (2002) *Twice Dead: Organ Transplants and the Reinvention of Death*. Berkeley: University of California Press.

Lopez, J. (2004) How Sociology Can Save Bioethics. Maybe. *Sociology of Health & illness*, 26 (7): 875–896.

Low, Lawrence, King, Suzanne, & Wilkie, Tom (1998) Genetic Discrimination in Life Insurance: Empirical Evidence from a Cross Sectional Survey of Genetic Support Groups in the United Kingdom. *British Medical Journal*, 317, 1632–1635.

Ludmerer, Kenneth M. (1972) *Genetics and American Society: A Historical Appraisal*. Baltimore, Md./London: Johns Hopkins University Press.

Lynch, Michael (1998) The Discursive Construction of Uncertainty: The O. J. Simpson "Dream Team" and the Sociology of Knowledge Machine. *Social Studies of Science*, 28 (5 / 6): 829–868.

Lyon, David (1994) *The Electronic Eye: The Rise of Surveillance Society*. Minneapolis: University of Minnesota Press.

M'Charek, Amade (2005) *The Human Genome Diversity Project: An Ethnography of Scientific Practice*. Cambridge: Cambridge University Press.

Manji, H. K., Drevets, W. C., & Charney, D. S. (2001a) The Cellular Neurobiology of Depression. *Nature Medicine*, 7 (5): 541–547.

Manji, H. K., Moore, G. J., & Chen, G. (2001b) Bipolar Disorder: Leads from the Molecular and Cellular Mechanisms of Action of Mood Stabilisers. *British Journal of Psychiatry*, 178, S107–S119.

Mann, Thomas (1960) *The Magic Mountain*. Harmondsworth: Penguin.

Manuck, Stephen B., et al. (2000) A Regulatory Polymorphism of the Monoamine Oxidase-a Gene May Be Associated with Variability in Aggression, Impulsivity, and Central Nervous System Serotonergic Responsivity. *Psychiatry Research*, 95 (1): 9–23.

Marcus, Steven, & Charles A. Dana Foundation (2002) *Neuroethics: Mapping the Field: Conference Proceedings, May 13–14, 2002, San Francisco, California*. New York: Dana Press.

Marks, John (1998) *Gilles Deleuze: Vitalism and Multiplicity*. London/Sterling, Va.: Pluto Press.

Marks, Jonathan (2002) *What It Means to Be 98% Chimpanzee: Apes, People, and Their Genes*. Berkeley: University of California Press.

Marshall, Elliot (1993) NIH Told to Reconsider Crime Meeting. *Science*, 262, 23–24.

Marshall, Thomas H. (1950) *Citizenship and Social Class: and Other Essays*. Cambridge: Cambridge University Press.

Marsit, C. J., et al. (2005) The Race Associated Allele of Semaphorin 3b (Sema3b) T415i and Its Role in Lung Cancer in African-Americans and Latino-Americans. *Carcinogenesis*, 26 (8): 1446–1449.

Martin, Emily (1994) *Flexible Bodies: Tracking Immunity in American Culture from the Days of Polio to the Age of Aids*. Boston: Beacon Press.

——— (Forthcoming) *Bipolar Explorations: Toward an Anthropology of Moods*.

Masoudi, F. A., & Havranek, E. P. (2001) Race and Responsiveness to Drugs for Heart Failure. *New England Journal of Medicine*, 345 (10): 767–767.

Mattick, J. S. (2004) The Hidden Genetic Program of Complex Organisms. *Scientific American*, 291 (4): 60–67.

McGuffin, Peter, Riley, B., & Plomin, R (2001) Toward Behavioral Genomics. *Science*, 16, 1232–1249.

McKibben, Bill (2003) *Enough: Staying Human in an Engineered Age*. London: Times Books.

McLeod, H. L. (2001) Race and Responsiveness to Drugs for Heart Failure. *New England Journal of Medicine*, 345 (10): 766–767.

Meadows, Donella H. (1972) *The Limits to Growth: A Report for the Club of Rome's Project on the Predicament of Mankind*. London: Earth Island Ltd.

Medawar, Peter (1984) A View from the Left. *Nature*, 310: 255–256.

Medewar, Charles (1997) The Antidepressant Web—Marketing Depression and Making Medicine Work. *International Journal of Risk and Safety in Medicine*, 10, 75–126.

Mehta, M. A., et al. (2000) Methylphenidate Enhances Working Memory by Modulating Discrete Frontal and Parietal Lobe Regions in the Human Brain. *Journal of Neuroscience*, 20 (6): RC65 1–6.

Meskus, Mianna (2003) Eugenics and the New Genetics as Technologies of Governing Reproduction: The Finnish Case. *Conference on Vital Politics: Health, Medicine and Bioeconomics into the 21st Century*. London School of Economics and Political Science.

Metzl, Jonathan (2003) *Prozac on the Couch: Prescribing Gender in the Era of Wonder Drugs*. Durham, N.C.: Duke University Press.

Mikluscak-Copper, Cindy, & Miller, Emmett E. (1991) *Living in Hope: A 12 Step Approach for Persons at Risk or Infected with HIV*. Berkeley Calif.: Celestial Arts.

Miller, Anton R., et al. (2001) Prescription of Methylphenidate to Children and Youth, 1990–1996. *Cmaj*, 165 (11): 1489–1494.

Miller, Peter B., & Rose, Nikolas (1986) *The Power of Psychiatry*. Cambridge: Polity.

——— (1990) Governing Economic Life. *Economy and Society*, 19 (1): 1–31.

Moncrieff, Joanna, & Cohen, David (2005) Rethinking Models of Psychotropic Drug Action. *Psychotherapy and Psychosomatics*, 74 (3): 145–153.

Moncrieff, Joanna, & Kirsch, Irving (2005) Efficacy of Antidepressants in Adults. *British Medical Journal*, 331 (7509): 155–157.

Moniz, Egas António (1948) *How I Came to Perform Prefrontal Leucotomy. Congress of Psychosurgery*. Lisboa: Ediçoes Atica.

——— (1954) *A Leucotomia Está Em Causa. Liçao, Etc*: Lisboa: Ediçoes Atica.

Moore, R. J., Chamberlain, R. M., & Khuri, F. R. (2004) Apolipoprotein E and the Risk of Breast Cancer in African-American and Non-Hispanic White Women—a Review. *Oncology*, 66 (2): 79–93.

Moran, Richard (1991) The Insanity Defense: Five Years after Hinckley. In Kelley, R. J., & Macnamara, D. E. (Eds.) *Perspectives on Deviance: Domination, Degradation and Denigration*. Cincinatti, Ohio: Anderson.

Mori, M., et al. (2005) Ethnic Differences in Allele Frequency of Autoimmune-Disease-Associated SNPs. *Journal of Human Genetics*, 50 (5): 264–266.

Morley, Katharine I., & Hall, Wayne D (2003) Is There a Genetic Susceptibility to Engage in Criminal Acts. *Trends and Issues in Crime and Criminal Justice.* Canberra: Australian Institute of Criminology.

Morse, Stephen J. (2004) New Neuroscience, Old Problems. In Garland, B. (Ed.) *Neuroscience and the Law: Brain, Mind and the Scales of Justice.* New York: Dana Press.

Mosse, George L. (1978) *Toward the Final Solution: A History of European Racism.* London: Dent.

Mountain, J. L., & Risch, N. (2004) Assessing Genetic Contributions to Phenotypic Differences among "Racial" and "Ethnic" Groups. *Nature Genetics*, 36 (11): S48–S53.

Moynihan, R., Heath, I., & Henry, D. (2002) Selling Sickness: The Pharmaceutical Industry and Disease Mongering. *British Medical Journal*, 324 (7342) 886–891.

Munoz, C., & Hilgenberg, C. (2005) Ethnopharmacology. *American Journal of Nursing*, 105 (8): 40–48.

Nash, Catherine (2004) Genetic Kinship. *Cultural Studies*, 18 (1): 1–33.

National Center for Human Genome Research (1993) *Genetic Information and Health Insurance: Report of the Task Force.* Bethesda Md: National Institute of Health.

National Health Service Quality Improvement Scotland (2004) *Health Indicators Report—December 2004: A Focus on Children.* NHS Quality Improvement Scotland.

Nazroo, J. Y. (1998) Genetic, Cultural or Socio-Economic Vulnerability? Explaining Ethnic Inequalities in Health. *Sociology of Health & Illness*, 20 (5): 710–730.

——— (2003) The Structuring of Ethnic Inequalities in Health: Economic Position, Racial Discrimination, and Racism. *American Journal of Public Health*, 93 (2): 277–284.

Nelkin, Dorothy, & Lindee, M. Susan (1995) *The DNA Mystique: The Gene as Cultural Icon.* New York: Freeman.

Nelkin, Dorothy, & Tancredi, Laurence R. (1989) *Dangerous Diagnostics: The Social Power of Biological Information.* New York: Basic Books.

Nguyen, Vhin-kim (2005) Therapeutic Citizenship. In Ong, A., & Collier, S. (Eds.) *Global Assemblages: Technology, Politics, and Ethics as Anthropological Problems.* Malden, Mass.: Blackwell Publishing.

Nightingale, Paul, & Martin, Paul A. (2004) The Myth of the Biotech Revolution. *Trends in Biotechnology*, 22 (11): 564–569.

Nilsson, Annika, & Rose, Joanna (1999) Sweden Takes Steps to Protect Tissue Banks. *Science*, 286, 894.

Novas, Carlos (2001) The Political Economy of Hope: Patients' Organisations, Science and Biovalue. *Paper presented at the Postgraduate Forum on Genetics and Society, University of Nottingham, June 21–22, 2001.*

——— (2003) Governing Risky Genes. PhD Thesis. London: University of London.

Novas, Carlos, & Rose, Nikolas (2000) Genetic Risk and the Birth of the Somatic Individual. *Economy and Society*, 29 (4): 485–513.

Nuffield Council on Bioethics (2003) *Pharmacogenetics: The Ethical Context*. London: Nuffield Council on Bioethics.

O'Malley, Pat (1996) Risk and Responsibility. In Barry, A., Osborne, T., & Rose, N. (Eds.) *Foucault and Political Reason*. London: UCL Press.

——— (1998) *Crime and the Risk Society*. Aldershot: Dartmouth.

——— (1999) Volatile and Contradictory Punishments. *Theoretical Criminology*, 3 (2): 175–196.

O'Neill, Onora (1998) Insurance and Genetics: The Current State of Play. *Modern Law Review*, 61 (5): 716–723.

Ofili, E., Flack, J., & Gibbons, G. (2001) Race and Responsiveness to Drugs for Heart Failure. *New England Journal of Medicine*, 345 (10): 767–767.

Ogden, Jane (1995) Psychosocial Theory and the Creation of the Risky Self. *Social Science & Medicine*, 40 (3): 409–415.

Ong, Aihwa, & Collier, Stephen J. (2005) *Global Assemblages: Technology, Politics, and Ethics as Anthropological Problems*, Malden, Mass.: Blackwell Publishing.

Organisation for Economic Co-operation and Development (1996) *The Knowledge Based Economy*. Paris: Organisation for Economic Co-operation and Development.

——— (2004) *Biotechnology for Sustainable Growth and Development*. Organisation for Economic Co-operation and Development.

Osborne, Thomas (1996) Security and Vitality: Drains, Liberalism and Power in the Nineteenth Century. In Barry, A., Osborne, T., & Rose, N. (Eds.) *Foucault and Political Reason*. London: UCL Press.

——— (1997) Of Health and Statecraft. In Petersen, A., & Bunton, R. (Eds.) *Foucault, Health and Medicine*. London: Routledge.

Palsson, Gisli, & Rabinow, Paul (1999) Iceland: The Case of a National Human Genome Project. *Anthropology Today*, 15 (5): 14.

Parens, Erik (Ed.) (1998) *Enhancing Human Traits: Ethical and Social Implications*. Washington, D.C.: Georgetown University Press.

Parton, Nigel (1991) *Governing the Family: Child Care, Child Protection and the State*. London: Macmillan Education.

Paul, Diane B. (1998a) Genetic Services, Economics and Eugenics. *Science in Context*, 11 (3–4): 481–491.

——— (1998b) *The Politics of Heredity: Essays on Eugenics, Biomedicine and the Nature-Nurture Debate*. Albany: State University of New York Press.

Pearson, Karl (1911) *The Academic Aspect of the Science of National Eugenics: A Lecture Delivered to Undergraduates*. London: Dulau and Co. Ltd.

——— (1912) *Darwinism, Medical Progress and Eugenics: The Cavendish Lecture, 1912, an Address to the Medical Profession*. London: Dulau & Co. Ltd.

Peräkyla, Anssi (1991) Hope Work in the Care of Seriously Ill Patients. *Qualitative Health Research*, 1 (4): 407–433.

Persson, I., et al. (1999) Genetic Polymorphism of Xenobiotic Metabolizing Enzymes among Chinese Lung Cancer Patients. *International Journal of Cancer*, 81 (3): 325–329.

Peterson, Alan, & Bunton, Robin (2002) *The New Genetics and the Public's Health*. London: Routledge.

Petryna, Adriana (2002) *Life Exposed: Biological Citizens after Chernobyl*. Princeton, N.J.: Princeton University Press.

——— (2005) Ethical Variability: Drug Development and Globalizing Clinical Trials. *American Ethnologist*, 32 (2): 183–197.

Pick, Daniel (1989) *Faces of Degeneration: A European Disorder C.1848-C.1918*. Cambridge: Cambridge University Press.

Plomin, Robert, & McGuffin, Peter (2003) Psychopathology in the Postgenomic Era. *Annual Review of Psychology*, 54, 205–228.

Pokorski, Robert J. (1997) Insurance Underwriting in the Genetic Era. *Cancer*, (Supplement) 80 (3): 587–599.

Pollock, Vikki, Mednick, Sarnoff A., & Gabrielli, WIlliam F. (1983) Crime Causation: Biological Theories. In Kadish, S. H. (Ed.) *Encyclopedia of Crime and Justice, Vol. 1*. New York: Free Press.

Porter, Roy (1997) *The Greatest Benefit to Mankind: A Medical History of Humanity From*. London: HarperCollins.

Porter, Theodore, M. (1995) *Trust in Numbers: The Pursuit of Objectivity in Science and Public Life*. Princeton, N.J.: Princeton University Press.

Potter, V. R. (1970) Bioethics, Science of Survival. *Perspectives in Biology and Medicine*, 14 (1): 127–153.

Pratt, John (1995) Dangerousness, Risk and Technologies of Power. *Australian and New Zealand Journal of Criminology*, 28 (1): 3–31.

——— (1998) The Rise and Fall of Homophobia and Sexual Psychopath Legislation in Postwar Society. *Psychology, Public Policy, and Law*, 4 (1–2): 25–49.

——— (2000) Dangerousness, Risk and Technologies of Power. *Australian and New Zealand Journal of Criminology*, 28 (1): 3–31.

President's Council on Bioethics (U.S.) & Kass, Leon (2003) *Beyond Therapy: Biotechnology and the Pursuit of Happiness*. New York: Regan Books.

Prewitt, Kenneth (1987) Public Statistics and Democratic Politics. In Alonso, W., & Starr, P. (Eds.) *The Politics of Numbers*. New York: Russell Sage Foundation.

Prior, Lindsay, et al. (2002) Making Risk Visible: The Role of Images in the Assessment of (Cancer) Genetic Risk. *Health Risk & Society*, 4 (3): 241–258.

Proctor, Robert (1988) *Racial Hygiene: Medicine under the Nazis*. Cambridge, Mass./London: Harvard University Press.

——— (1999) *The Nazi War on Cancer*. Princeton, N.J.: Princeton University Press.

Rabeharisoa, Vololona, & Callon, Michel (1998a) L'implication des Malades dans les Activités De Recherche Soutenues par L'Association Française Contre Les Myopathies. *Sciences sociales et santé*, 16 (3): 41–65.

Rabeharisoa, Vololona, & Callon, Michel (1998b) The Participation of Patients in the Process of Production of Knowledge: The Case of the French Muscular Dystrophies Association. *Sciences sociales et santé*, 16 (3): 41–66.

Rabinow, Paul (1994) The Third Culture. *History of the Human Sciences*, 7 (2): 53–64.

——— (1996a) Artificiality and Enlightenment: From Sociobiology to Biosociality. *Essays on the Anthropology of Reason*. Princeton, N.J.: Princeton University Press.

——— (1996b) *Making PCR: A Story of Biotechnology*. Chicago: University of Chicago Press.

——— (1999) *French DNA: Trouble in Purgatory*. Chicago: University of Chicago Press.

Rabinow, Paul, & Rose, Nikolas (2006) Biopower Today. *BioSocieties: An Interdisciplinary Journal for the Social Study of the Life Sciences*, 1 (2): 195–218.

Rafter, Nicole Hahn (1997) *Creating Born Criminals*. Urbana: University of Illinois Press.

Raine, A., Buchsbaum, M., & LaCasse, L. (1997) Brain Abnormalities in Murderers Indicated by Positron Emission Tomography. *Biological Psychiatry*, 42 (6): 495–508.

Ramchandani, Paul (2004) Treatment of Major Depressive Disorder in Children and Adolescents. *British Medical Journal*, 328 (7430): 3–4.

Rapp, Rayna (1999) *Testing Women, Testing the Fetus: The Social Impact of Amniocentesis in America*. New York: Routledge.

Reardon, Jennifer (2001) The Human Genome Diversity Project: A Case Study in Coproduction. *Social Studies of Science*, 31 (3): 357–388.

——— (2005) *Race to the Finish: Identity and Governance in an Age of Genomics*. Princeton, N.J.: Princeton University Press.

Reed, Sheldon C. (1974) A Short History of Genetic Counseling. *Social biology*, 21 (4): 332–339.

Reilly, Phillip R. (1991) *The Surgical Solution: A History of Involuntary Sterilization in the United States*. Baltimore, Md.: Johns Hopkins University Press.

Reiss, Albert J., Roth, Jeffrey A., & Miczek, Klaus A. (1993–94) *Understanding and Preventing Violence: Report of the National Research Council Panel on the Understanding and Control of Violent Behavior*. Washington, D.C.: National Academy Press.

Relman, A. S., & Angell, M. (2002) Americas' Other Drug Problem. *New Republic*, 227 (25): 27–41.

Rheinberger, Hans-Jörg (2000) Beyond Nature and Culture: Modes of Reasoning in the Age of Molecular Biology and Medicine. In Lock, M., Young, Allan and Cambrosio, Alberto (Ed.) *Living and Working with the New Medical Technologies*. Cambridge: Cambridge University Press.

Rifkin, Jeremy (1998) *The Biotech Century: Harnessing the Gene and Remaking the World*. New York: Jeremy P. Tarcher/Putnam.

Risch, Neil J. (2000) Searching for Genetic Determinants in the New Millennium. *Nature*, 405, 847–856.

Risch, N., et al. (2002) Categorization of Humans in Biomedical Research: Genes, Race and Disease. *Genome Biology*. 3 (7): 2007.

Roberton, John (1812) *Medical Police: Or the Causes of Diseases with the Means of Prevention*. London: Routledge.

Roberts, J. A. Fraser (1961) Genetic Advice to Patients. In Jones, F. A. (Ed.) *Clinical Aspects of Genetics*. London: Pitman Medical Publishing.

Robertson, Ann (2000) Embodying Risk, Embodying Political Rationality: Women's Accounts of Risks for Breast Cancer. *Health Risk & Society*, 2 (2): 219–235.

Robertson, John A. (1993) Procreative Liberty and the Control of Conception, Pregnancy and Childbirth. *Virginia Law Review*, 69: 405–464.

Robey, B., Rutstein S. O., Morris L. and Blackburn, R. (1992) *The Reproductive Revolution: New Survey Findings*. Baltimore, Md.: Population Information Program.

Roh, H. K., et al. (1996) Debrisoquine and S-Mephenytoin Hydroxylation Phenotypes and Genotypes in a Korean Population. *Pharmacogenetics*, 6 (5): 441–447.

Rose, Hilary (2003) *The Commodification of Bioinformation: The Icelandic Health Sector Database*. London: Wellcome Trust.

Rose, Nikolas (1985) *The Psychological Complex: Psychology, Politics and Society in England, 1869–1939*. London/Boston: Routledge & Kegan Paul.

——— (1986) Psychiatry: The Discipline of Mental Health. In Miller, P., & Rose, N. (Eds.) *The Power of Psychiatry*. Cambridge: Polity Press.

——— (1989) *Governing the Soul: The Shaping of the Private Self*. London/New York: Routledge.

——— (1991) Governing by Numbers. *Accounting Organizations and Society*, 16 (7): 673–692.

——— (1992) Governing the Enterprising Self. In Heelas, P., & Morris, P. (Eds.) *The Values of the Enterprise Culture: The Moral Debate*. London: Routledge.

——— (1994) Medicine, History and the Present. In Jones, C. & Porter, R. (Eds.) *Reassessing Foucault: Power, Medicine and the Body*. London: Routledge.

——— (1996a) The Death of the Social? Re-Figuring the Territory of Government. *Economy and Society*, 25 (3): 327–356.

——— (1996b) *Inventing Our Selves: Psychology, Power, and Personhood*. New York: Cambridge University Press.

——— (1996c) Psychiatry as a Political Science: Advanced Liberalism and the Administration of Risk. *History of the Human Sciences*, 9 (2): 1–23.

——— (1998) Governing Risky Individuals: The Role of Psychiatry in New Regimes of Control. *Psychiatry, Psychology & Law*, 5 (2): 177–195.

——— (1999) *Powers of Freedom: Reframing Political Thought*. Cambridge/New York: Cambridge University Press.

——— (2000a) Biological Psychiatry as a Style of Thought. Unpublished manuscript.

——— (2000b) The Biology of Culpability: Pathological Identity and Crime Control in a Biological Culture. *Theoretical Criminology*, 4 (1): 5–43.

——— (2000c) Government and Control. *British Journal of Criminology*, 40 (2): 321–339.

——— (2001) The Politics of Life Itself. *Theory, Culture & Society*, 18 (6): 1–30.

——— (2002) The Politics of Bioethics Today. Conference on Biomedicalization, Social Conflicts and the New Politics of Bioethics. Vienna.

——— (2003) Neurochemical Selves. *Society*, 41 (1): 46–59.

——— (2004) Becoming Neurochemical Selves. In Stehr, N. (Ed.) *Biotechnology, Commerce and Civil Society*. New York: Transaction Press.

Rose, Nikolas, & Miller, Peter (1992) Political Power Beyond the State—Problematics of Government. *British Journal of Sociology*, 43 (2): 173–205.

Rose, Nikolas, & Novas, Carlos (2004) Biological Citizenship. In Ong, A., & Collier, S. (Eds.) *Blackwell Companion to Global Anthropology*. Oxford: Blackwell.

Rose, Nikolas, & Singh, Ilina (2006) Neuroforum. *BioSocieties: An Interdisciplinary Journal for the Social Study of the Life Sciences*, 1 (1): 97–102.

Rose, Steven P. R. (1995) The Rise of Neurogenetic Determinism. *Nature*, 373 (6513): 380–382.

——— (1998b) *Lifelines: Biology Beyond Determinism*. Oxford/New York: Oxford University Press.

——— (2005) *The 21st Century Brain: Explaining, Mending and Manipulating the Mind*. London: Jonathan Cape.

Rosell, Sune (1991) Sweden's Answer to Genomics Ethics (Letter). *Nature*, 401 (September 16).

Rosen, George (1958) *A History of Public Health. Foreword by Félix Martí-Ibánez*. New York: MD Publications.

Rosenberg, Charles (2003) What Is Disease? in Memory of Owsei Temkin. *Bulletin of the History of Medicine*, 77 491–505.

Rosenberg, N. A., et al. (2002) Genetic Structure of Human Populations. *Science*, 298 (5602): 2381–2385.

Rosenhan, David L. (1973) On Being Sane in Insane Places. *Science*, 179, 250–258.

Roses, Allen D. (2002a) Genome-Based Pharmacogenetics and the Pharmaceutical Industry. *Nature Reviews Drug Discovery*, 1 (7): 541–549.

——— (2002b) Pharmacogenetics' Place in Modern Medical Science and Practice. *Life Sciences*, 70 (13): 1471–1480.

——— (2004) Pharmacogenetics and Drug Development: The Path to Safer and More Effective Drugs. *Nature Reviews Genetics*, 5 (9): 645–656.

Rotimi, Charles N. (2003) Genetic Ancestry Tracing and the African Identity: A Double Edged Sword? *Developing World Bioethics*, 3 (2): 151–158.

Royal, Charmaine D. M. & Dunston, Georgia M. (2004) Changing the Paradigm from "Race" to Human Genome Variation. *Nature Genetics*, 36 (Supplement): S5–S7.

Rubinsztein, D. C., et al. (1996) Phenotypic Characterization of Individuals with 30–40 Cag Repeats in the Huntington Disease (Hd) Gene Reveals Hd Cases with 36 Repeats and Apparently Normal Elderly Individuals with 36–39 Repeats. *American Journal of Human Genetics*, 59 (1): 16–22.

Ruddick, William (1999) Hope and Deception. *Bioethics*, 13 (3–4): 343–357.

Saleeby, Caleb Williams (1914) *The Progress of Eugenics*. London.

Salter, Brian, & Jones, Mavis (2002) Human Genetic Technologies, European Governance and the Politics of Bioethics. *Nature Reviews Genetics*, 3 (10): 808–814.

——— (2005) Biobanks and Bioethics: The Politics of Legitimation. *Journal of European Public Policy*, 12 (4): 710–732.

Sanders, T., et al. (2003) Risk Constructions among People Who Have a First-Degree Relative with Cancer. *Health Risk & Society*, 5 (1): 53–69.

Santos, S., & Bizzo, N. (2005) From "New Genetics" to Everyday Knowledge: Ideas About How Genetic Diseases Are Transmitted in Two Large Brazilian Families. *Science Education*, 89 (4): 564–576.

Saulitis, Andrew (1979) Chromosomes and Criminality: The Legal Implications of Xyy Syndrome. *Journal of Legal Medicine*, 1 (3): 269–291.

Scheingold, Stuart A., Olson, T., & Pershing, J. (1994) Sexual Violence, Victim Advocacy, and Republican Criminology—Washington States Community Protection Act. Law *& Society Review*, 28 (4): 729–763.

Scheper-Hughes, Nancy (2000) The Global Traffic in Human Organs. *Current Anthropology*, 41 (2): 191–224.

——— (2003a) Review of "The Twice Dead: Organ Transplants and the Reinvention of Death" by Margaret Lock. *American Anthropologist*, 105 (1): 172–174.

——— (2003b) Scarce Goods: Justice, Fairness, and Organ Transplantation. *American Anthropologist*, 105 (1): 172–174.

Scheper Hughes, Nancy, & Waquant, Loic (Eds.) (2002) *Commodifying Bodies*. London: Sage.

Schwartz, R. S. (2001a) Race and Responsiveness to Drugs for Heart Failure. Reply. *New England Journal of Medicine*, 345 (10): 768–768.

——— (2001b) Racial Profiling in Medical Research. [Comment]. *New England Journal of Medicine*, 344 (18): 1392–1393.

Scott, Randy (2001) Genomics: The Forces of Acceleration Are Upon Us. In Ernst & Young, *Focus on Fundamentals: The Biotechnology Report*. Ernst & Young.

Scott, Russell (1981) *The Body as Property*. London: Allen Lane.

Scott, Susie et al. (2005) Repositioning the Patient: The Implications of Being "at Risk." *Social Science & Medicine*, 60 (8): 1869–1879.

Shaw, Margery W. (1987) Testing for the Huntington Gene: A Right to Know, a Right Not to Know, or a Duty to Know. *American Journal of Medical Genetics*, 26 (2): 243–246.

Sherrington, R., et al. (1988) Localization of a Susceptibility Locus for Schizophrenia on Chromosome-5. *Nature*, 336 (6195): 164–167.

Shockley, William, & Pearson, Roger (1992) *Shockley on Eugenics and Race: The Application of Science to the Solution of Human Problems*. Washington, D.C.: Scott-Townsend Publishers.

Shriver, M. D., et al. (2003) Skin Pigmentation, Biogeographical Ancestry and Admixture Mapping. *Human Genetics*, 112 (4): 387–399.

Silver, Lee M. (1998) *Remaking Eden: Cloning and Beyond in a Brave New World*. London: Weidenfeld and Nicholson.

Silverman, Chloe (2003) Brains, Pedigrees, and Promises: The Material Politics of Autism Research. *Conference on Vital Politics: Health, Medicine and Bioeconomics into the Twenty First Century*. London School of Economics and Political Science.

——— (2004) A Disorder of Affect: Love, Tragedy, Biomedicine and Citizenship in American Autism Research, 1943–2003. Ph.D. Thesis, University of Pennsylvania.

Simon, Jonathan (2000) Megan's Law: Crime and Democracy in Late Modern America. *Law and Social Inquiry*, 25 (4): 1111–1150.

Simpson, Robert (2000) Imagined Genetic Communities: Ethnicity and Essentialism in the Twenty-First Century. *Anthropology Today*, 16 (3): 3–6.
Singh, Ilina (2002) Bad Boys, Good Mothers, and the "Miracle" of Ritalin. *Science in Context*, 15 (4): 577–603.
——— (2003) Boys Will Be Boys: Fathers' Perspectives on ADHD Symptoms, Diagnosis, and Drug Treatment. *Harvard Review of Psychiatry*, 11 (6): 308–316.
——— (2004) Doing Their Jobs: Mothering with Ritalin in a Culture of Mother-Blame. *Social Science & Medicine*, 59 (6): 1193–1205.
Slater, Lauren (1999) *Prozac Diary*. London: Hamish Hamilton.
Sleeboom, Margaret (2005) The Harvard Case of Xu Xiping: Exploitation of the People, Scientific Advance, or Genetic Theft? *New Genetics and Society*, 24 (1): 57–78.
Smith, H. O., et al. (2003) Generating a Synthetic Genome by Whole Genome Assembly: Phi X174 Bacteriophage from Synthetic Oligonucleotides. *Proceedings of the National Academy of Sciences of the United States of America*, 100 (26): 15440–15445.
Snipp, Matthew C. (2003) Racial Measurement in the American Census: Past Practices and Implications for the Future. *Annual Review of Sociology*, 29: 563–588.
Snyder, S. H., & Ferris, C. D. (2000) Novel Neurotransmitters and Their Neuropsychiatric Relevance. *American Journal of Psychiatry*, 157 (11): 1738–1751.
Solomon, Andrew (2001) *The Noonday Demon: An Anatomy of Depression*. London: Chatto & Windus.
Soodyall, Himla (2003) Reflections and Prospects for Anthropological Genetics in South Africa. In Goodman, A. H., Heath, D., & Lindee, M. S. (Eds.) *Genetic Nature/Culture*. Berkeley: University of California Press.
Spallone, Pat (1998) The New Biology of Violence: New Geneticisms for the Old? *Body and Society*, 4 (4): 47–65.
Speybroeck, Linda van, Vijver, Gertrudis van de, & Waele, Dani de (Eds.) (2002) *From Epigenesis to Epigenetics: The Genome in Context*. New York: New York Academy of Sciences.
Stahl, Steven M. (1996) *Essential Psychopharmacology: Neuroscientific Basis and Practical Applications*. Cambridge: Cambridge University Press.
Staley, L. W., et al. (1995) Identification of Cytogenetic Abnormalities as a Consequence of FMRI Testing in Schools. *Developmental Brain Dysfunction*, 8 (4–6): 310–318.
Starr, Douglas (2002) *Blood: An Epic History of Medicine and Commerce*. New York: HarperCollins.
Starr, Paul (1982) *The Social Transformation of American Medicine*. New York: Basic Books.
Stepan, Nancy (1991) *"The Hour of Eugenics": Race, Gender and Nation in Latin America*, Ithaca, N.Y./London: Cornell University Press.
Stephens, Joe, et al. (2000) The Body Hunters. *The Washington Post* (Six Part Series: December 17–22, 2000).
Stock, Gregory (2003) *Redesigning Humans: Choosing Our Children's Genes*. London: Profile.

Stoler, Ann Laura (1995) *Race and the Education of Desire: Foucault's History of Sexuality and the Colonial Order of Things*. Durham, N.C./London: Duke University Press.

Stone, Emma (1996) A Law to Protect, a Law to Prevent: Contextualising Disability Legislation in China. *Disability & Society*, 11 (4): 469–483.

Strathern, Marilyn (1992) *Reproducing the Future: Essays on Anthropology, Kinship and the New Reproductive Technologies*. Manchester: Manchester University Press.

——— (1999) *Property, Substance and Effect*. London: Athlone.

Strous, Rael D., et al. (1997) Analysis of a Functional Catechol-O-Methyltransferase Gene Polymorphism in Schizophrenia: Evidence for Association with Aggressive and Antisocial Behavior. *Psychiatry Research*, 69, 71–77.

Styron, William (1990) *Darkness Visible: A Memoir of Madness*. New York: Random House.

Su, Xiaokang (1991) River Elegy. *Chinese Sociology and Anthropology*, 24 (2): 9.

Sung, Wen-Ching (2005) Chinese DNA: How China Uses Genome Projects to Construct Chineseness. Paper delivered to the April 2005 meeting of the 2005 Meeting of the American Ethnological Society on "Anxious Borders: Traversing Anthropological Divides."

Sutherland, Edwin H. (1931) Mental Deficiency and Crime. In Young, K. (Ed.) *Social Attitudes*. New York: Holt.

Suzman, A. (1960) Race Classification and Definition in the Legislation of the Union of South Africa. *Acta Juridica*, 339–367.

Tapper, Melbourne (1999) in *The Blood: Sickle Cell Anemia and the Politics of Race*. Philadelphia: Pennsylvania University Press.

Taussig, Karen Sue (2005) The Molecular Revolution in Medicine: Promise, Reality, and Social Organization. In McKinnon, S., & Silverman, S. (Eds.) *Complexities: Beyond Nature & Nurture*. Chicago: Chicago University Press.

Taussig, Karen Sue, Heath, Deborah & Rapp, Rayna (2003) Flexible Eugenics: Technologies of the Self in the Age of Genetics. In Goodman, A. H., Heath, D., & Lindee, M. S. (Eds.) *Genetic Nature/Culture*. Berkeley: University of California Press.

Tecott, L. H., & Barondes, S. H. (1996) Behavioral Genetics: Genes and Aggressiveness. *Current Biology*, 6 (3): 238–240.

Thapar, A., et al. (1999) Genetic Basis of Attention Deficit and Hyperactivity. *British Medical Journal*, 174: 105–111.

Thompson, Charis (2005) *Making Parents: The Ontological Choreography of Reproductive Technologies*. Cambridge, Mass/London: MIT.

Thompson, L. A., Detterman, D. K., & Plomin, R. (1993) Differences in Heritability across Groups Differing in Ability, Revisited. *Behavior Genetics*, 23 (4): 331–336.

Trimble, Michael R. (1996) *Biological Psychiatry*. Chichester: Wiley.

Trouiller, P., et al. (2002) Drug Development for Neglected Diseases: A Deficient Market and a Public-Health Policy Failure. *The Lancet*, 359 (9324): 2188–2194.

Tsuang, M. T., et al. (2005) Assessing the Validity of Blood-Based Gene Expression Profiles for the Classification of Schizophrenia and Bipolar Disorder: A Preliminary Report. *American Journal of Medical Genetics Part B-Neuropsychiatric Genetics*, 133B (1): 1–5.

Valenstein, Elliot, S. (1986) *Great and Desperate Cures: The Rise and Decline of Psychosurgery*. New York: Basic Books.

Valverde, Mariana (1991) *The Age of Light, Soap, and Water: Moral Reform in English Canada, 1885–1925*, Toronto: McClelland & Stewart.

——— (1998) *Diseases of the Will: Alcohol and the Dilemmas of Freedom*. Cambridge: Cambridge University Press.

van Dijk, S., et al. (2004) Feeling at Risk: How Women Interpret Their Familial Breast Cancer Risk. *American Journal of Medical Genetics Part A*, 131A (1): 42–49.

Vartiainen, H. (1995) Free Will and 5-Hydroxytryptamine. *Journal of Forensic Psychiatry*, 6 (1): 6–9.

Vaughan, Megan (1998) Slavery and Colonial Identity in Eighteenth-Century Mauritius. *Transactions of the Royal Historical Society, Sixth Series*, 8: 189–214.

Venter, J. C., et al. (2001) The Sequence of the Human Genome. *Science*, 291 (5507): 1304–1351.

Virkkunen, M., et al. (1994) Csf Biochemistries, Glucose-Metabolism, and Diurnal Activity Rhythms in Alcoholic, Violent Offenders, Fire Setters, and Healthy-Volunteers. *Archives of General Psychiatry*, 51 (1): 20–27.

Waldby, Catherine (2000) *The Visible Human Project: Informatic Bodies and Posthuman Medicine*. London, New York: Routledge.

——— (2002) Stem Cells, Tissue Cultures and the Production of Biovalue. *Health*, 6 (3): 305–323.

Waldby, Catherine, & Mitchell Robert (2006) *Tissue Economies: Gifts, Commodities, and Bio-Value in Late Capitalism*. Durham, N.C.: Duke University Press.

Walker, Francis Amasa, & Dewy, Davis Rich (1899) *Discussions in Economics and Statistics*. New York: H. Holt and Company.

Walker, Gladstone (1924) *Practical Eugenics: A Study of Eugenic Principles in Their Application to Social Conditions*. London: Poor-Law Publications.

Walters, G. D. (1992) A Metaanalysis of the Gene-Crime Relationship. *Criminology*, 30 (4): 595–613.

Wang, Jiye, & Hull, Terence H. (1991) *Population and Development Planning in China*, North Sydney, NSW: Allen & Unwin.

Wasserman, David (1995) Science and Social Harm: Genetic Research into Crime and Violence. *Report from the Institute of Philosophy and Public Policy*, 15 (1): 14–19.

——— (1996) Research into Genetics and Crime: Consensus and Controversy. *Politics and the Life Sciences*, 15 (1): 107–109.

Wasserman, David, & Wachbroit, Robert (Eds.) (2001) *Genetics and Criminal Behavior*. Cambridge: Cambridge University Press.

Waterfield, Henry (1875) *Memorandum on the Census of British India of 1871–72: Presented to Both Houses of Parliament by Command of Her Majesty*. London: HMSO.

Weber, Max (1930) *The Protestant Ethic and the Spirit of Capitalism*. London: Allen & Unwin.

Weindling, Paul (1999) *Epidemics and Genocide in Eastern Europe, 1890–1945*. Oxford/New York: Oxford University Press.

Weir, Lorna (1996) Recent Developments in the Government of Pregnancy. *Economy and Society*, 25 (3): 372–392.

Weismann, August (1893 [1982]) *The Germ-Plasm: A Theory of Heredity. Translated by W. N. Parker and Harriet Ronnfeldt*. London: Walter Scott.

Wexler, Alice (1996) *Mapping Fate: A Memoir of Family, Risk and Genetic Research*. Berkeley, Calif.: University of California Press.

Wheeler, Leonard Richmond (1939) *Vitalism: Its History and Validity*. London: H.F. & G. Witherby Ltd.

Wheelwright, Jett (2005) Finland's Fascinating Genes. *Discover*, 26 (4): April 2005.

White House (2000) President Clinton Takes Historic Action to Ban Genetic Discrimination in the Federal Workplace. February 8: *http://clinton4.nara.gov/WH/New/html/20000208.html*.

Wilson, Elizabeth A. (2004a) The Brain in the Gut. In Wilson, E. A. (Ed.) *Psychosomatic*. Durham, N.C.: Duke University Press.

——— (2004b) *Psychosomatic: Feminism and the Neurological Body*. Durham, N.C.: Duke University Press.

Wilson, J. F., et al. (2001) Population Genetic Structure of Variable Drug Response. *Nature Genetics*. 29 (3): 265–269.

Wilson, James Q., & Herrnstein, Richard J. (1985) *Crime and Human Nature*. New York: Simon and Schuster.

Wittgenstein, Ludwig (1958) *Philosophical Investigations*. Oxford: Basil Blackwell.

Wolpe, Paul Root (2003) Neuroethics of Enhancement. *Brain and Cognition*, 50, 387–395.

——— (2002) Treatment, Enhancement, and the Ethics of Neurotherapeutics. *Brain and Cognition*, 50 (3): 387–395.

Wolpert, Louis (1999) *Malignant Sadness: The Anatomy of Depression*. London: Faber and Faber.

World Health Organization (2002) *Genomics and World Health*. Geneva: World Health Organization.

——— (2004) *Mental Health of Children and Adolescents: Briefing Paper for Who European Ministerial Conference on Mental Health*. Finland (January 12–15, 2005). Geneva: World Health Organization.

Wright, R. A., & Miller, J. M. (1998) Taboo until Today? The Coverage of Biological Arguments in Criminology Textbooks, 1961 to 1970 and 1987 to 1996. *Journal of Criminal Justice*, 26 (1): 1–19.

Wurtzel, Elizabeth (1995) *Prozac Nation: Young and Depressed in America*. New York: Riverhead.

Xie, H. G., et al. (2001) Molecular Basis of Ethnic Differences in Drug Disposition and Response. *Annual Review of Pharmacology and Toxicology*, 41, 815–850.

Yancy, C. W., et al. (2001) Race and the Response to Adrenergic Blockade with Carvedilol in Patients with Chronic Heart Failure. *New England Journal of Medicine*, 344 (18): 1358–1365.

Yasar, U., et al. (1999) Validation of Methods for Cyp2c9 Genotyping: Frequencies of Mutant Alleles in a Swedish Population [Erratum Appears in *Biochemical & Biophysical Research Communications*, 1999, 258 (1): 227]. *Biochemical & Biophysical Research Communications*, 254 (3): 628–631.

Yoxen, Edward (1981) Life as a Productive Force: Capitalising the Science and Technology of Molecular Biology. In Levidow, L., & Young, B. (Eds.) *Science, Technology and the Labour Process: Marxist Studies*. London: Blackrose Press.

Zimring, F. E. (1996) The Genetics of Crime: A Skeptic's Vision of the Future. *Politics and the Life Sciences*, 15 (1): 105–106.

Zournazi, Mary (Ed.) (2002) *Hope: New Philosophies for Change*. London: Lawrence and Wishart

Index

The Shadows and Lights of Waco: Millennialism Today by James D. Faubion

Life Exposed: Biological Citizens after Chernobyl by Adriana Petryna

Anthropos Today: Reflections on Modern Equipment by Paul Rabinow

When Nature Goes Public: The Making and Unmaking of Bioprospecting in Mexico by Cori Hayden

Picturing Personhood: Brain Scans and Biomedical Identity by Joseph Dumit

Fiscal Disobedience: An Anthropology of Economic Regulation in Central Africa by Janet Roitman

Race to the Finish: Identity and Governance in an Age of Genomics by Jenny Reardon

Everything Was Forever, Until It Was No More: The Last Soviet Generation by Alexei Yurchak

Wild Profusion: Biodiversity Conservation in an Indonesian Archipelago by Celia Lowe

The Politics of Life Itself by Nikolas Rose